Test Item File for

Precalculus

SIXTH EDITION

Larson/Hostetler

HOUGHTON MIFFLIN COMPANY **Boston New York**

Publisher: Jack Shira
Managing Editor: Cathy Cantin
Development Manager: Maureen Ross
Development Editor: Laura Wheel
Assistant Editor: James Cohen
Supervising Editor: Karen Carter
Senior Project Editor: Patty Bergin
Editorial Assistant: Allison Seymour
Production Technology Supervisor: Gary Crespo
Executive Marketing Manager: Michael Busnach
Senior Marketing Manager: Danielle Potvin
Marketing Associate: Nicole Mollica
Senior Manufacturing Coordinator: Jane Spelman

Printed in the United States of America

ISBN 0-618-31439-3

2 3 4 5 6 7 8 9 – EB – 07 06 05 04

GENERAL TABLE OF CONTENTS

Chapter 1 Functions and Their Graphs

Section 1.1: Graphs of Equations

Objective 1: Sketch graphs of equations

Match the equation with its graph.

1. $y = x^3 - x^2 - 2x$

(A)

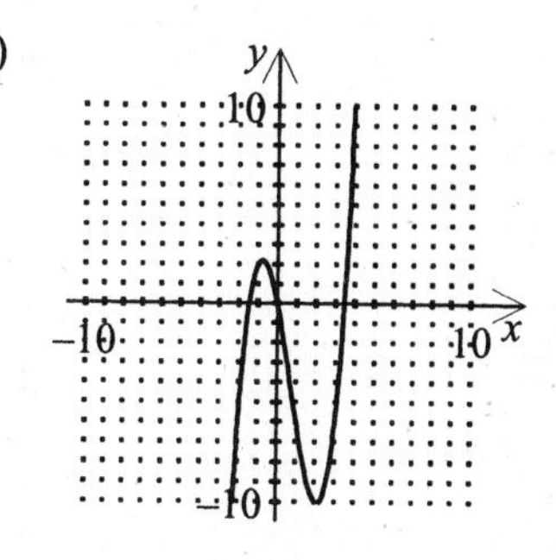

(B)

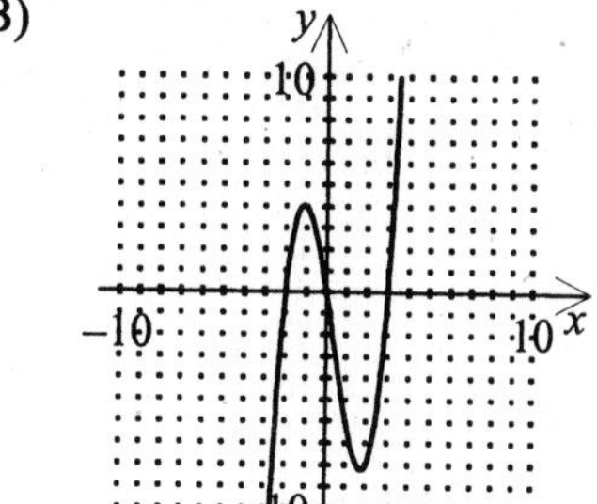

(C)

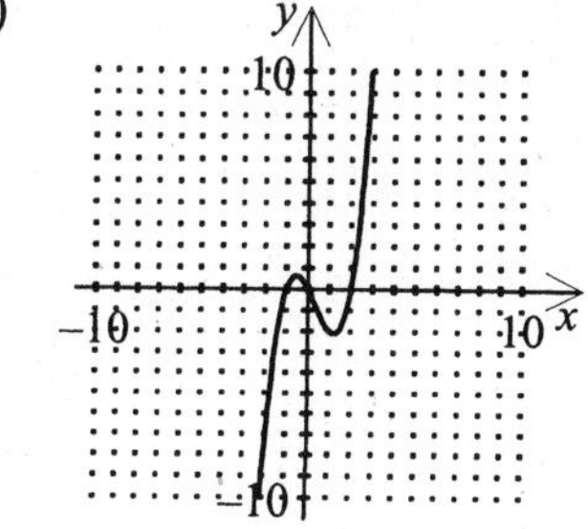

(D) None of these

Match the equation with its graph.

2. $y = 4\sqrt{x} - 3$

(A)

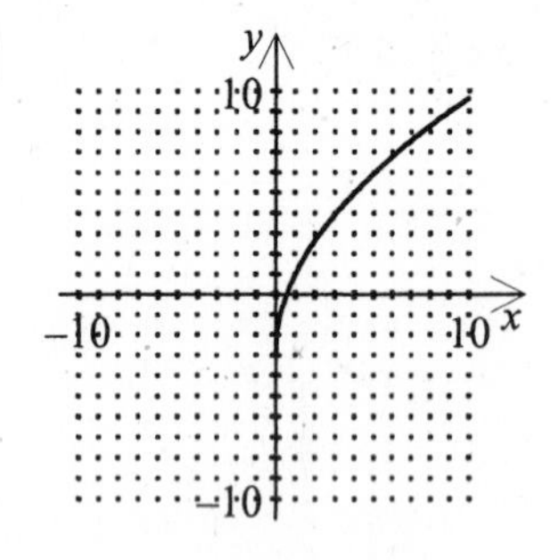

(B)

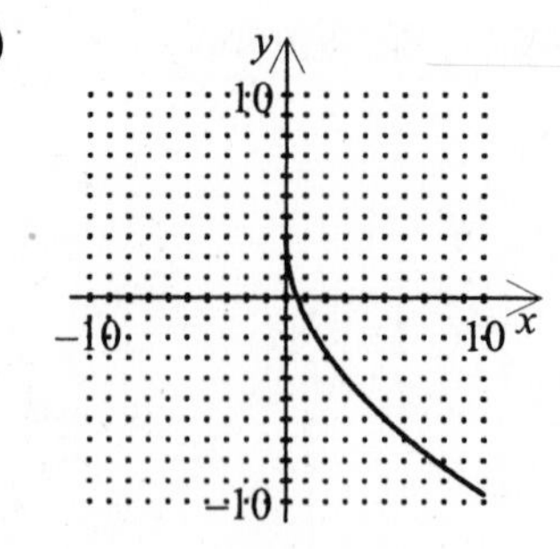

(C)

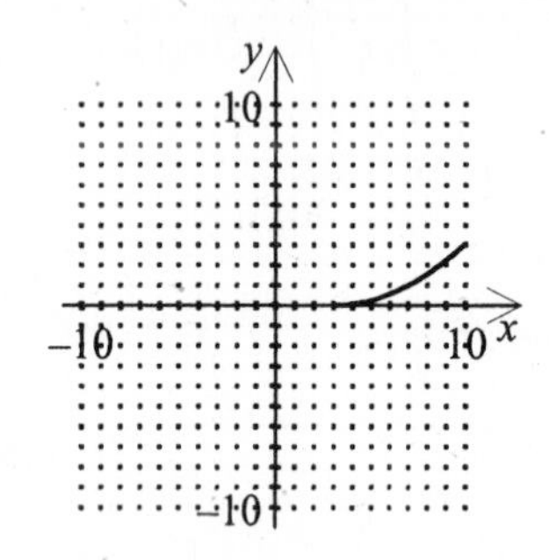

(D)

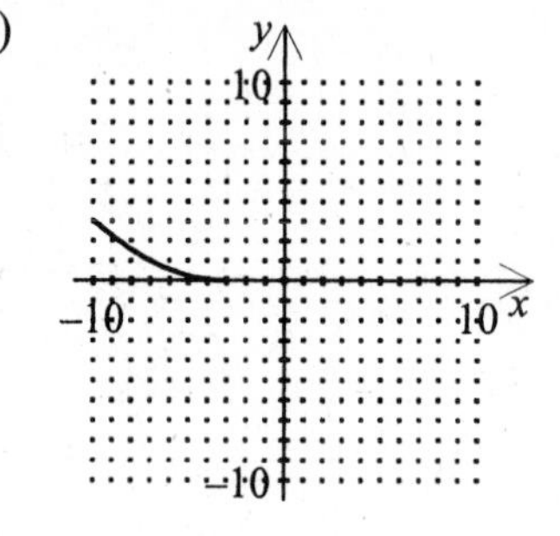

3. Sketch the graph of $y = |x - 2| + 2$

4. Complete the table, then sketch the graph of the equation by plotting points.
$5x + y = -3$

x	-2	-1	0	1	2
y					

Objective 2: Find x- and y-intercepts of graphs of equations

5. Identify the x-intercept and the y-intercept.

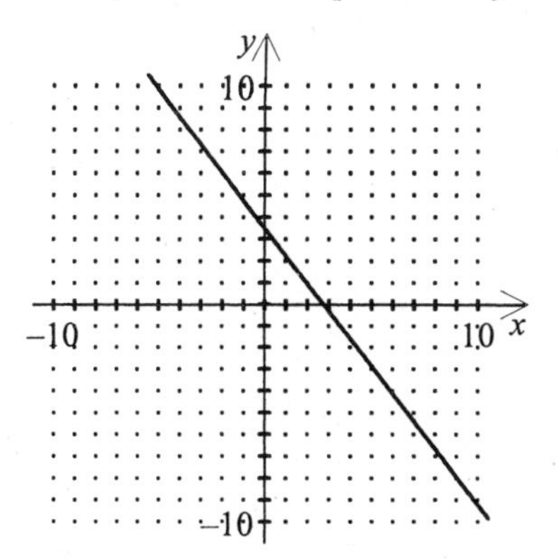

(A) $x\text{-intercept} = \frac{11}{4}$, $y\text{-intercept} = \frac{5}{3}$

(B) $x\text{-intercept} = \frac{11}{4}$, $y\text{-intercept} = \frac{7}{2}$

(C) $x\text{-intercept} = 4$, $y\text{-intercept} = \frac{5}{3}$

(D) $x\text{-intercept} = 4$, $y\text{-intercept} = \frac{7}{2}$

6. Identify the x- and y-intercepts.

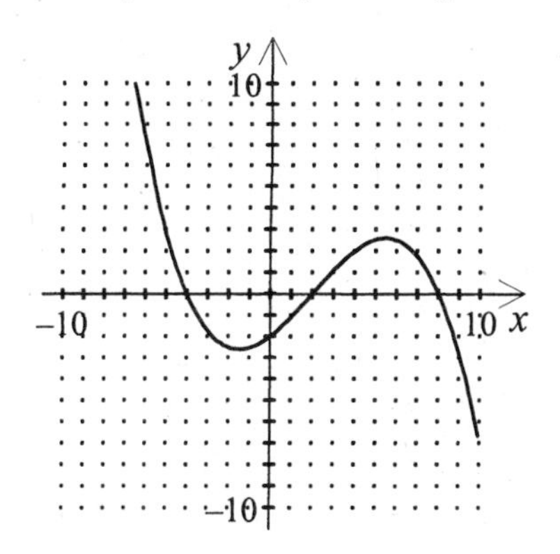

(A) x-intercept: $(0,\ 2)$
y-intercepts: $(-4,\ 0)$, $(2,\ 0)$, $(8,\ 0)$

(B) x-intercepts: $(-4,\ 0)$, $(2,\ 0)$, $(-8,\ 0)$
y-intercept: $(0,\ 2)$

(C) x-intercept: $(0,\ -2)$
y-intercepts: $(-4,\ 0)$, $(2,\ 0)$, $(8,\ 0)$

(D) x-intercepts: $(-4,\ 0)$, $(2,\ 0)$, $(8,\ 0)$
y-intercept: $(0,\ -2)$

7. Find the x- and y-intercepts of the graph of the equation $y = -6x + x^2$.

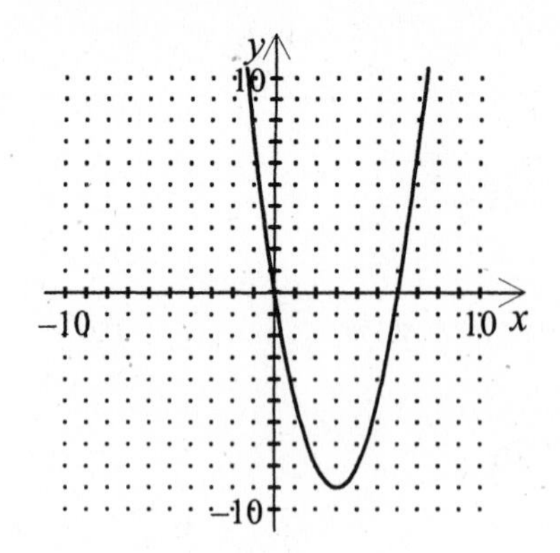

8. Identify the x- and y-intercepts and the intercept points.

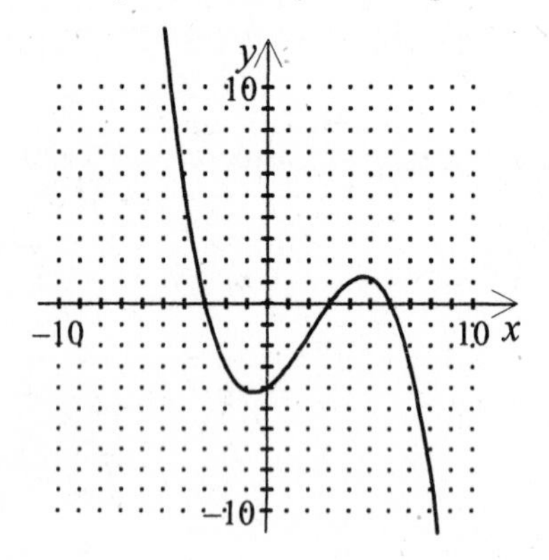

Objective 3: Use symmetry to sketch graphs of equations

9.

Assume that the graph above is a portion of a complete graph with y-axis symmetry. Which graph below is the complete graph?

(A)

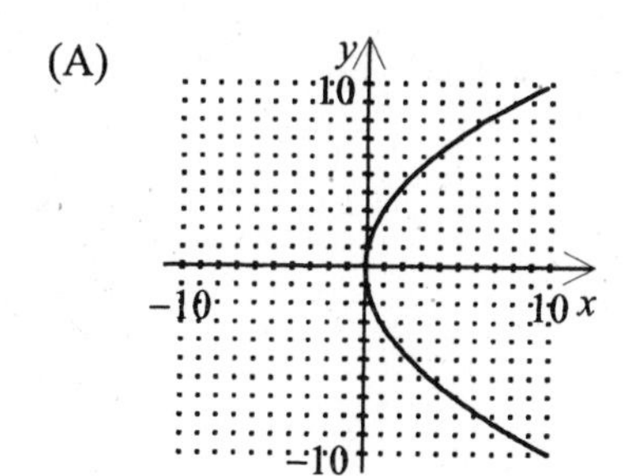

(B)

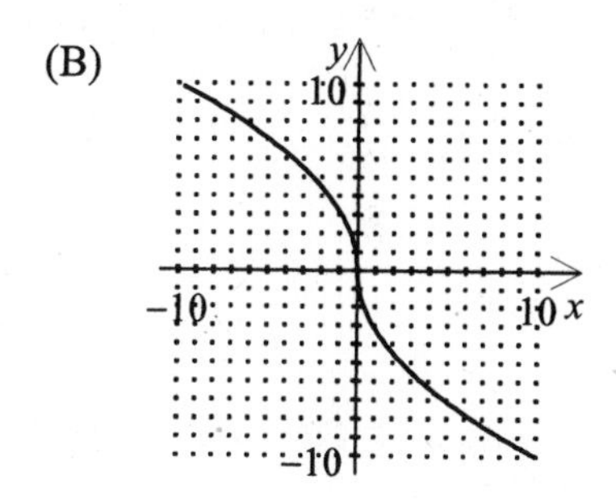

(C)

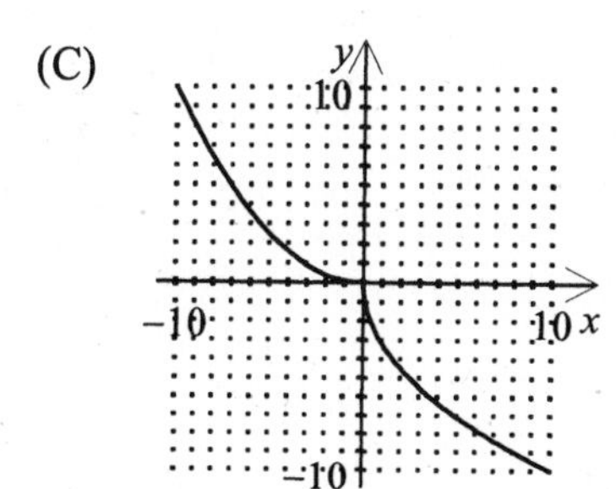

(D)

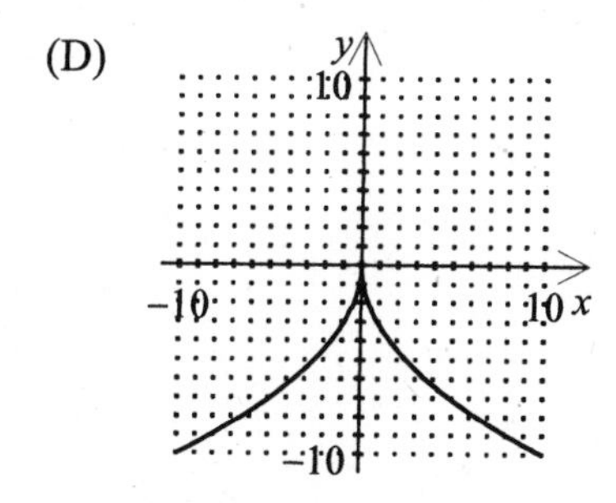

(9.)

10. Identify the equation whose graph is symmetric with respect to the the origin.

(A) $25x^2 + 4y^2 = 100$ (B) $y = x^3 - 4$ (C) $x = |y - 4|$ (D) $y^3 = 25x^4 - 5$

11. The graph below is a portion of a complete graph. Sketch the complete graph assuming it is symmetric with respect to the x-axis.

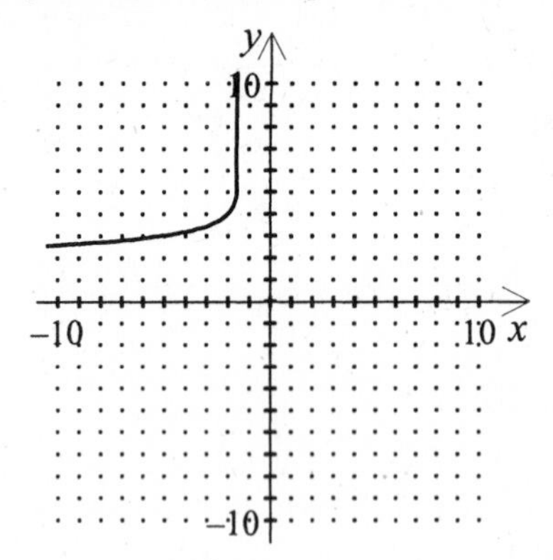

12. Determine whether the graph of $y = 5x^2 - 5$ is symmetric with respect to the x-axis, y-axis, and/or origin.

Objective 4: Find equations and sketch graphs of circles

13. Which equation does this graph represent?

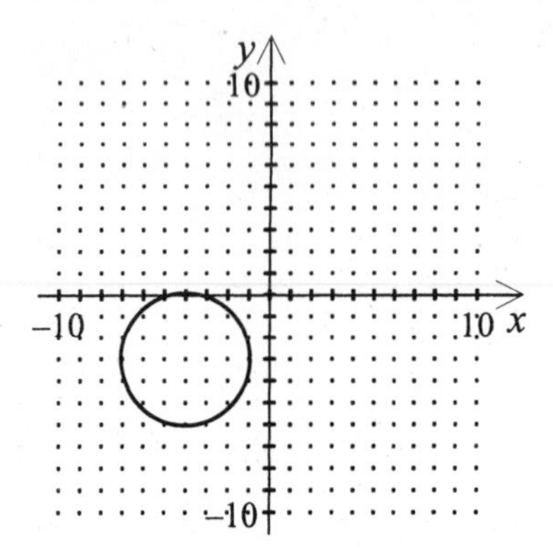

(A) $(x-4)^2 + (y-3)^2 = 3$ (B) $(x+3)^2 + (y-3)^2 = 9$

(C) $(x-3)^2 + (y+4)^2 = 3$ (D) $(x+4)^2 + (y+3)^2 = 9$

14. Find the center and radius of the circle with the given equation.
$(x+8)^2 + (y+4)^2 = 144$

(A) $(-8, -4)$; 12 (B) $(-4, -8)$; 144 (C) $(-4, 8)$; 12 (D) $(8, 4)$; 12

15. Sketch a graph of the equation $(x-1)^2+(y+3)^2=9$.

16. Find the equation in standard form of the circle where $C(-2,\ -5)$ and $D(4,\ 7)$ are endpoints of a diameter.

Objective 5: Use graphs of equations in solving real-life problems

17. The area of a rectangle with a perimeter of 240 units is given by
$A=x(120-x)$
where x represents the width of the rectangle. A farmer has available 240 yards of fencing and wishes to enclose a rectangular area. Use a graphing utility to graph the equation for the area of the rectangle and find the value of x that gives the largest area.

(A) 65 yd (B) 60 yd (C) 55 yd (D) 70 yd

18. A model for the demand for motors is
$d=-2p^2+108p-120$
where d is the number of motors a manufacturer can sell at a price of p dollars each. Use a graphing utility to graph the equation. Then find the price that results in the maximum demand for motors.

(A) \$27 (B) \$30 (C) \$54 (D) None of these

19. You own a silk-screening business that prints designs on T-shirts. The model for the average cost per T-shirt is
$\overline{A}=\frac{4.25x+175}{x}$
where x is the number of shirts in the production run, \$175 is the one time charge for creating the design and purchasing the supplies, and \$4.25 is the cost of each plain T-shirt. Sketch a graph of the equation and find the average cost per T-shirt for a production run of 300 shirts.

20. The height of a diver jumping from a diving platform is about
$h=-16.1t^2+12t+31.1$
where h is the height of the diver in feet above the water and t is the time measured in seconds, when diving from a platform about 31.1 feet above the water with an initial upward velocity of 12 ft/sec.
(a) Sketch a graph the equation from $t=0.0$ to $t=2.0$.
(b) After how many seconds is the diver's height above the water 30 feet? Round your answer to the nearest tenth of a second.
(c) After how many seconds is the diver's height above the water 39 feet? Round your answer to the nearest tenth of a second.

Section 1.2: Linear Equations in Two Variables

Objective 1: Use slope to graph linear equations in two variables

21. Identify the graph of the line that has the given slope and y-intercept.
$m = -1$; y-intercept: $(0, -7)$

(A)

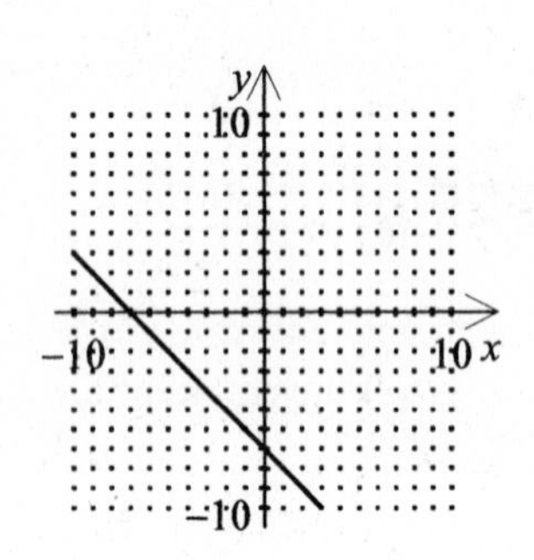

(B)

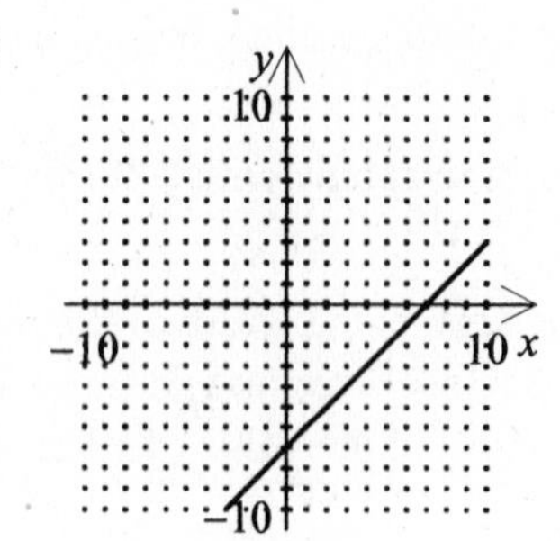

(C)

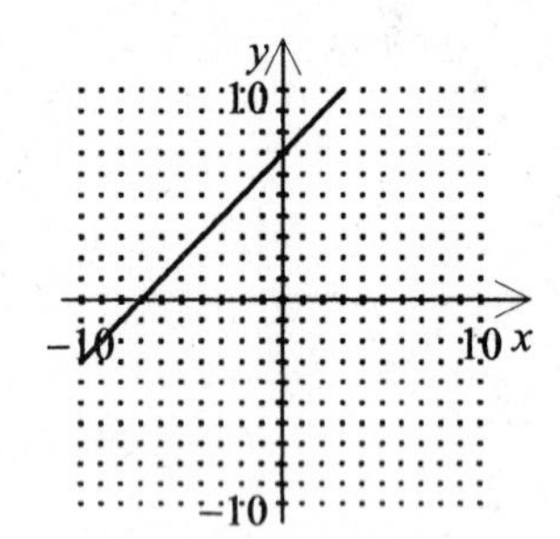

(D)

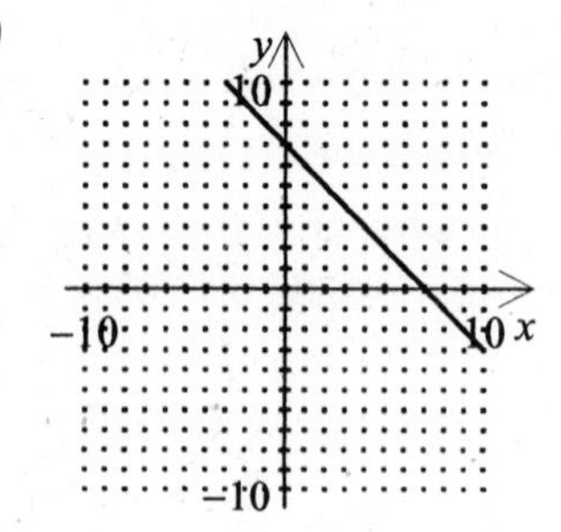

22. Identify the graph of the linear equation.

$-3x = 6$

(A)

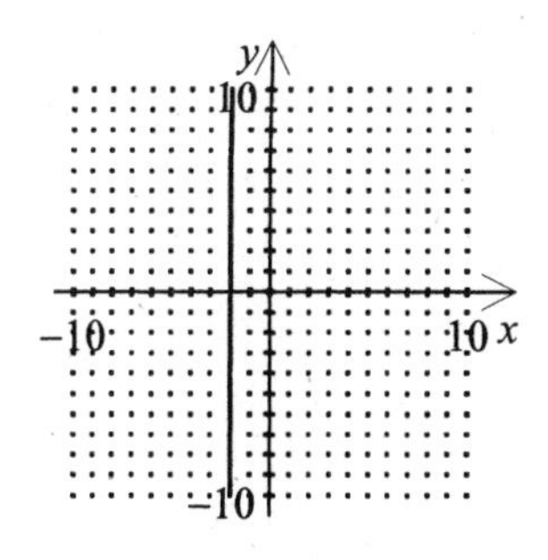

(B)

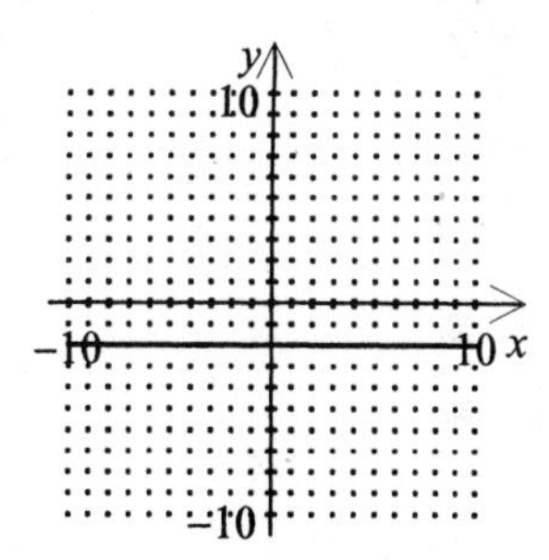

(C)

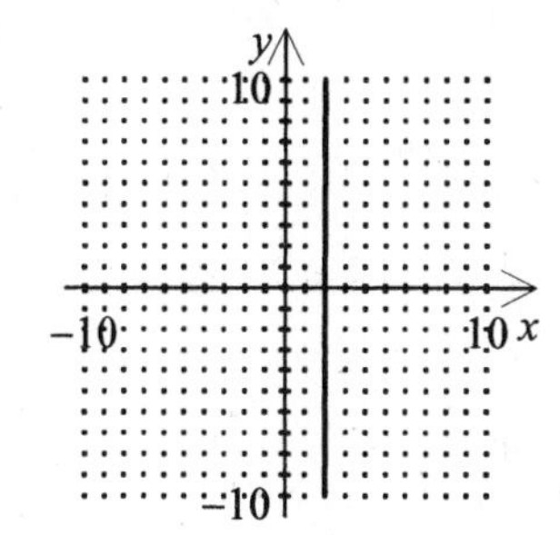

(D)

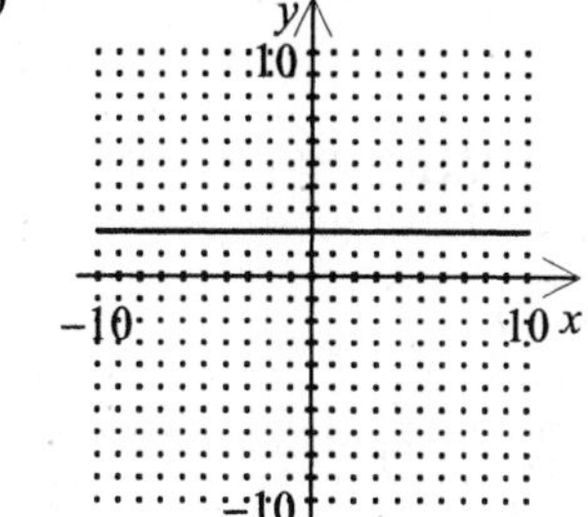

23. Use the point on the line and the slope of the line to find three additional points through which the line passes. (There are many correct answers.)

$(4, -7), \; m = 1$

24. Sketch the graph of the line through the given point with the indicated slope.

Point $(2, -4)$ with slope $\frac{1}{3}$

Objective 2: Find slopes of lines

25. Find the slope of the line passing through the pair of points.

$(-1, 9), \; (9, 9)$

(A) $\frac{9}{5}$ (B) 0 (C) $\frac{5}{4}$ (D) Undefined

26. Estimate the slope of the line.

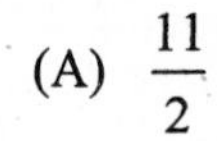
(A) $\frac{11}{2}$ (B) $-\frac{2}{11}$ (C) $-\frac{11}{2}$ (D) $\frac{2}{11}$

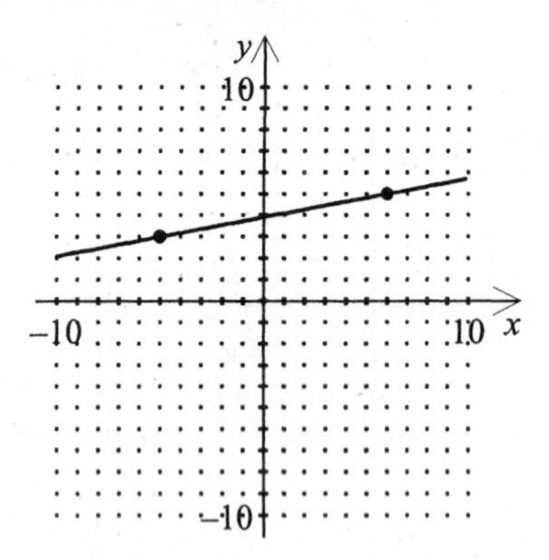

27. Find the slope of the line passing through the pair of points.
$(-5,\ 2),\ (-2,\ -7)$

28. Plot the points and find the slope of the line passing through the pair of points.
$\left(\frac{4}{5},\ -\frac{10}{3}\right),\ \left(-\frac{4}{7},\ -\frac{8}{9}\right)$

Objective 3: Write linear equations in two variables

29. Find the slope-intercept form of the equation of the line that passes through the given point and has the indicated slope.
$(-2,\ -1),\ m=2$

(A) $y=2x+3$ (B) $y=-2x-3$ (C) $y=2x-3$ (D) $y=-2x+3$

Find the slope-intercept form of the equation of the line passing through the points.

30. $(-6,\ -2),\ (-1,\ -4)$

(A) $y=\frac{5}{2}x+\frac{22}{5}$ (B) $y=\frac{5}{2}x+\frac{5}{22}$ (C) $y=-\frac{2}{5}x-\frac{22}{5}$ (D) $y=\frac{2}{5}x-\frac{5}{22}$

31. $(2,\ -6),\ (2,\ 5)$

32. Use the *intercept form* to find the equation of the line with the given intercepts. The intercept form of the equation of a line with intercepts $(a,\ 0)$ and $(0,\ b)$ is $\frac{x}{a}+\frac{y}{b}=1,\ a \neq 0,\ b \neq 0$.

x-intercept: $(-2,\ 0)$

y-intercept: $(0,\ 6)$

Objective 4: Use slope to identify parallel and perpendicular lines

33. Find the slope-intercept form of the equation of the line through the point $(-8,\ 2)$, parallel to the line $-7x-4y=-6$.

(A) $y=-\frac{4}{7}x+\frac{1}{12}$ (B) $y=-\frac{7}{4}x+\frac{1}{12}$ (C) $y=-\frac{7}{4}x-12$ (D) $y=\frac{7}{4}x-12$

34. Determine whether the graphs of the equations are parallel, perpendicular, or neither.

$7x+7y=0$

$-7x+7y=2$

(A) Parallel (B) Perpendicular (C) Neither

35. Find the slope-intercept form of the equation of the line that passes through the point $(-5,\ 1)$ and is perpendicular to the line $-7x-8y=-3$.

36. Determine whether the lines L_1 and L_2 passing through the pair of points are parallel, perpendicular, or neither.

$L_1:(-3,\ -10),\ (-9,\ -14)$

$L_2:(-8,\ 5),\ (-14,\ 1)$

Objective 5: Use linear equations in two variables to model and solve real-life problems

37. In 1980, the average price of a home in Lake County was $92,000. By 1987, the average price of a home was $127,000. Which of the following is a linear model for the price *P* of a home in Lake County, in terms of the year *t*? Let $t=0$ correspond to 1980.

(A) $P=5000t+92{,}000$ (B) $P=127{,}000-35{,}000t$

(C) $P=35{,}000t+92{,}000$ (D) $P=127{,}000-5000t$

38. The graph for a stable that charges a \$20 flat fee plus \$10 per hour for horseback riding is shown below. How will the graph change if the stable changes its charges to a flat fee of \$40 plus \$15 per hour?

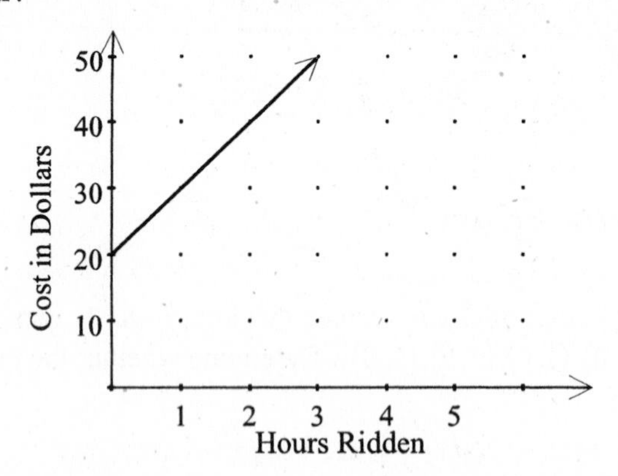

(A) The slope will be 15 and the y-intercept will be 20.

(B) The slope will be 15 and the y-intercept will be 40.

(C) The slope will be 10 and the y-intercept will be 40.

(D) The slope will be 40 and the y-intercept will be 15.

39. The graph below shows the weights of several kittens at various ages.

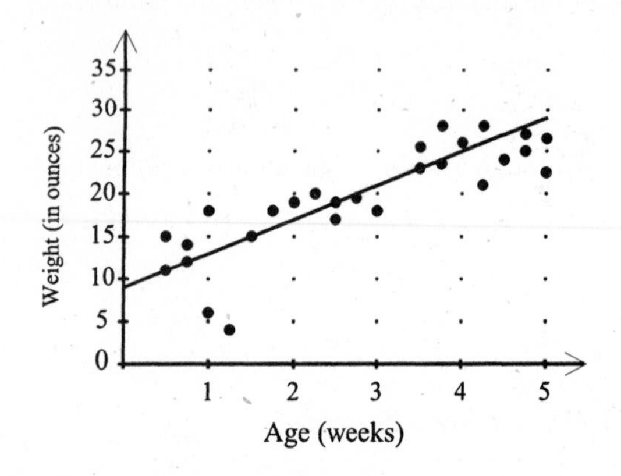

(a) What is an equation of the line of best fit?
(b) What is the meaning of the slope and y-intercept of the graph of the line of best fit?

40. Write an equation for the following situation in slope-intercept form.

The gas tank in a truck holds 13 gallons. The truck uses $\frac{1}{4}$ gallon per mile.

Section 1.3: Functions

Objective 1: Determine whether relations between two variables are functions

41. Determine which set of ordered pairs (x, y) represents y as a function of x.

(A) $\{(-4, 0), (0, -7), (-4, -6)\}$ (B) $\{(-4, 0), (-7, -6), (-7, -4), (-6, -7)\}$

(C) $\{-4, 0, -7, -6\}$ (D) $\{(-4, 0), (0, -4), (-6, -6)\}$

42. Identify the model of the relation $\{(-1, 0), (1, 0), (5, 0), (7, 0)\}$. Determine whether the relation is a function.

(A)

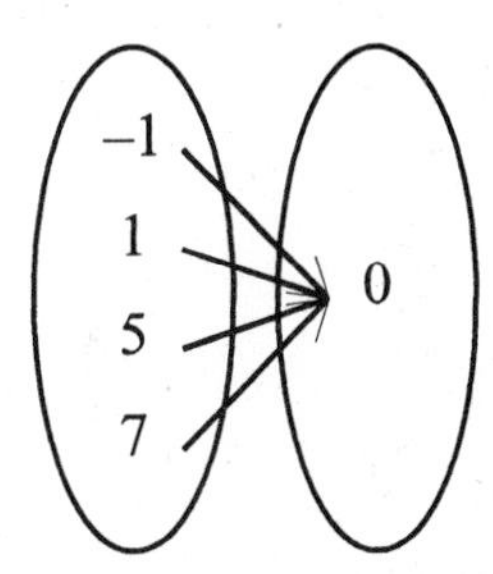

Not a function

(B)

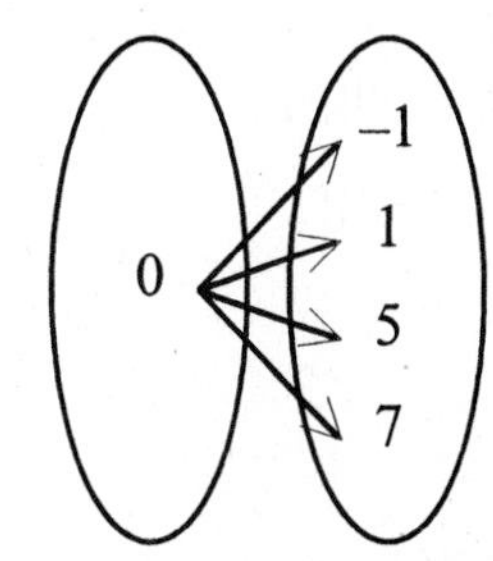

Function

(C)

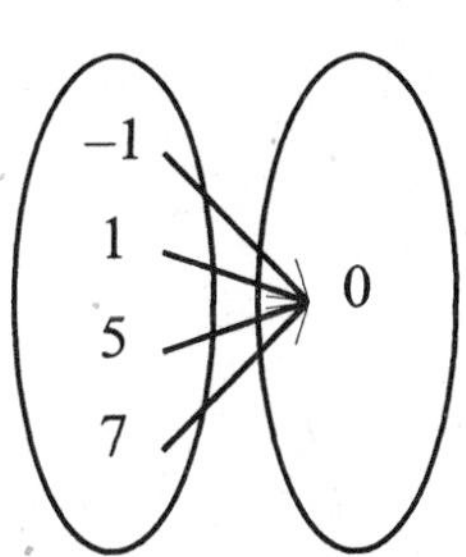

Function

(D)

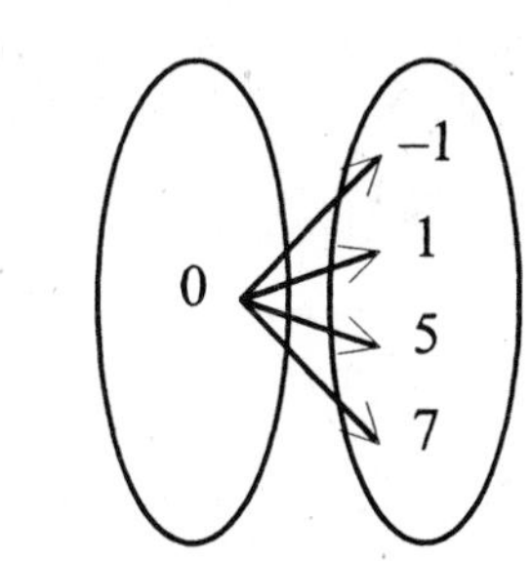

Not a function

43. Does the table describe y as a function of x? Explain your reasoning.

Input x	1	2	3	4
Output y	1	4	9	16

44. Determine whether the equation represents y as a function of x.

$x = y$

Objective 2: Use function notation and evaluate functions

Evaluate the function at the specified value(s) of the independent variable and simplify.

45. $f(x)=2x^2-\sqrt{-8x}$; $f(-7)$ (A) 90.517 (B) 42 (C) 119.166 (D) 117.799

46. $g(x)=\dfrac{x^2-5}{2x}$; $g(n-2)$

(A) $\dfrac{n^2-5}{2n}-2$ (B) $\dfrac{n^2-4n+-1}{2n-4}$ (C) $\dfrac{n^2-7}{2n-4}$ (D) $\dfrac{n^2-4n+-1}{2n-2}$

47. $f(x)=\begin{cases}\dfrac{3}{5}x & \text{if } x<-3\\ -3+8x & \text{if } x\geq -3\end{cases}$

(a) $f(-3)$ (b) $f(0)$
(c) $f(4)$ (d) $f(-2.8)$

48. $f(x)=\dfrac{3}{7}x-5$; $f(-28)$, $f(21)$

Objective 3: Find domains of functions

Find the domain of the function.

49. $h(x)=\dfrac{7x}{x(x^2-36)}$

(A) All real numbers $x\neq\pm 36,\ 0$ (B) All real numbers $x\neq\pm 6$

(C) All real numbers $x\neq\pm 6,\ 0$ (D) All real numbers $x\neq 6$

50. $f(x)=\sqrt{-5x+10}$ (A) $x\leq 2$ (B) $x\geq 2$ (C) $x\leq 0$ (D) $x\geq 0$

51. $f(x)=\dfrac{\sqrt{x+6}}{x^2-4x+3}$

Find the domain of the function.

52. $f(x) = \sqrt[7]{x^4}$

Objective 4: Use functions to model and solve real-life problems

53. A slug travels at a rate of 0.3 meter per minute. Let t be the number of minutes spent traveling.
(a) Find the distance d traveled by the slug as a function of t.
(b) Find the distance traveled by the slug in 2 minutes.
(c) Identify the graph illustrating the distance traveled by the slug versus time.

(A) (a) $d(t) = t + 0.3$
(b) 0.3 m
(c)

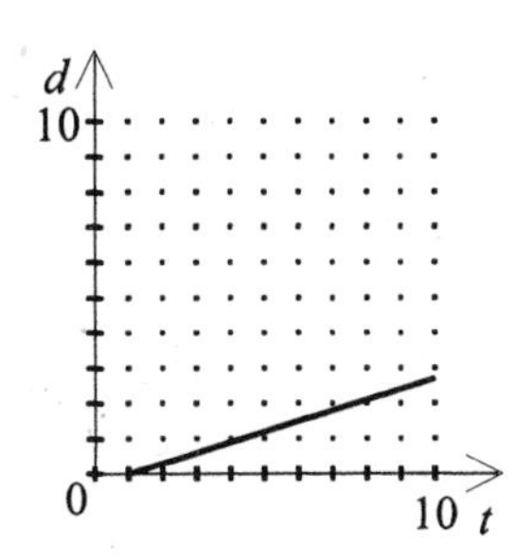

(B) (a) $d(t) = 0.3t + 1$
(b) 0.5 m
(c)

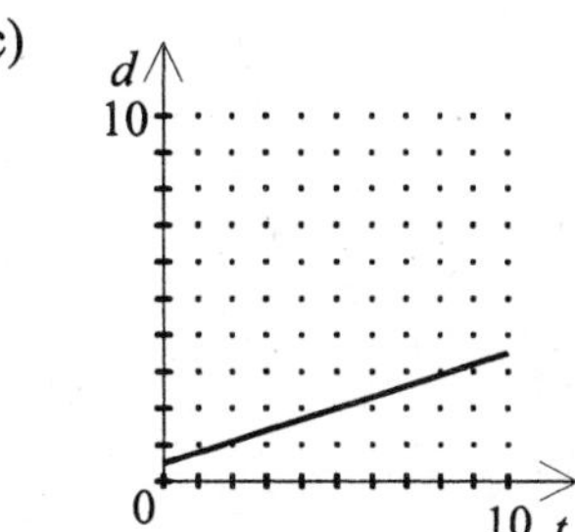

(C) (a) $d(t) = 0.3t$
(b) 0.6 m
(c)

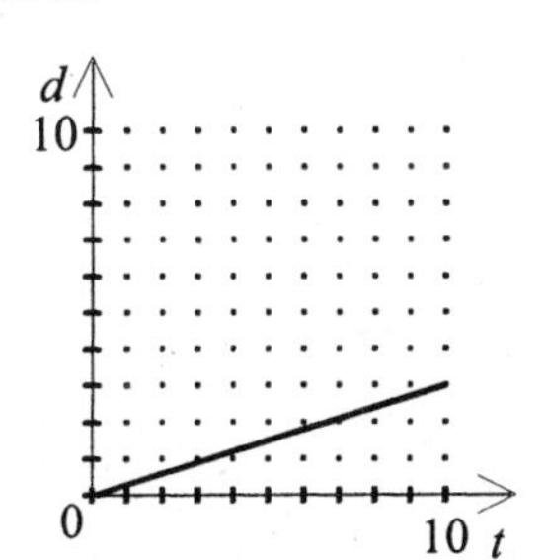

(D) (a) $d(t) = 0.3d$
(b) 0.8 m
(c)

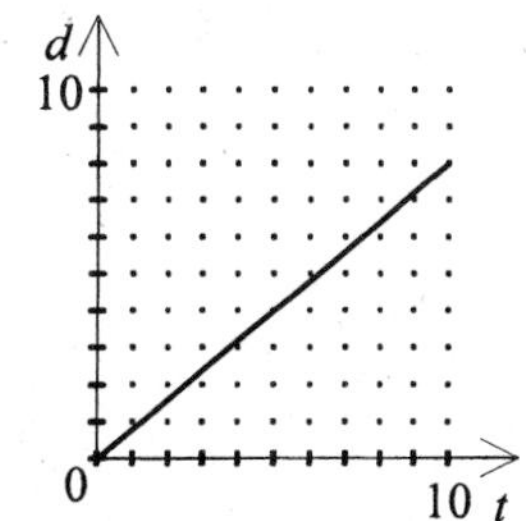

54. A publishing company estimates that the average cost (in dollars) for one copy of a new scenic calendar it plans to produce can be approximated by the function

$$C(x) = \frac{1.75x + 550}{x}$$

where x is the number of calendars printed. Find the average cost per calendar when the company prints 10,000 calendars.

(A) \$1.81 (B) \$56.75 (C) \$7.25 (D) \$0.06

55. Since 1993, Raul Perez has owned a bookstore called Basically Books. The number of books B, in thousands, that Basically Books has sold each year can be modeled by the function

$B(t) = t^2 + 27t + 300$

where t is the number of years after 1993. Using this model, estimate the number of books sold in 1995.

56. A tomato plant in Jeremy's garden was 11 centimeters tall when it was first planted. Since then, it has grown approximately 1.2 centimeters per day. Write a function expressing the tomato plant's height H in terms of the number of days d since it was planted.

Section 1.4: Analyzing Graphs of Functions

Objective 1: Use the Vertical Line Test for functions

57. Use the Vertical Line Test to determine which graph represents y as a function of x.

(A)

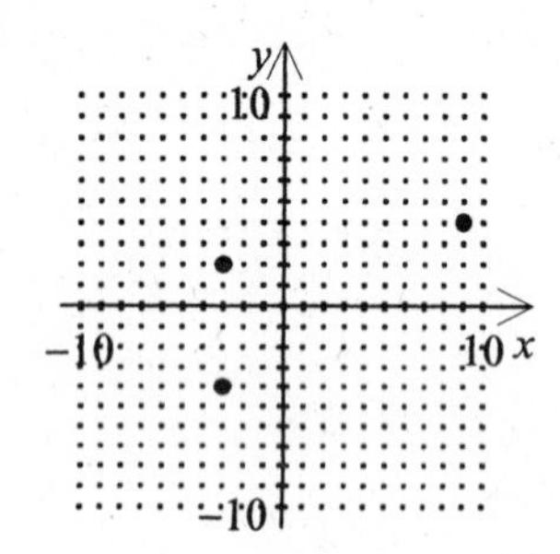

(B)

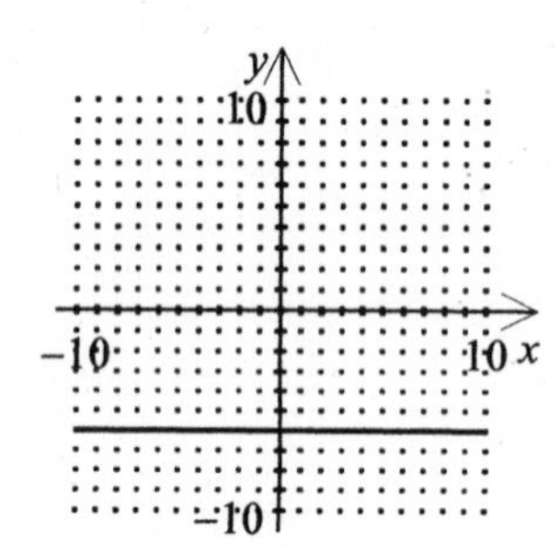

(C)

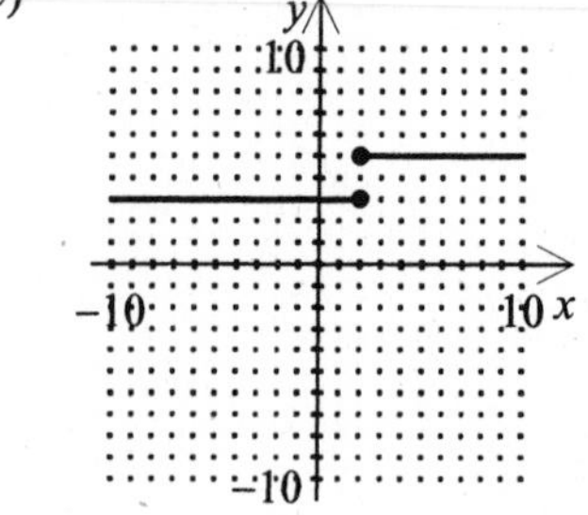

(D)

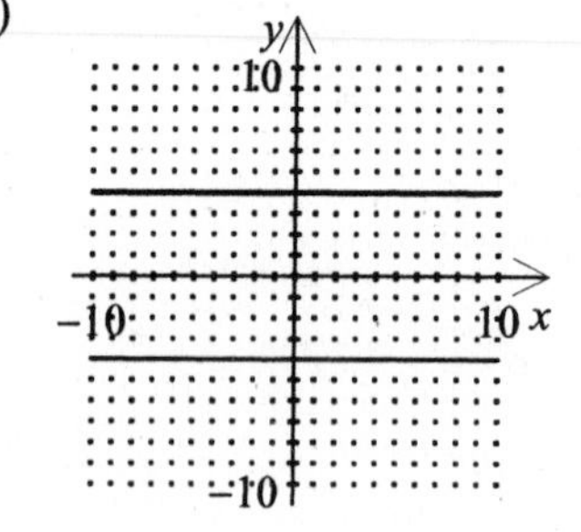

58. Use the Vertical Line Test to determine which graph represents y as a function of x.

(A)

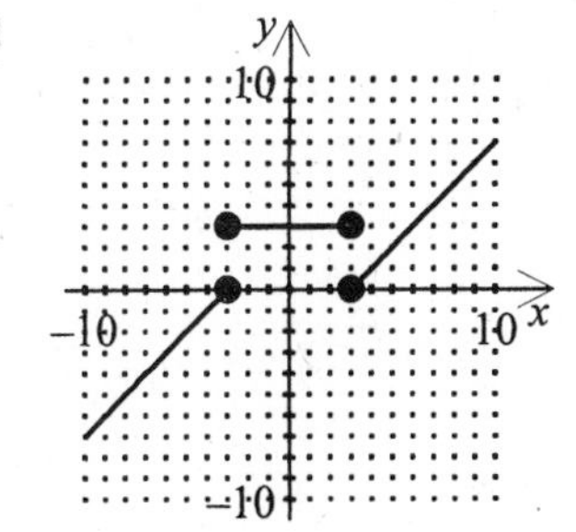

(B)

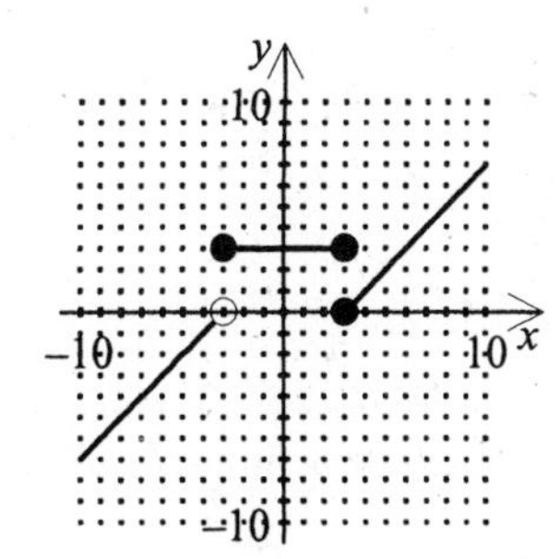

(C)

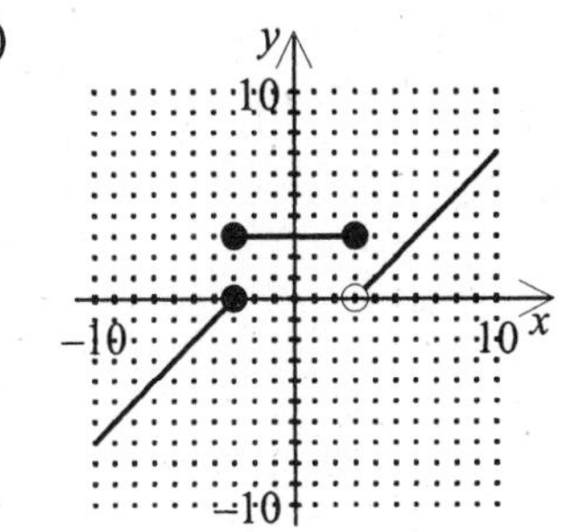

(D)

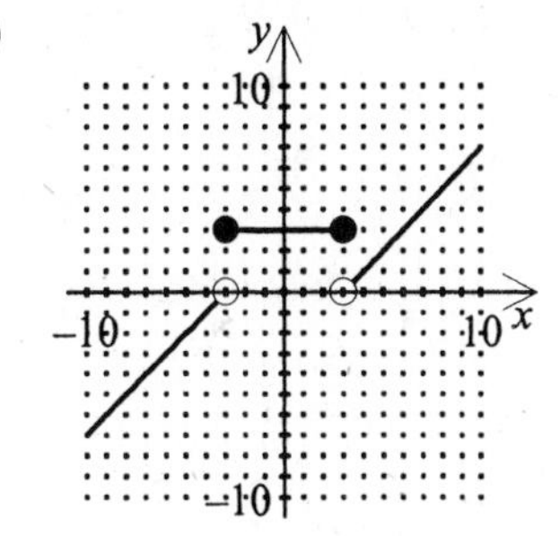

Use the Vertical Line Test to determine if the graph represents y as a function of x.

59.

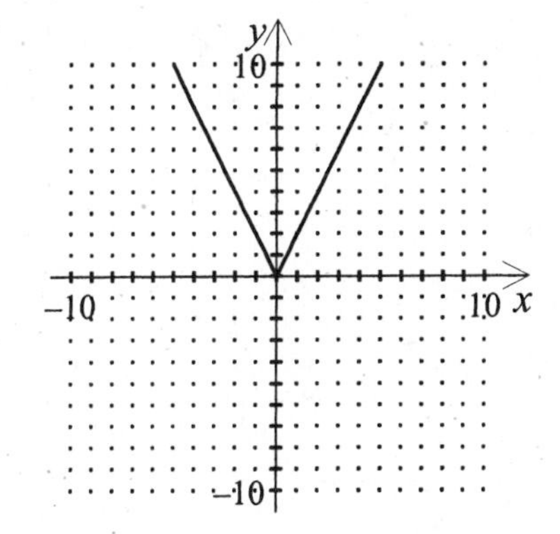

Use the Vertical Line Test to determine if the graph represents y as a function of x.

60.

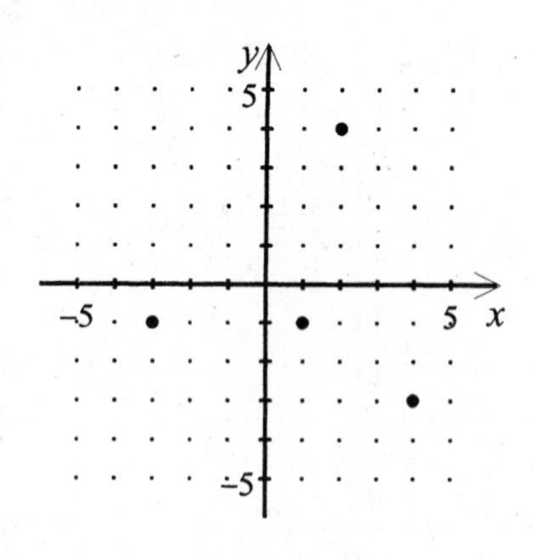

Objective 2: Find the zeros of functions

Find the zeros of the function by factoring or algebraically.

61. $f(x) = -x^2 + 3x + 6$

(A) –1.37, 4.37 (B) –0.44, 3.44 (C) –3.44, 0.44 (D) –4.37, 1.37

62. $f(x) = \dfrac{x^3 - 9x}{(2x+3)^2}$ (A) $0, \pm 3$ (B) $0, -\frac{3}{2}, \pm 3$ (C) $9, \pm 3$ (D) $0, \pm\frac{3}{2}$

63. $f(x) = x - 4\sqrt{x} - 32$

64. $f(x) = -2x^4 + 20x^3 - 42x^2$

Objective 3: Determine intervals on which functions are increasing or decreasing

65. Use a graphing utility to graph the function and visually determine the intervals on which the function is increasing, decreasing, or constant.

$$f(x)=\begin{cases}\sqrt{-x}-2, & x\le -1\\ -x^2, & x>-1\end{cases}$$

(A)

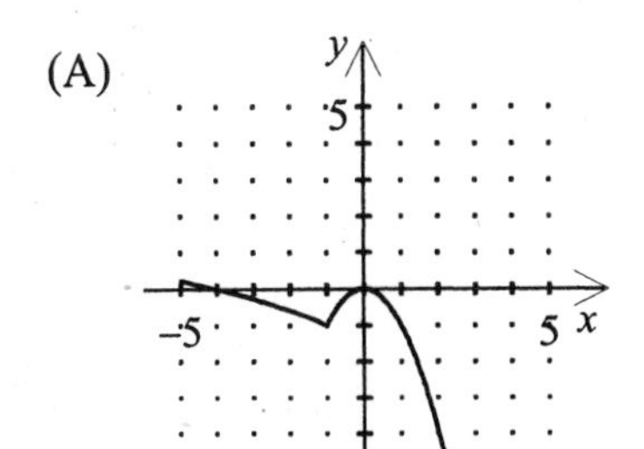

Decreasing on $(-\infty,\ -1)$ and $(0,\ \infty)$

Increasing on $(-1,\ 0)$

(B)

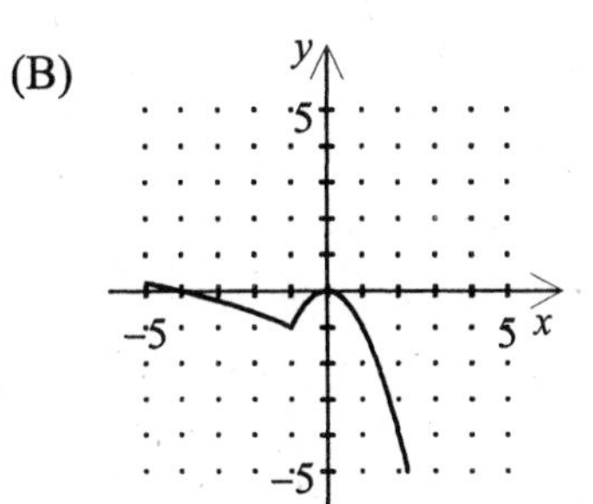

Increasing on $(-\infty,\ -1)$ and $(0,\ \infty)$

Decreasing on $(-1,\ 0)$

(C)

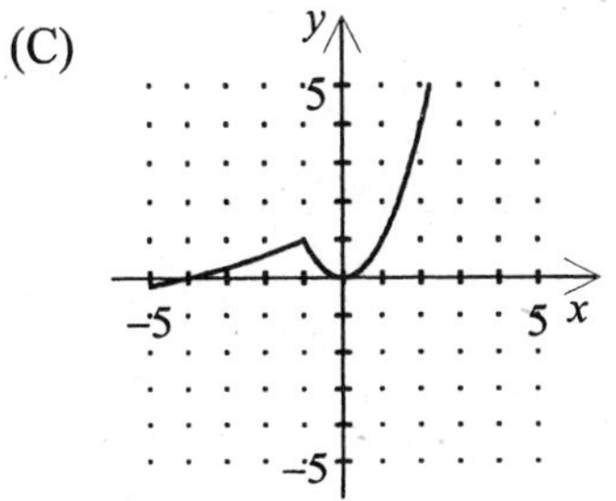

Increasing on $(-\infty,\ -1)$ and $(0,\ \infty)$

Decreasing on $(-1,\ 0)$

(D)

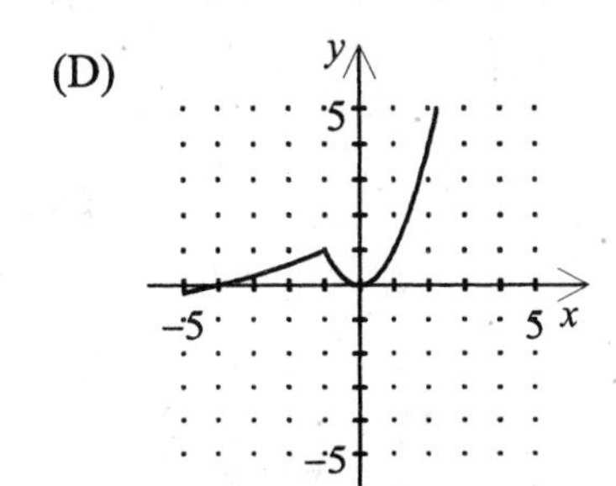

Decreasing on $(-\infty,\ -1)$ and $(0,\ \infty)$

Increasing on $(-1,\ 0)$

66. Determine the intervals on which the function is increasing, decreasing, or constant.

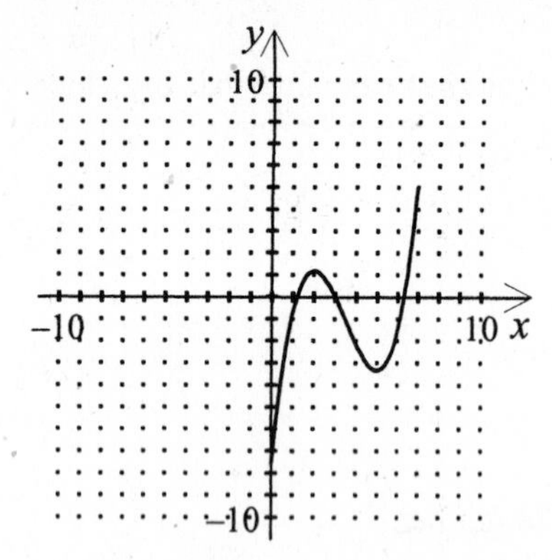

(A) Increasing on $(0, 5)$; Decreasing on $(5, 7)$

(B) Decreasing on $(0, 5)$; Increasing on $(5, 7)$

(C) Decreasing on $(0, 2)$ and $(5, 7)$; Increasing on $(2, 5)$

(D) Increasing on $(0, 2)$ and $(5, 7)$; Decreasing on $(2, 5)$

67. Use a graphing utility to graph the function and visually determine the intervals on which the function is increasing, decreasing, or constant.

$$f(x)=\begin{cases} \frac{5}{3}x, & x \le 3 \\ 5, & 3 < x \le 7 \\ -\frac{3}{7}x+8, & x > 7 \end{cases}$$

68. (a) Use a graphing utility to graph the function and visually determine the intervals on which the function is increasing, decreasing, or constant.
(b) Make a table of values to verify whether the function is increasing, decreasing, or constant.
$f(x)=\sqrt[4]{x-1}$

Objective 4: Identify even and odd functions

69. Identify the function that is *neither* odd nor even.

(A) $f(x)=4x^2-|x^2|+8$

(B) $f(x)=x^3-7x$

(C) $f(x)=x^4-7x+2+|x-2|$

(D) $f(x)=x^5-7x^3+4x$

70. Identify the function that is odd.

(A) $g(x) = -2x^4 + 2x^2 - 4$

(B) $p(x) = \dfrac{-2x^3}{-2x^2 + 2}$

(C) $h(x) = |-2x + 2| - 2$

(D) $f(x) = -4x^3 + 4x^2 - 4$

71. Determine whether the function is even, odd, or neither.
$f(x) = 3x^5 + 6x$

72. Use the graph to determine if the function is even, odd, or neither.

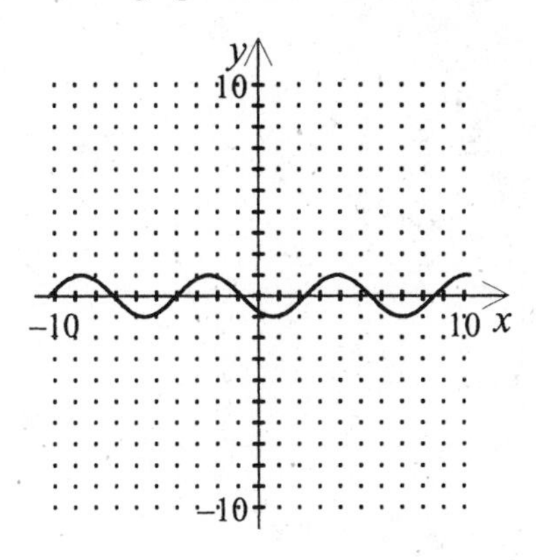

Section 1.5: A Library of Functions

Objective 1: Identify and graph linear and squaring functions

73. Identify the graph of the linear function.
$f(x) = -2x + 2$

(A)

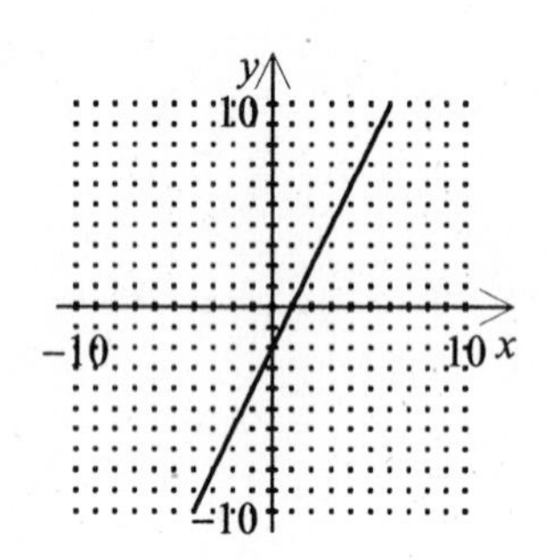

(B)

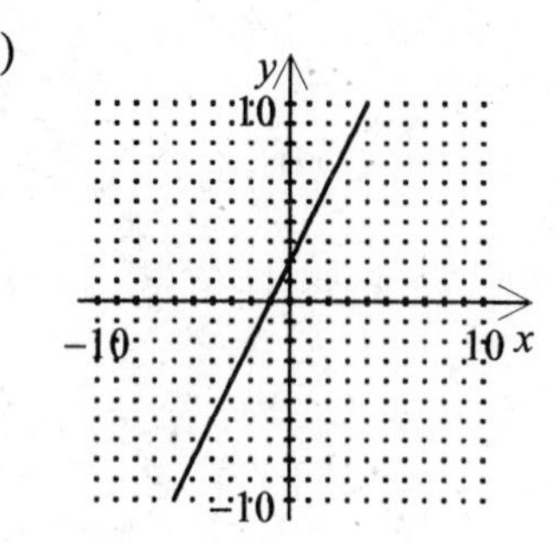

(C)

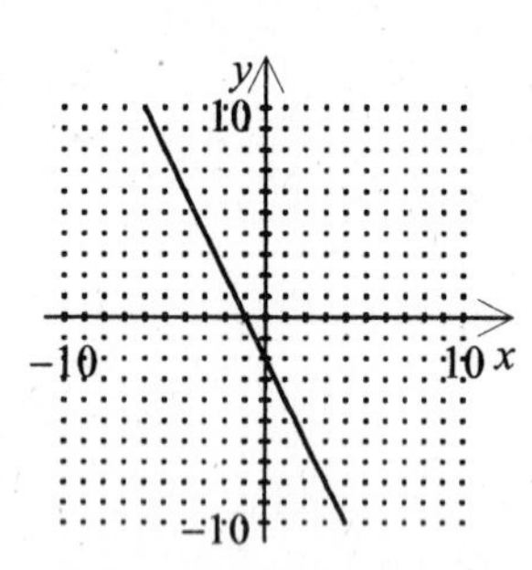

(D)

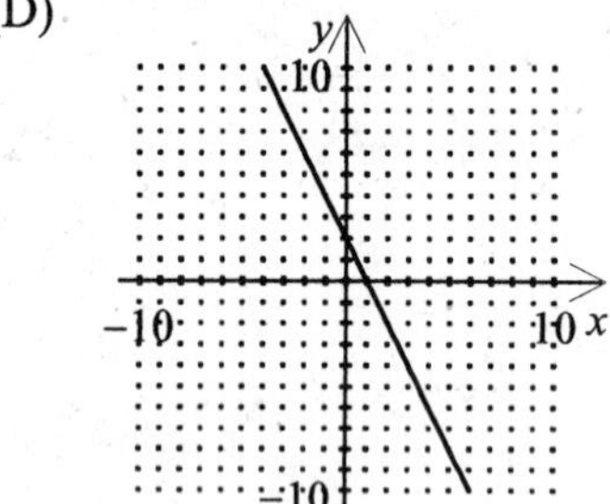

74. Identify the graph of the function and find the y-intercept.
$f(x) = 0.9x - 4$

(A)

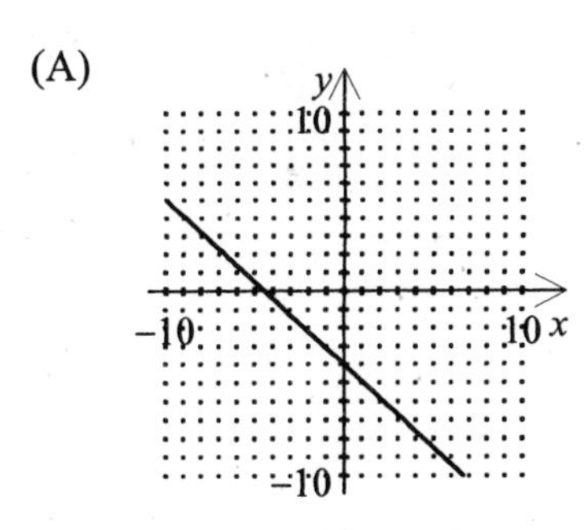

y-intercept: $(0, -4)$

(B)

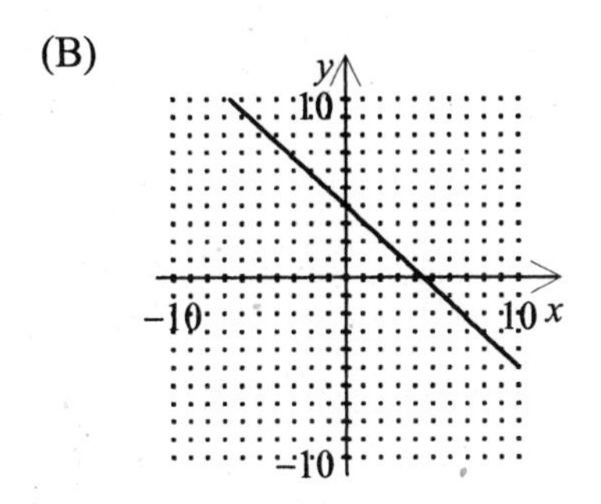

y-intercept: $(0, 4)$

(C)

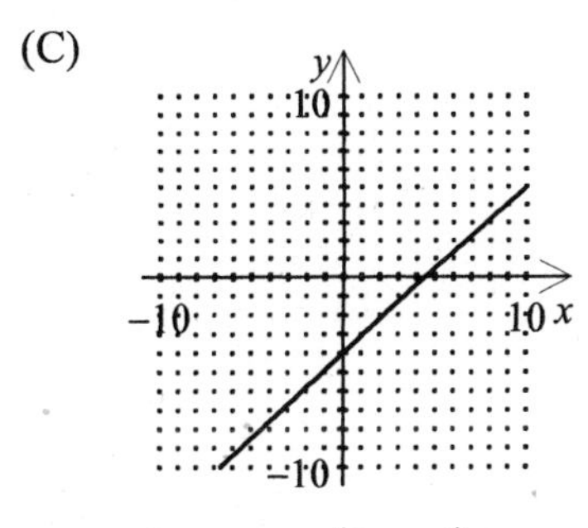

y-intercept: $(0, -4)$

(D)

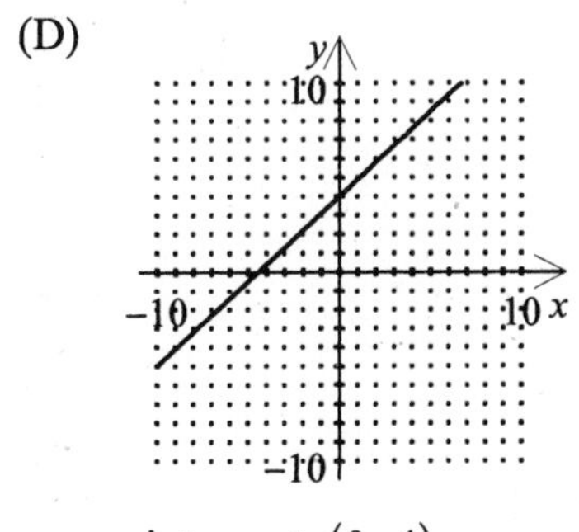

y-intercept: $(0, 4)$

(74.)

75. Sketch the graph of the linear function.

$$f(x) = -\frac{1}{3}x - 2$$

76. Write the linear function that has the indicated function values.

$f(4) = 5$, $f(6) = 9$

Objective 2: Identify and graph cubic, square root, and reciprocal functions

77. Which could be the graph of $y = 5x^3$?

(A)

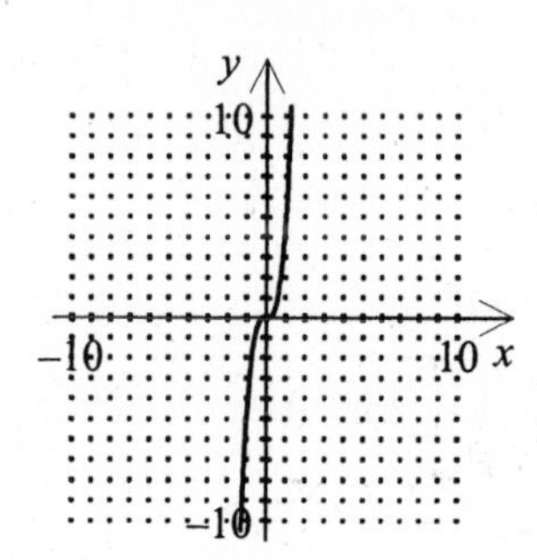

(B)

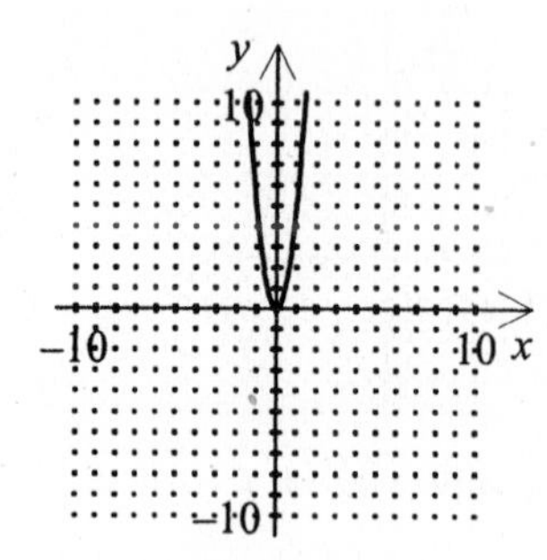

(C)

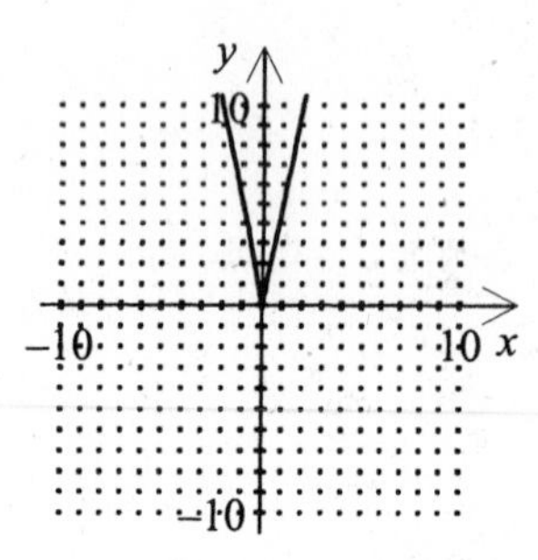

(D)

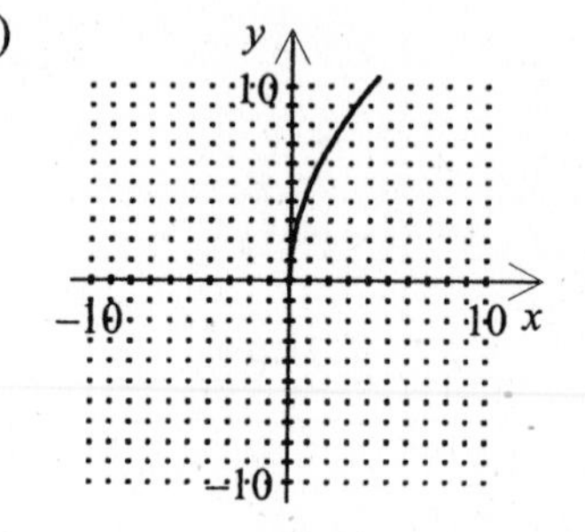

78. Which could be the graph of $y = \dfrac{1}{x-5}$?

(A)

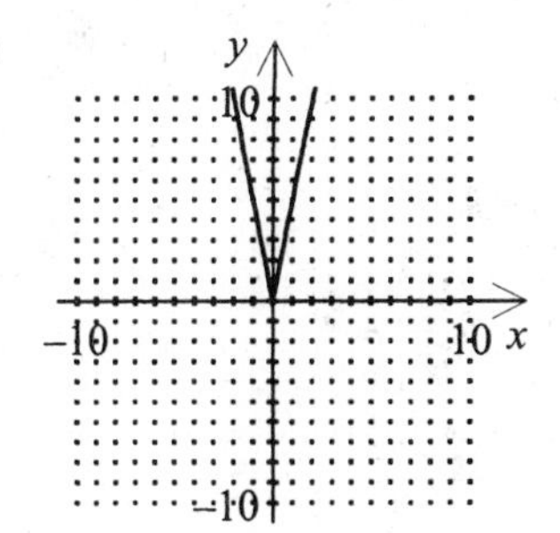

(B)

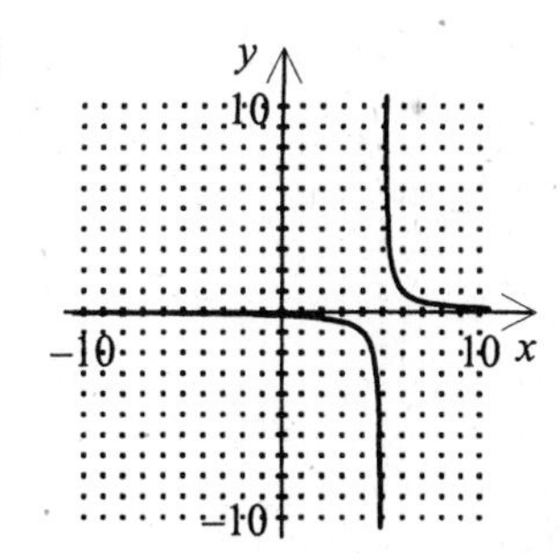

(C)

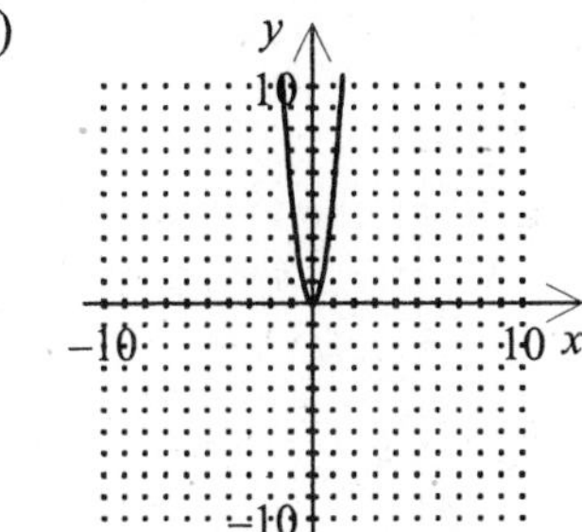

(D)

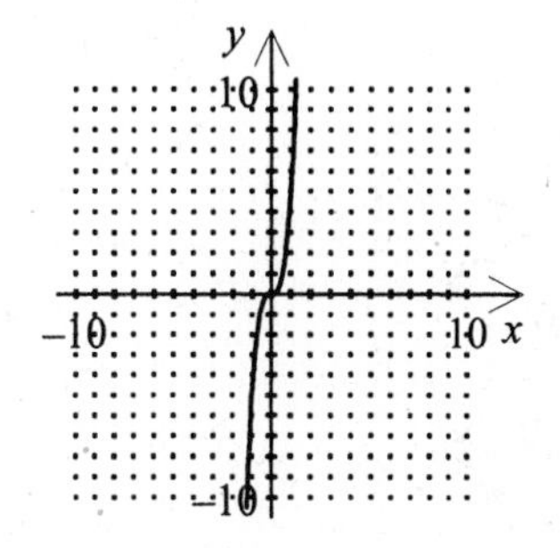

79. List an intercept of the graph of the cubic function.

80. What is the domain of the reciprocal function?

Objective 3: Identify and graph step and other piecewise-defined functions

81. Find the function that represents the graph.

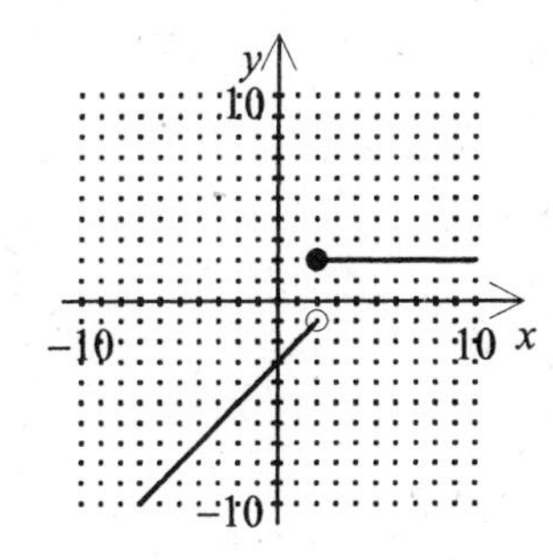

(A) $f(x) = \begin{cases} 2, & x \geq 2 \\ x-3, & x > 2 \end{cases}$

(B) $f(x) = \begin{cases} 2, & x \geq 2 \\ x-3, & x < 2 \end{cases}$

(C) $f(x) = \begin{cases} 2, & x \leq 2 \\ x-3, & x > 2 \end{cases}$

(D) $f(x) = \begin{cases} 2, & x \leq 2 \\ x-3, & x < 2 \end{cases}$

82. Find the graph of the function.

$$f(x)=\begin{cases}\sqrt{x}+3, & x\geq 0\\ -3x+1, & x<0\end{cases}$$

(A)

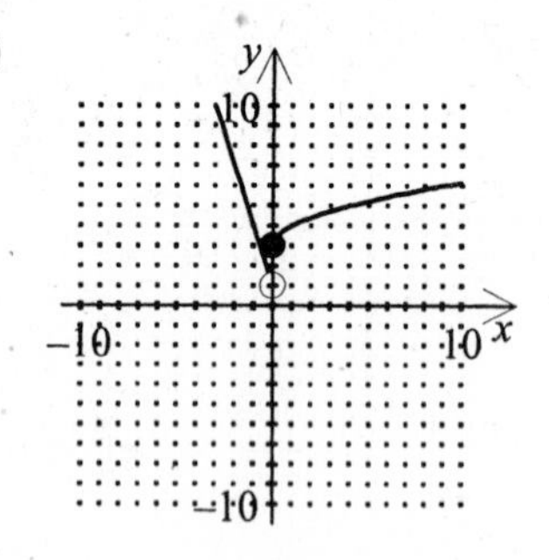

(B)

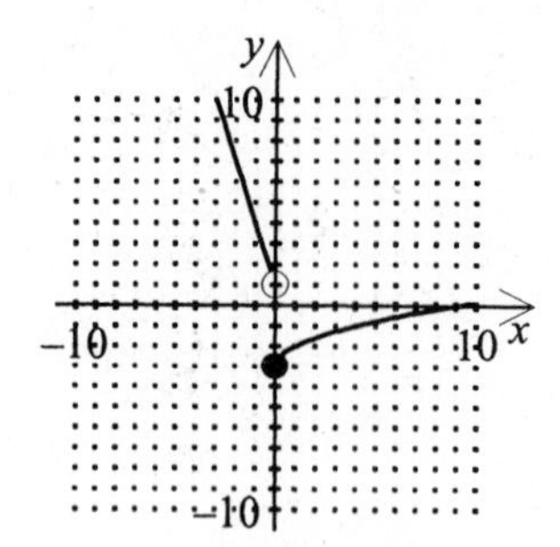

(C)

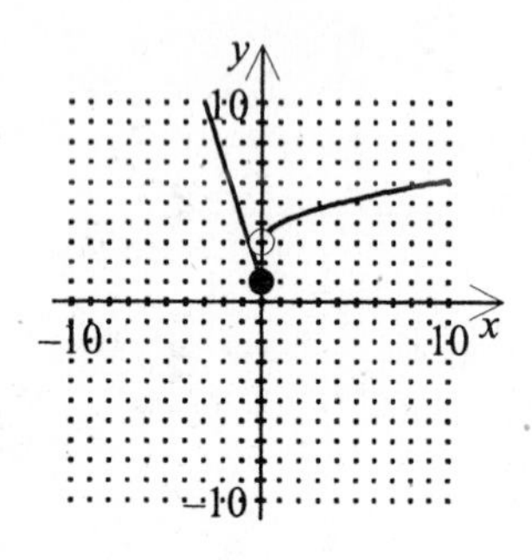

(D)

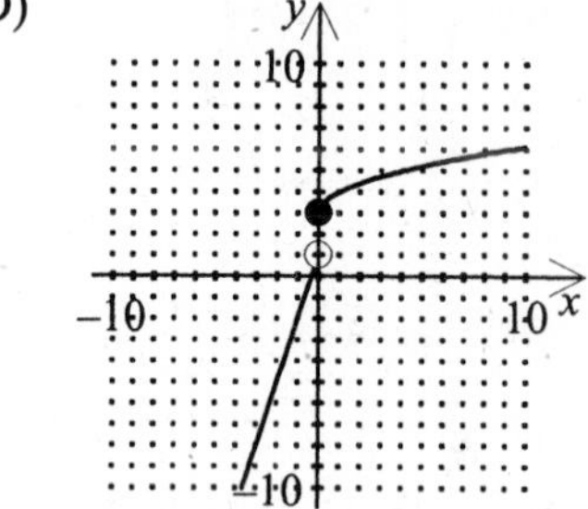

83. Graph the function.

$$f(x)=\begin{cases}2, & x<-2\\ x^2+4x+4, & x\geq -2\end{cases}$$

84. Sketch the graph of the function. (The symbol [[ ]] represents the greatest integer function.)

$f(x)=[[2x-4]]$

Objective 4: Recognize graphs of common functions

85. Identify the common function type shown in the graph.

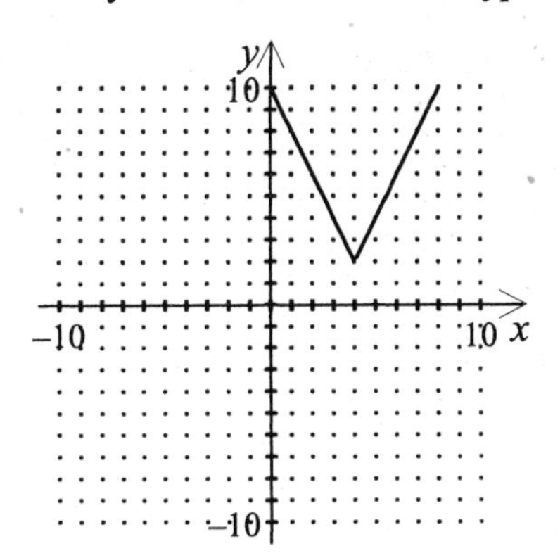

(A) Cubic Function

(B) Quadratic Function

(C) Absolute Value Function

(D) Identity Function

86. Find the equation for the common function shown in the graph.

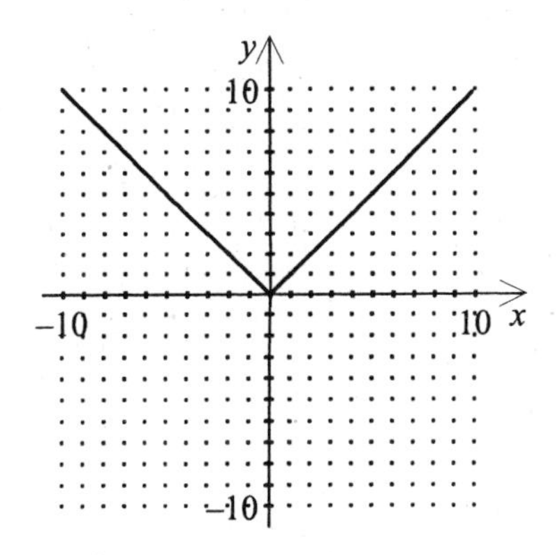

(A) $f(x)=\sqrt{x}$ (B) $f(x)=x^3$ (C) $f(x)=x$ (D) $f(x)=|x|$

87. Identify the common function type shown in the graph.

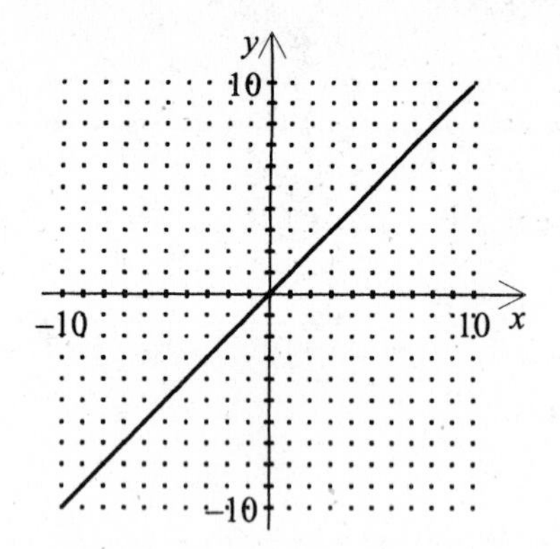

88. Identify the common function shown in the graph.

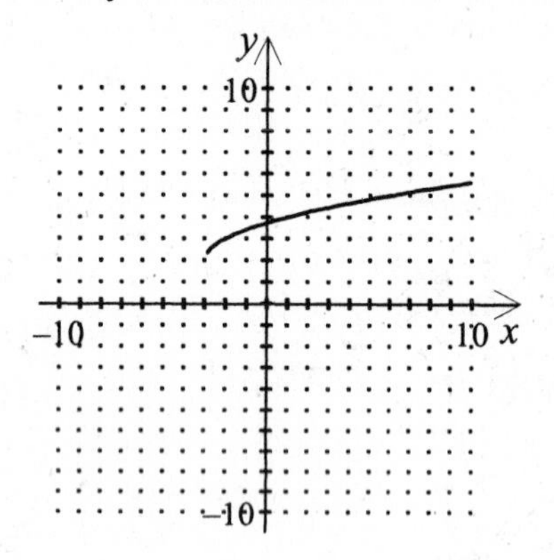

Section 1.6: Shifting, Reflecting, and Stretching Graphs

Objective 1: Use vertical and horizontal shifts to sketch graphs of functions

89. Find an equation and identify the graph of the function whose graph has the shape of $f(x) = x^2$ but has moved 4 units to the left and 3 units downward.

(A) $f(x)=(x+4)^2-3$

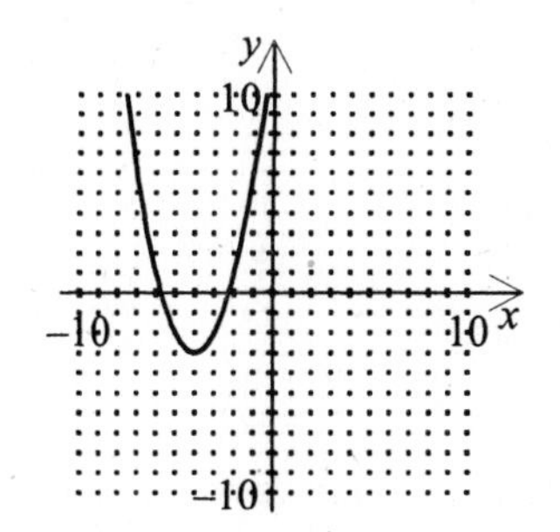

(B) $f(x)=(x+4)^2+3$

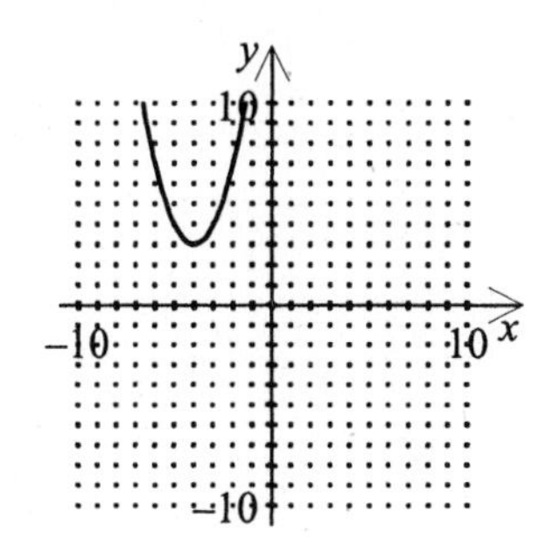

(C) $f(x)=(x-4)^2-3$

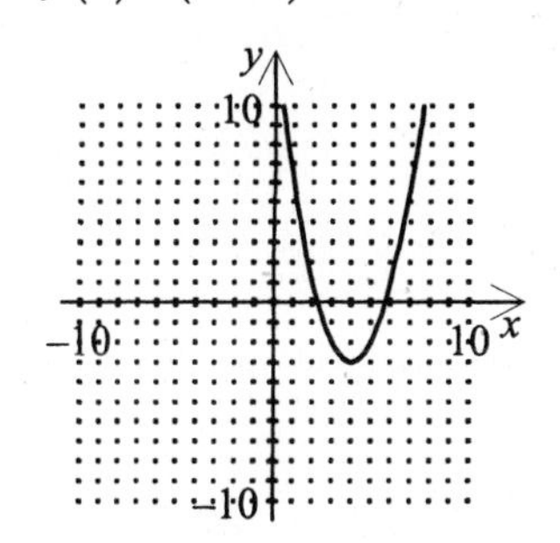

(D) $f(x)=(x-4)^2+3$

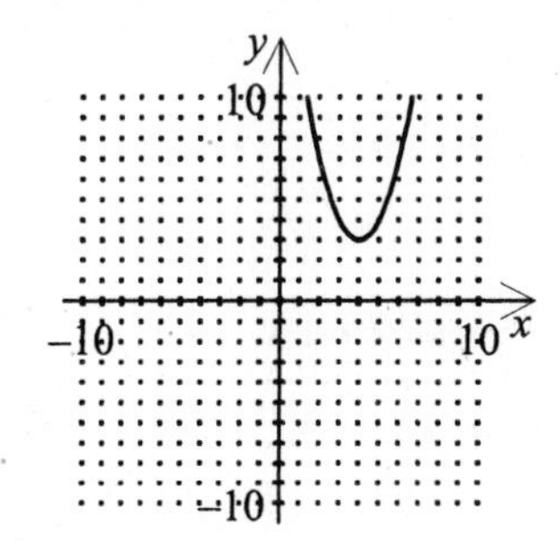

(89.)

90. Suppose $f(x) = |\,x\,|$ and $h(x) = |\,x + 2\,|$. Determine which of the following statements is true.

(A) The graph *h* is the graph *f* shifted horizontally right 2 units.

(B) The graph *h* is the graph *f* shifted vertically up 2 units.

(C) The graph *h* is the graph *f* shifted vertically down 2 units.

(D) The graph *h* is the graph *f* shifted horizontally left 2 units.

91. Given $f(x) = x^2$, consider the translation 3 units to the left and 2 units downward. Give the translated function and its graph.

92. Sketch the graphs of the following functions.
(a) $f(x) = x^3$
(b) $h(x) = x^3 + 3$
(c) $k(x) = (x - 3)^3$

Objective 2: Use reflections to sketch graphs of functions

93. Identify the function $g(x)$ that is the reflection of $f(x) = \dfrac{1}{x}$ in the x -axis.

(A) $g(x) = -\left(\dfrac{1}{x}\right)$ (B) $g(x) = x$ (C) $g(x) = \dfrac{-1}{-x}$ (D) $g(x) = -x$

94. The graph of the function $f(x) = |\,x\,|$ is shown below. Find the equation and the graph of the function $g(x)$ which is the reflection of $f(x)$ in the y -axis.

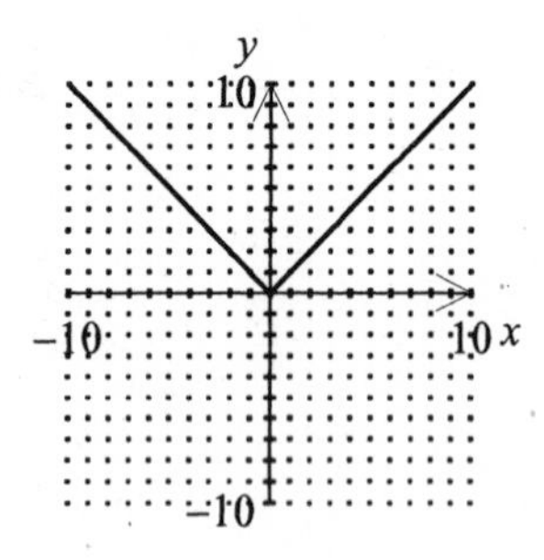

(A) $g(x) = |-x+1|$

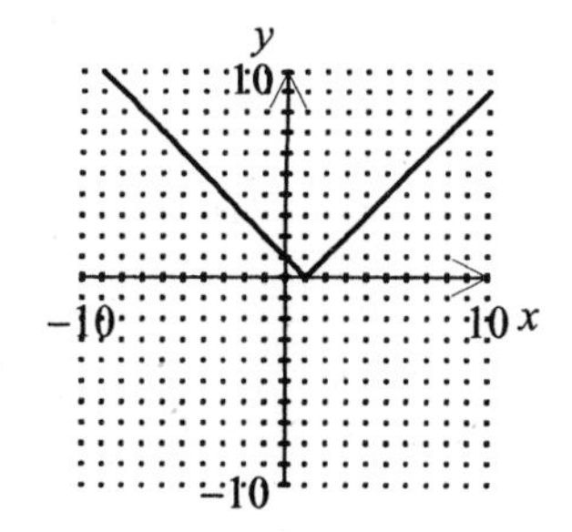

(B) $g(x) = -|x+1|$

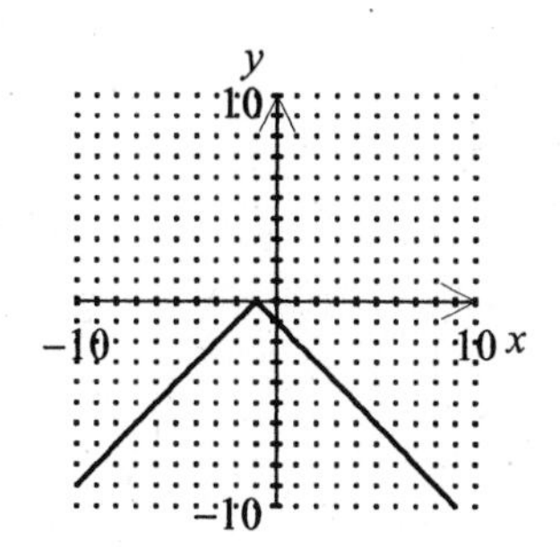

(C) $g(x) = -|x|$

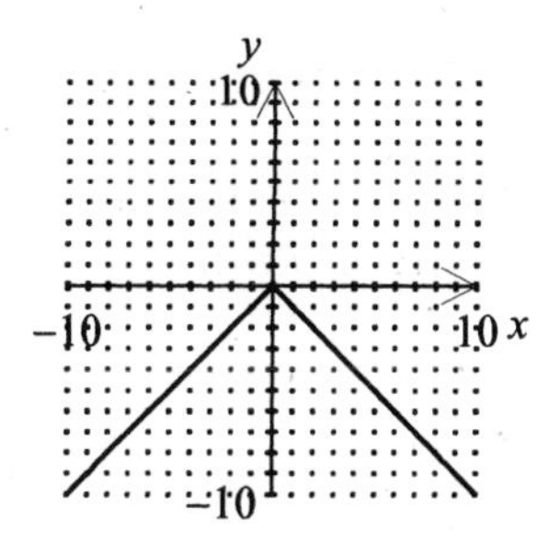

(D) $g(x) = |-x|$

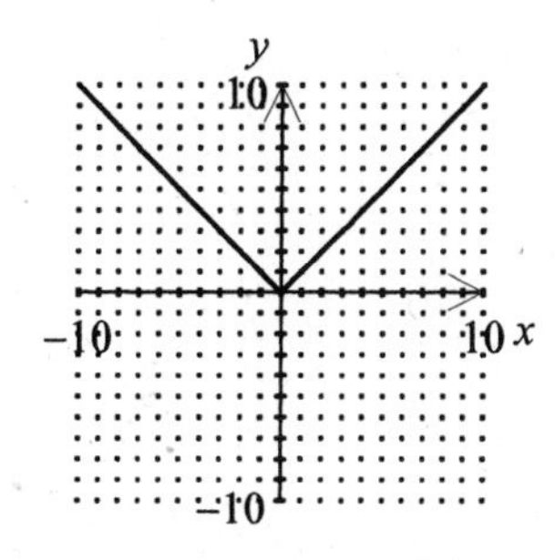

(94.)

95. Graph the pair of functions and identify the transformation. $f(x) = |x| - 2$, $g(x) = |-x| - 2$

96. Given below is the graph of a function $f(x)$. Graph $-f(x)$.

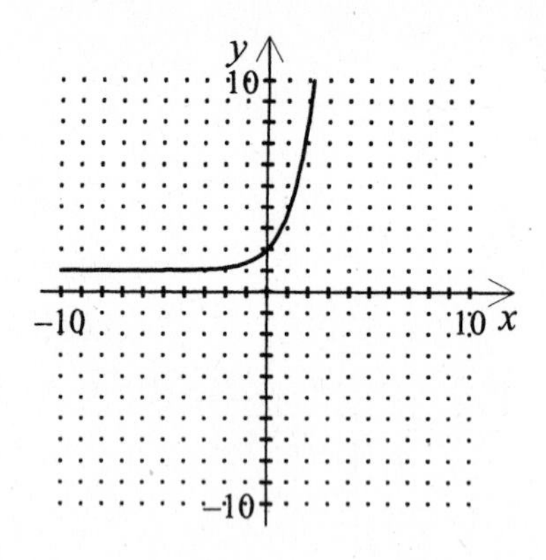

Objective 3: Use nonrigid transformations to sketch graphs of functions

97. The graph of the function $v(x)$ is given. Identify the graph of $3v(x)$.

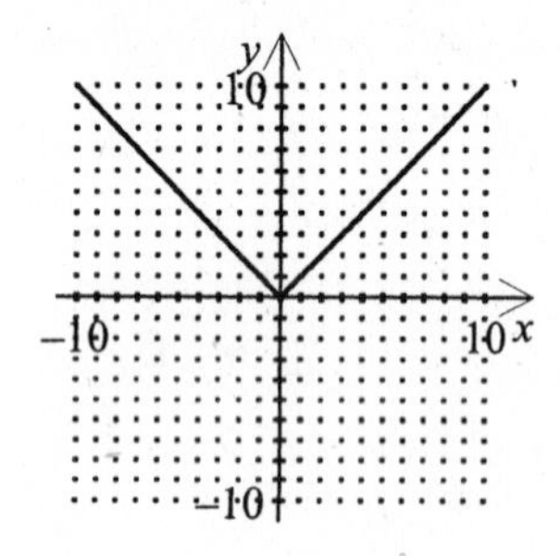

(A)

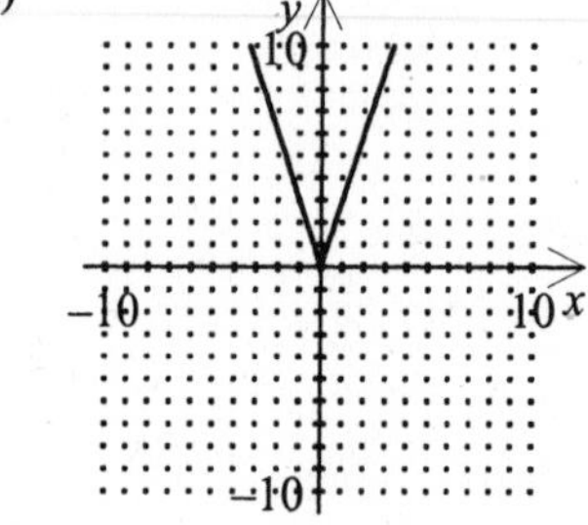

(B)

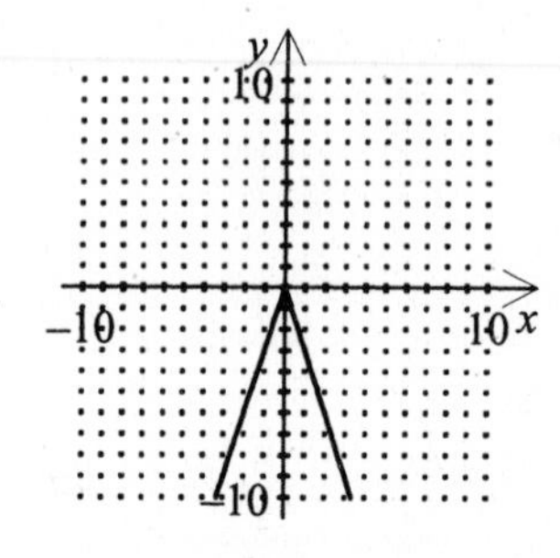

(C)

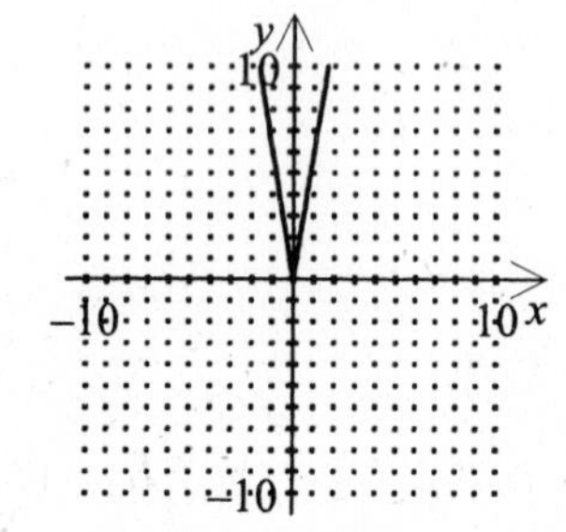

(D)

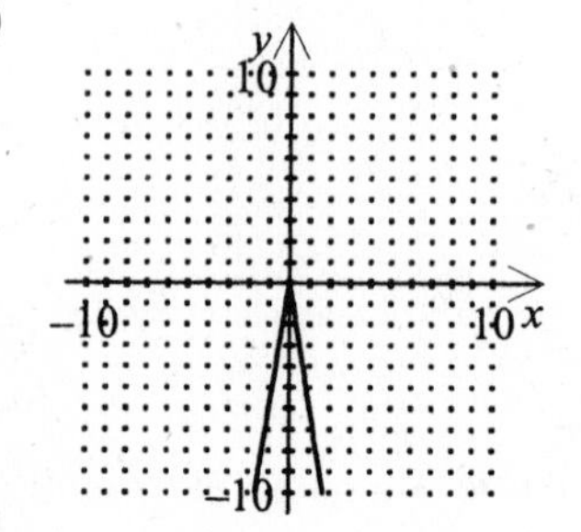

98. Find an equation and graph for the function that is described by the given characteristics.
The shape of $f(x) = x^2$, with a vertical stretch of 2.

(A) $g(x) = -2x^2$

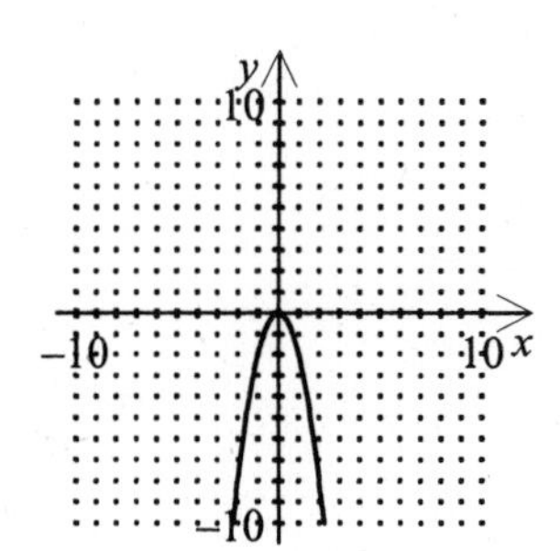

(B) $g(x) = 2x^2$

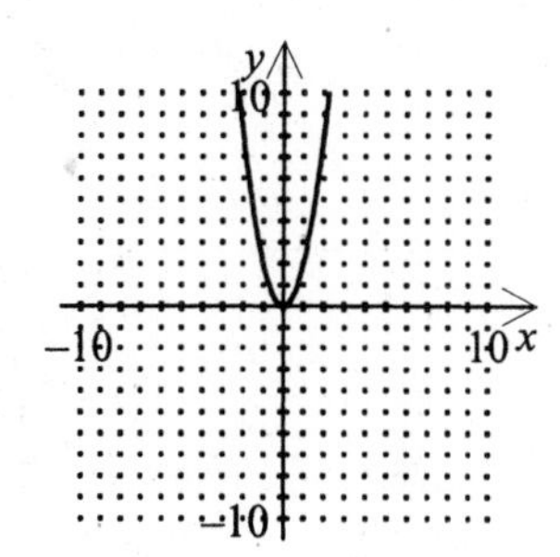

(C) $g(x) = -\frac{1}{2}x^2$

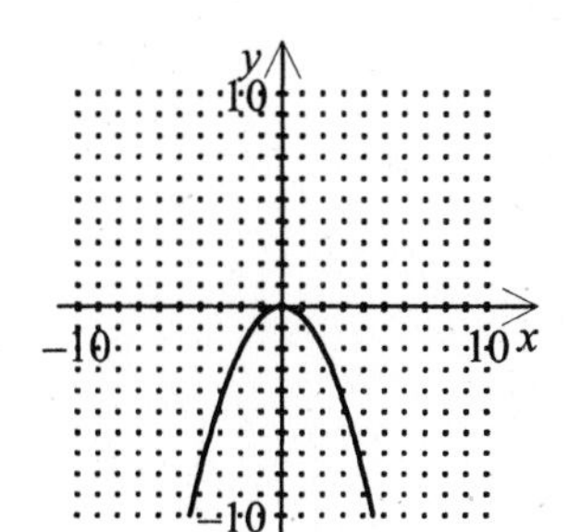

(D) $g(x) = \frac{1}{2}x^2$

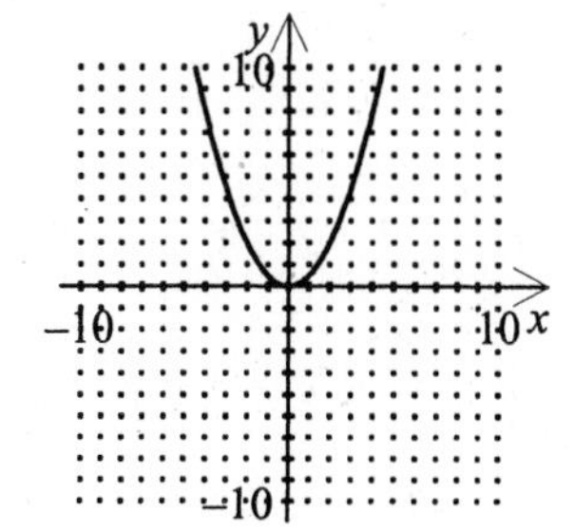

99. Write an equation for the function whose graph is shown.

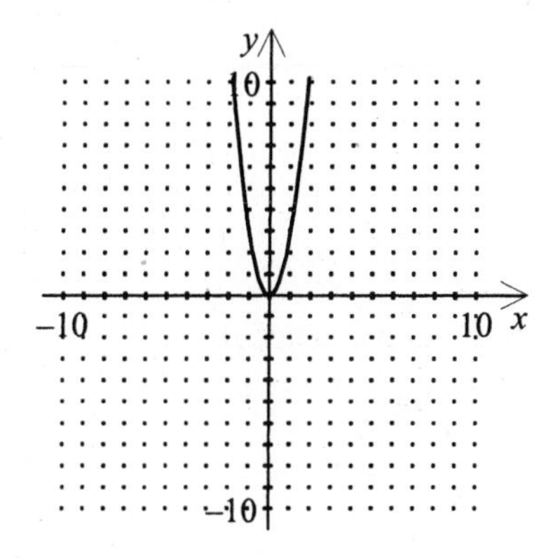

100. The graph of a common function is on the left and the graph of a transformation of that function is on the right. Find the equation of the transformed graph on the right.

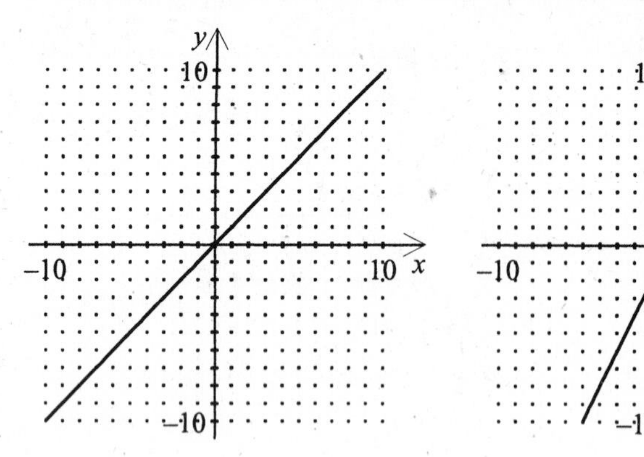

Section 1.7: Combinations of Functions

Objective 1: Add, subtract, multiply, and divide functions

101. Find $(fg)(x)$ for $f(x)=16-x^2$ and $g(x)=4-x$.

(A) $4+x$ (B) $x^3-4x^2-16x+64$ (C) $-x^2+x+12$ (D) $-x^2-x+20$

102. Find $(f-g)(x)$ for $f(x)=\frac{1}{x^2}$ and $g(x)=x$.

(A) $\frac{1}{x^3}$ (B) $\frac{1}{x^2}$ (C) $\frac{1-x^3}{x^2}$ (D) $\frac{x^2}{1-x^3}$

103. Evaluate $(f+g)(x)$ for $f(x)=4x+5$ and $g(x)=x-1$.

104. If $f(x)=-2\sqrt{x}-3$, and $g(x)=-5\sqrt{x}+5$, find $(f-g)(x)$. What is the domain of $(f-g)(x)$?

Objective 2: Find the compositions of one function with another function

105. If $f(x)=|x|$ and $g(x)=-3x$, find $(g\circ f)(x)$.

(A) $-3x+|x|$ (B) $|x|+3x$ (C) $-3|x|$ (D) $3|x|$

106. If $f(x)=2x-4$ and $g(x)=3x+8$, find $(f\circ g)(-7)$.

(A) –30 (B) –29 (C) –46 (D) –45

107. If $f(x) = x + 6$ and $g(x) = \sqrt{x-3}$, find $(f \circ g)(x)$.

108. If $f(x) = 5 - 6x$ and $g(x) = x^2 - 3$, find (a) $(g \circ f)(x)$ and (b) $(f \circ g)(x)$.

Objective 3: Use combinations of functions to model and solve real-life problems

109. The number of pounds of apples a cannery can process and the processing cost are
$P(h) = 325h$ and $C(n) = 0.4n + 600$
where $P(h)$ is the number of pounds of apples that can be processed in h hours and $C(n)$ is the cost of processing n pounds of apples. Use composition functions to find the cost of operating the cannery 40 hours.

(A) \$6100 (B) \$5800 (C) \$5440 (D) \$13,240

110. The selling price of x number of a certain stereo can be modeled by the function $R(x) = 120x$.
The total cost of making x stereos is
$C(x) = 75x - 0.04x^2$.
What is the percent markup for 41 stereos?

(A) 16% (B) 64% (C) 164% (D) 191%

111. The production in board feet and the cost of manufacturing lumber at a sawmill are
$P(h) = 250h$ and $C(n) = 0.4n + 12{,}000$
where $P(h)$ is the number of board feet that can be produced in h hours and $C(n)$ is the cost of producing n board feet. Find $(C \circ P)(h)$ and interpret its meaning.

112. Spheres are being packed into a square box (see figure).
(a) Express the radius r of each sphere as a function of the length x of the sides of the square.
(b) Express the volume V of a sphere as a function of the radius r.
(c) Find $(V \circ r)(x)$.
(d) Find and interpret $(V \circ r)(5)$.

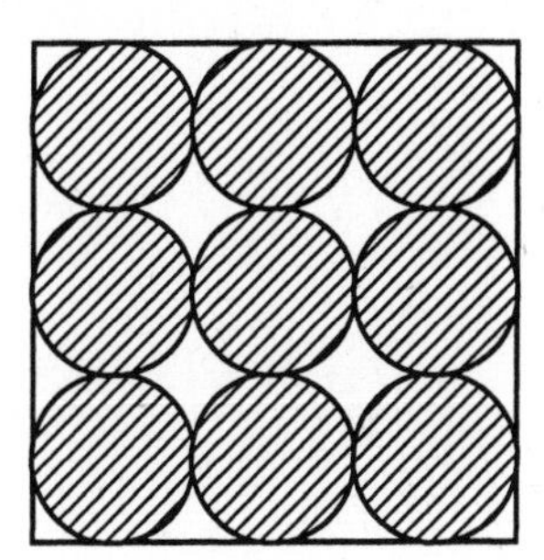

Section 1.8: Inverse Functions

Objective 1: Find inverse functions informally and verify that two functions are inverse functions of each other

113. Find the inverse of the function.

$f(x) = x - \frac{3}{4}$

(A) $f^{-1}(x) = \frac{1}{4}x - \frac{3}{4}$ (B) $f^{-1}(x) = \frac{1}{4}x + 3$ (C) $f^{-1}(x) = 4x - 3$ (D) $f^{-1}(x) = x + \frac{3}{4}$

114. Are the following functions inverses? If not, rewrite the second function so that it is an inverse of the first.

$f(x) = 2x - \frac{3}{4}$

$g(x) = 2x - \frac{3}{2}$

(A) Yes (B) No; $g(x) = \frac{1}{2}x + \frac{3}{2}$ (C) No; $g(x) = \frac{1}{2}x + \frac{3}{8}$ (D) No; $g(x) = \frac{1}{4}x - \frac{3}{8}$

Find the inverse of *f* informally. Verify that $f(f^{-1}(x)) = x$ and $f^{-1}(f(x)) = x$.

115. $f(x) = 8x$

116. $f(x) = \sqrt[5]{x}$

Objective 2: Use graphs of functions to determine whether functions have inverse functions

117. Use the Horizontal and Vertical Line Tests to determine which of the following is a function that has an inverse.

(i)

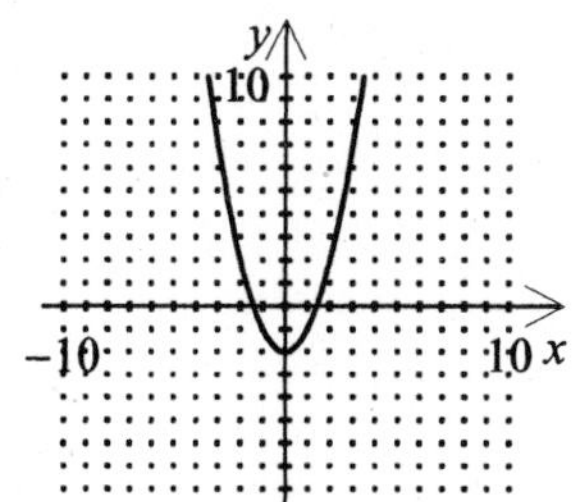

(ii)

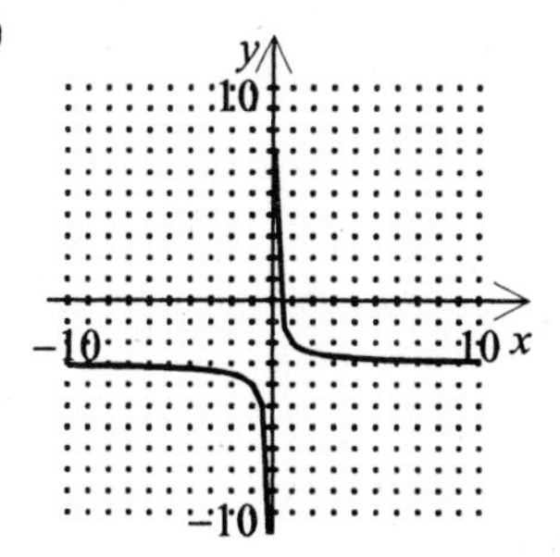

(iii)

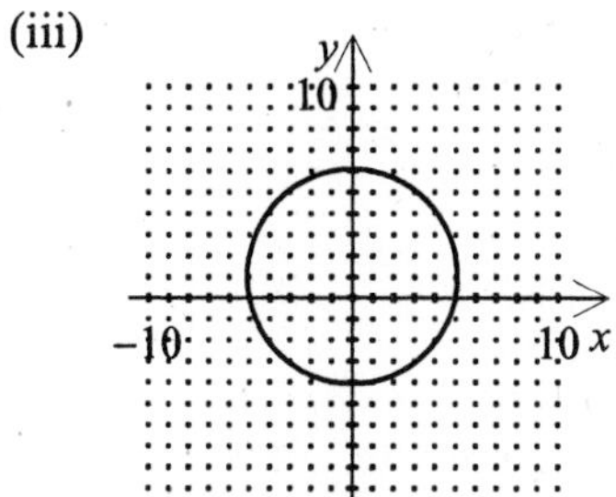

(iv)

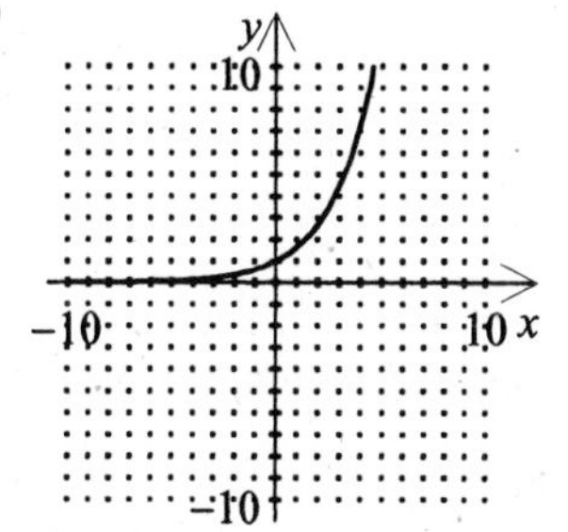

(A) i, ii, and iv only (B) ii and iv only (C) iv only (D) i and iv only

118. Let $f(x) = 2x - 1$. Identify the graph of f and f^{-1}.

(A)

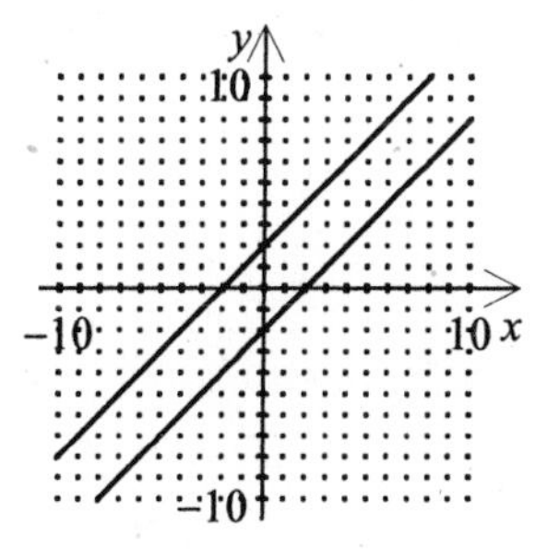

(B)

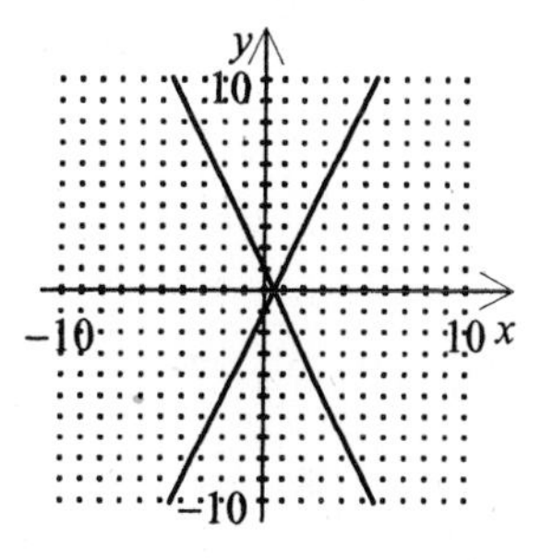

(C)

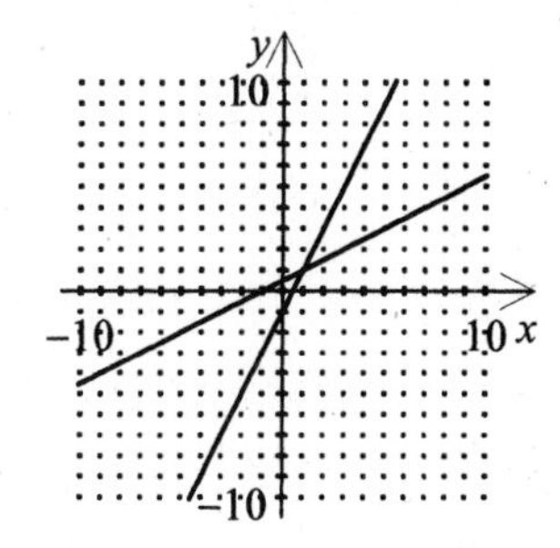

(D)

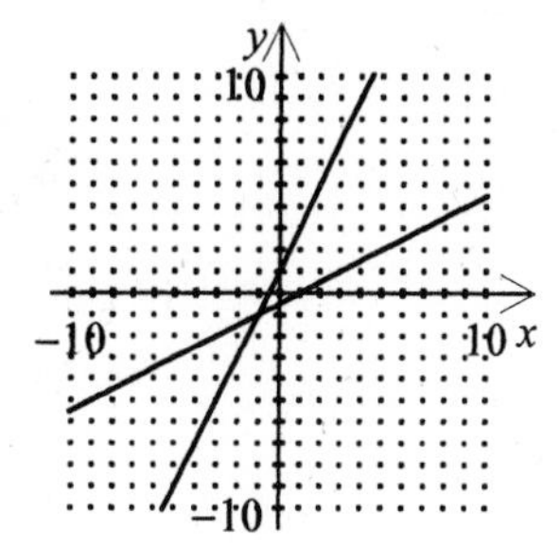

119. Use a graphing utility to graph the function $f(x) = \sqrt{x-1}$ and use the Horizontal Line Test to determine whether the function has an inverse.

120. The graph of a function $f(x)$ is illustrated below. Graph the inverse function $f^{-1}(x)$.

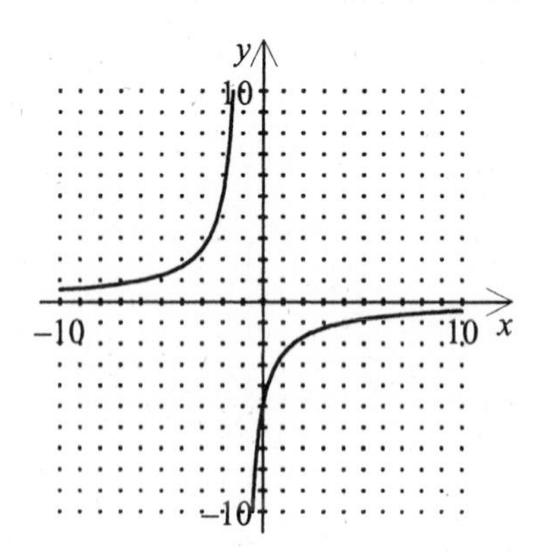

Objective 3: Use the Horizontal Line Test to determine if functions are one-to-one

121. Determine which of the following are one-to-one functions.

i.

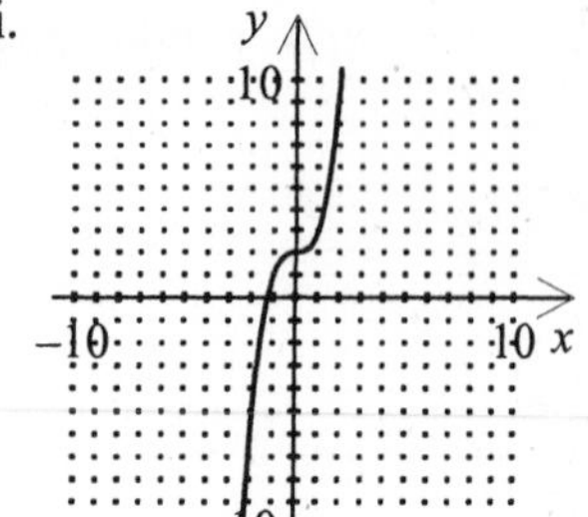

ii.

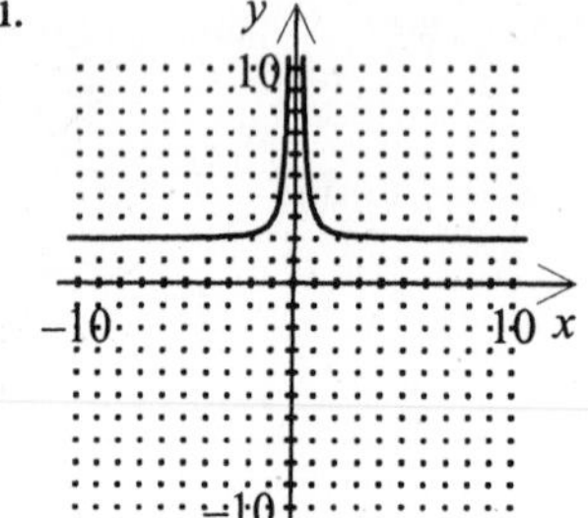

iii.

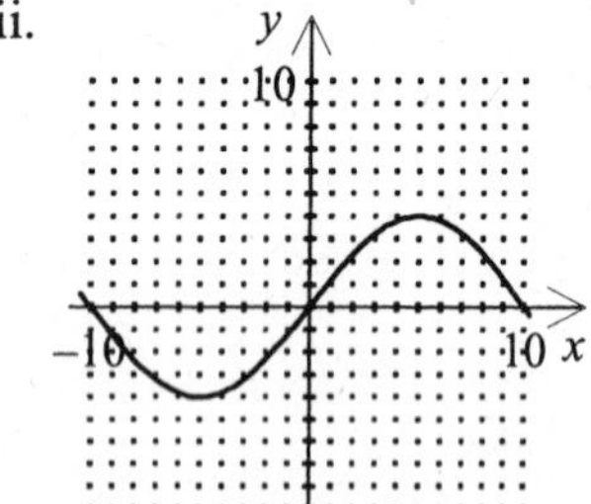

iv.

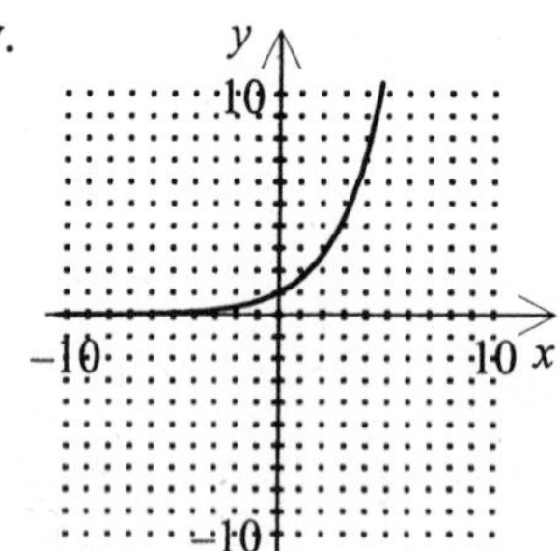

(A) iv only (B) i, ii, and iv only (C) ii and iv only (D) i and iv only

122. Which of the following is a one-to-one function?

(A) $\{(-2, 5), (-4, 1), (1, -4), (4, 4)\}$

(B) $\{(-2, 5), (-4, 9), (1, 1), (-4, -3)\}$

(C) $\{(-2, 5), (-4, 9), (1, 2), (4, 5)\}$

(D) $\{(-2, 5), (-4, 9), (-2, 2), (9, -4)\}$

123. Is $3x + 7y = -3$ a one-to-one function?

124. Which of the following are one-to-one functions?

i.

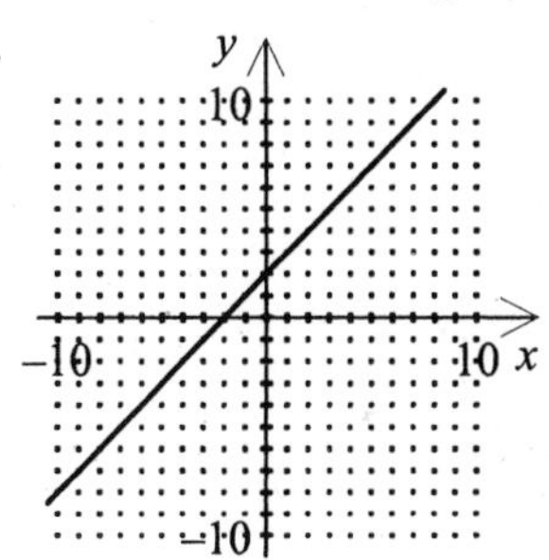

ii.

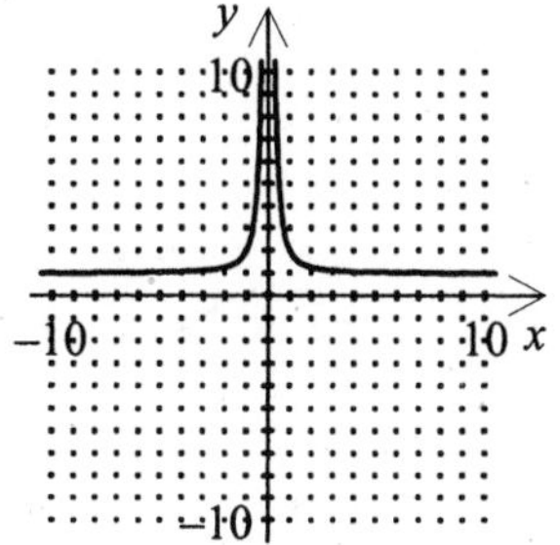

iii.

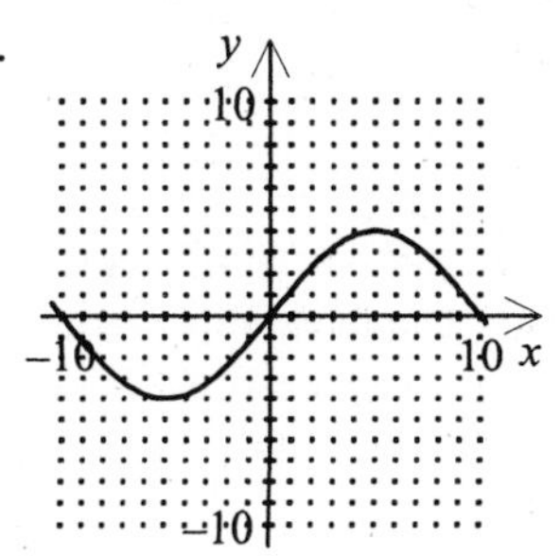

iv.

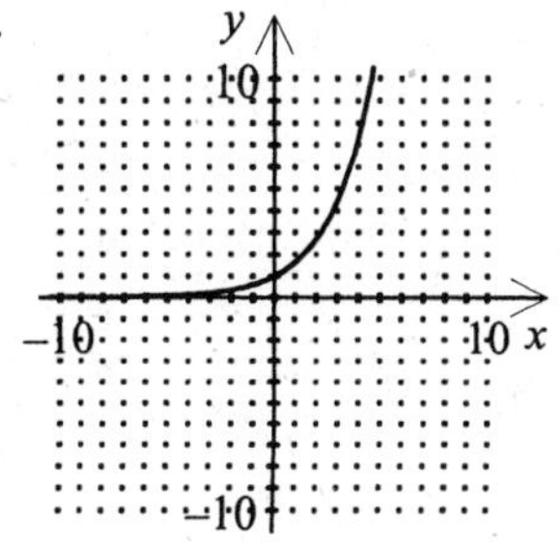

Objective 4: Find inverse functions algebraically

Find the inverse of the function.

125. $f(x) = \dfrac{3-2x}{4-2x}$

(A) $f^{-1}(x) = \dfrac{-2+4x}{-2+3x}$

(B) $f^{-1}(x) = \dfrac{-4x+3}{-2x+2}$

(C) $f^{-1}(x) = \dfrac{-2x+4}{-2x+3}$

(D) $f^{-1}(x) = \dfrac{-2+3x}{-2+4x}$

Find the inverse of the function.

126. $f(x) = -7\sqrt[3]{7x-4} - 5$

(A) $f^{-1}(x) = \dfrac{x^3 + 15x^2 + 75x - 1247}{-2401}$

(B) $f^{-1}(x) = \dfrac{x^3 - 15x^2 + 75x - 1247}{-2401}$

(C) $f^{-1}(x) = \dfrac{x^3 - 15x^2 + 75x + 1497}{-2401}$

(D) $f^{-1}(x) = \dfrac{x^3 + 15x^2 + 75x + 1497}{-2401}$

127. $f(x) = \dfrac{x-9}{6}$

128. $f(x) = 3\sqrt{4x-2} - 1$

Section 1.9: Mathematical Modeling

Objective 1: Use mathematical models to approximate sets of data points

129. Wouldn't it be nice to win a million dollars? Most people think so, but one thing many forget to consider is the amount of taxes that must be paid on the winnings. The table shows the relationship between the winnings and the approximate amount of taxes to be paid to the IRS. Draw a scatter plot to model the data and determine what relationship exists in the data. Find an equation for the best-fit line and use the equation to estimate the amount of taxes to be paid on winnings of $8,200,000.

Amount Won	Taxes Due
$380,000	$150,000
$540,000	$250,000
$940,000	$500,000
$1,468,000	$830,000
$1,884,000	$1,090,000

(A) $5,125,000 (B) $8,200,000 (C) $3,947,500 (D) $5,037,500

130. In a study of the relationship between the number of anteaters A and the number of anthills C in an area, the following data was collected. Graph the relationship that expresses C as a function of A. Use the regression feature of a graphing utility to find the least squares regression line that fits this data.

Anteaters (A)	34	49	63	77	91	106	120
Anthills (C)	710	890	1080	960	1150	1330	1220

(A)

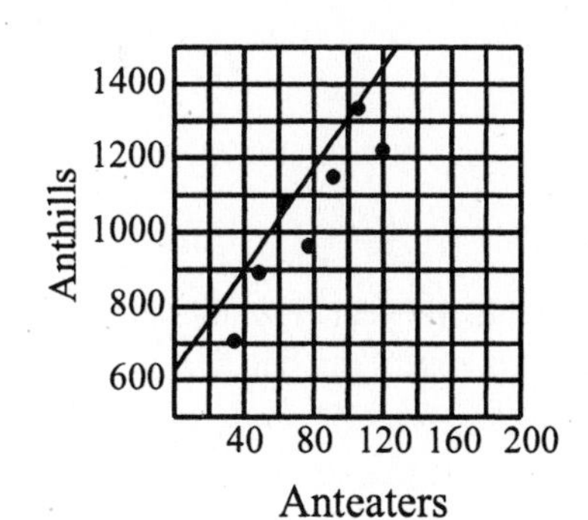

$C = 6.84A + 626.14$

(B)

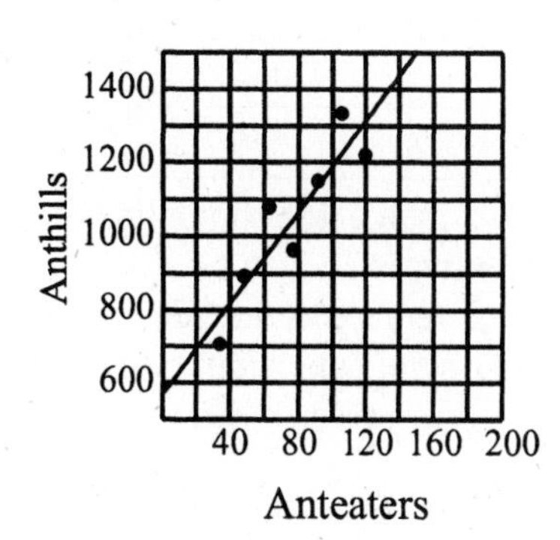

$C = 6.21A + 569.22$

(C)

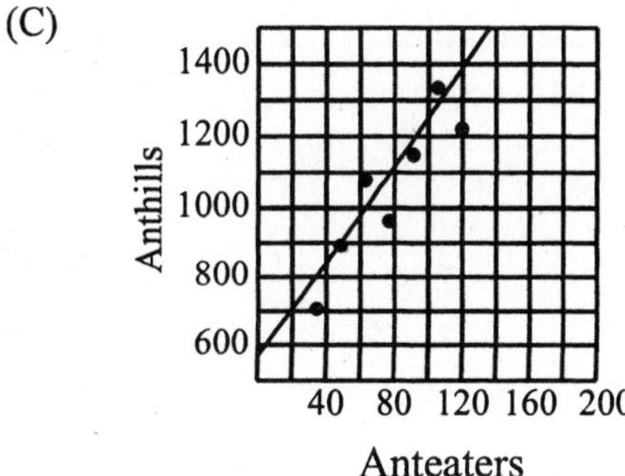

$C = 6.84A + 569.22$

(D)

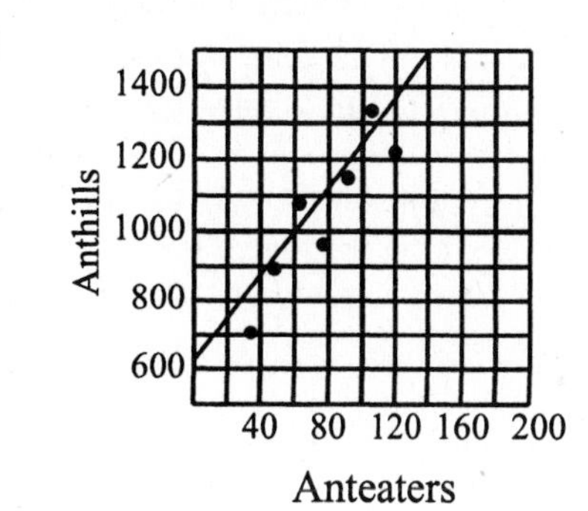

$C = 6.21A + 626.14$

131. The table shows the number of times a cricket chirped in 1 minute and the temperature at the time of the chirps.

Temperature (°C)	18	8	15	14	10	7	14	6	14	9	14	13	15
Number of chirps	20	17	18	18	16	16	17	15	16	15	17	16	16

A linear model that approximates this data is
$N = 0.203t + 11.5$
where N represents the number of chirps and t represents the temperature. Plot the actual data and the model on the same graph. How closely does the model represent the data?

132. The population of a once-endangered animal species has been increasing.

Year (t)	Population (P)
1	292
2	320
3	442
4	1560
5	2380
6	3660

A linear model that approximates this data is
$P = 690t - 971, \quad 0 \le t \le 6$
where P represents the number of animals and t represents the year. Plot the actual data and the model on the same graph. How closely does the model represent the data?

Objective 2: Write mathematical models for direct variation

133. An enclosed gas exerts a pressure P on the walls of its container. This pressure is directly proportional to the temperature T of the gas. If the pressure is 4 pounds per square inch when the temperature is 440 K, identify the appropriate direct variation equation.

(A) $P = 16T$ (B) $P = 110T$ (C) $P = \frac{1}{16}T$ (D) $P = \frac{1}{110}T$

134. Identify a direct variation model.

(A) $y = -5x$ (B) $y = \frac{7}{x}$ (C) $x = -\frac{7}{y}$ (D) $y = x + 2$

135. A new Internet company is selling more software packages each day it is in business. From the data in the table, write a model relating the number of days in business to the total units sold. Estimate the number of units sold after 70 days in business.

Days in business, x	30	40	50	60
Units sold, y	225	300	375	450

136. If w varies directly as x, and $w = -12$ when $x = 3$, find
(a) the constant of variation.
(b) the direct variation equation.

Objective 3: Write mathematical models for direct variation as an nth power

137. On planet *X*, an object falls 17 feet in 3 seconds. Knowing that the distance it falls varies directly with the square of the time of the fall, how long does it take an object to fall 87 feet?

(A) 27.982 sec (B) 36.547 sec (C) 6.787 sec (D) 1.326 sec

138. Find a mathematical model representing the statement.
y varies directly as the cube of x and $y = 14$ when $x = 10$.

(A) $y = \dfrac{500x^3}{7}$ (B) $y = \dfrac{7x^3}{500}$ (C) $y = \dfrac{7x^3}{5000}$ (D) $y = 1000x^3$

139. Does the equation $x^2y = 12.5$ represent direct variation of one variable with the square of another variable? Write Yes or No.

140. Write a sentence using variation terminology to describe the equation.
$y = 7x^3$

Objective 4: Write mathematical models for inverse variation

141. Find an equation of variation if y varies inversely as x and $y = 3$ when $x = 9$.

(A) $y = \dfrac{27}{x}$ (B) $\dfrac{9}{3} = \dfrac{x}{y}$ (C) $\dfrac{y}{3} = \dfrac{x}{9}$ (D) $\dfrac{y}{9} = \dfrac{x}{3}$

142. The intensity I of light received from a source varies inversely as the square of the distance d from the source. If the light intensity is 5 foot-candles at 17 feet, find the light intensity at 13 feet.

(A) 0.5 foot-candles (B) 111.15 foot-candles

(C) 8.55 foot-candles (D) 1.71 foot-candles

143. Find a mathematical model representing the statement and determine the indicated quantity.
y varies inversely as x, and $y = \dfrac{4}{9}$ when $x = 3$. Find y when $x = 4$.

144. A drama club is planning a bus trip to New York City to see a Broadway play. The cost per person for the bus rental varies inversely as the number of people going on the trip. It will cost \$23 per person if 66 people go on the trip. How much will it cost per person if 87 people go on the trip?

Objective 5: Write mathematical models for joint variation

145. The horsepower that a rotating shaft can safely generate varies jointly with the cube of its diameter and its speed in revolutions per minute. A shaft with a 3-inch diameter turning at a speed of 1500 revolutions per minute can safely transmit 12 horsepower. Find the horsepower that a shaft with a 4-inch diameter can safely transmit at a speed of 2000 revolutions per minute.

(A) 28 hp (B) 38 hp (C) 114 hp (D) 9 hp

146. The wattage rating of an appliance W varies jointly as the square of the current I and the resistance R. If the wattage is 3 watts when the current is 0.1 ampere and the resistance is 300 ohms, find the wattage when the current is 0.4 ampere and the resistance is 200 ohms.

(A) 16,000 W (B) 160 W (C) 32 W (D) 80 W

147. A variable z varies jointly with x and the square of y. If $z = 5$ when $x = 3$ and $y = 2$, write the equation of joint variation.

148. The variable R varies jointly as S and the square of T. If R is 43.2 when $S = 0.8$ and $T = 3$, find R when $S = 0.3$ and $T = 5$.

Objective 6: Use the regression feature of a graphing utility to find the equation of a least squares regression line

149. Use the regression feature of a graphing utility to find the least squares regression line that fits these data.

x	39	45	18	22	11	29	14	33	36	37	26	40	31
y	24	23	31	29	33	27	32	26	25	25	28	24	27

(A) $y = -0.3037x + 36.1303$ (B) $y = -0.3340x + 32.5173$

(C) $y = -0.3037x + 32.5173$ (D) $y = -0.3340x + 36.1303$

150. The table shows Christine's best javelin throws each year. Use a graphing utility to determine an equation for the line of best fit for the data. Use $x = 0$ for 1989.

Year	1989	1990	1991	1992	1993	1994	1995	1996
Distance (m)	42.1	42.7	41.55	44.65	44	45.85	49.7	47.55

(A) $y = 0.984x + 41.204$ (B) $y = 1.017x + 41.704$

(C) $y = 0.984x + 41.704$ (D) $y = 1.017x + 41.204$

151. State Fair

Winning Zucchini Lengths

Year	1986	1987	1988	1989	1990	1991	1992	1993	1994	1995
Length (in.)	31.8	33.1	29.5	32.8	34.2	33.9	34.5	36.7	34.4	34.6

(a) Make a scatter plot of the ten data points. Let x represent the number of years after 1985 and y represent the winning length that year.
(b) Use the regression feature of a graphing utility to find the least squares regression line that fits these data. What is the equation of the linear regression model? Graph the equation on the scatter plot for part (a).
(c) Predict the winning length for the year 2000.

152. This year the Wolverine football team scored the following number of points in its 10 games.

Game	1	2	3	4	5	6	7	8	9	10
Points	27	17	22	16	35	39	42	26	45	29

(a) Make a scatter plot of the ten data points. Let x represent the game number and y represent the number of points scored during the game.
(b) Use the regression feature of a graphing utility to find the least squares regression line that fits these data. Graph the line on the scatter plot in part (a).

Answer Key for Chapter 1 Functions and Their Graphs

Section 1.1: Graphs of Equations

Objective 1: Sketch graphs of equations

[1] (C)

[2] (A)

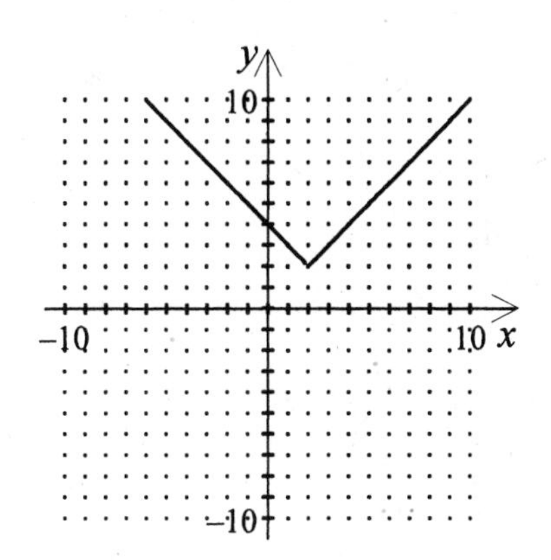

[3]

x	-2	-1	0	1	2
y	7	2	-3	-8	-13

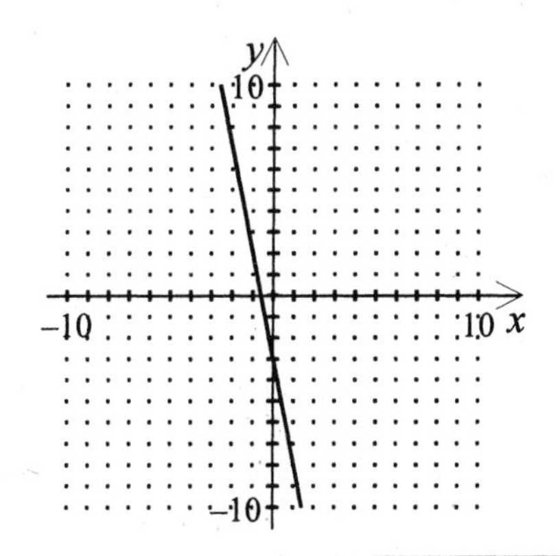

[4]

Objective 2: Find x- and y-intercepts of graphs of equations

[5] (B)

[6] (D)

x-intercept(s): $(0,\ 0)$, $(6,\ 0)$

[7] y-intercept: $(0,\ 0)$

x-intercepts are -3, 3, and 6

y-intercept is -4

[8] intercept points are $(-3,\ 0)$, $(3,\ 0)$, $(6,\ 0)$, and $(0,\ -4)$

Objective 3: Use symmetry to sketch graphs of equations

[9] (D)

[10] (A)

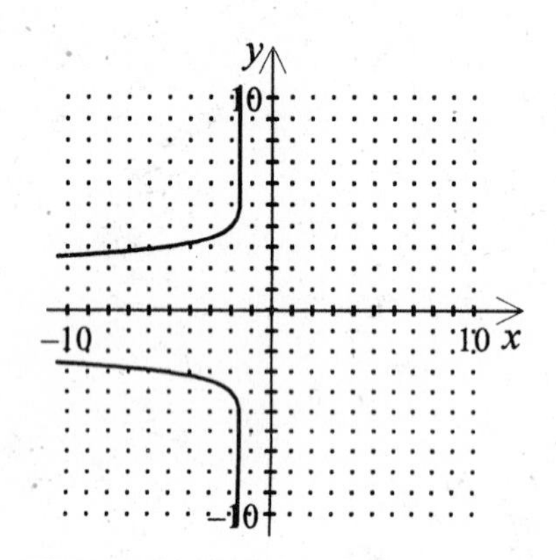

[11]

[12] The graph is symmetric with respect to the y-axis.

Objective 4: Find equations and sketch graphs of circles

[13] (D)

[14] (A)

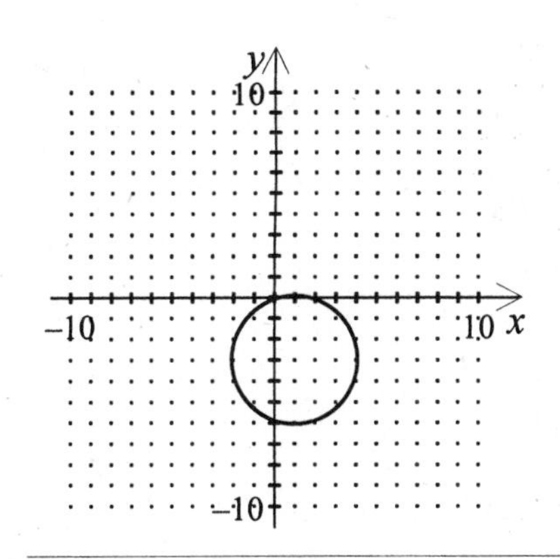

[15]

[16] $(x-1)^2+(y-1)^2=45$

Objective 5: Use graphs of equations in solving real-life problems

[17] (B)

[18] (A)

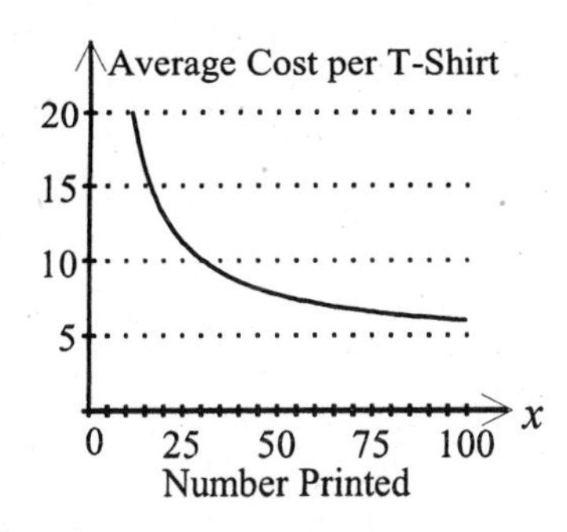

[19] The average cost per T-shirt for 300 shirts is \$4.83.

(a)

(b) Approximately 0.8 sec

[20] (c) Never

Section 1.2: Linear Equations in Two Variables

Objective 1: Use slope to graph linear equations in two variables

[21] (A)

[22] (A)

[23] Answers may vary. Sample answer: $(5,\ -6)$, $(6,\ -5)$, $(7,\ -4)$

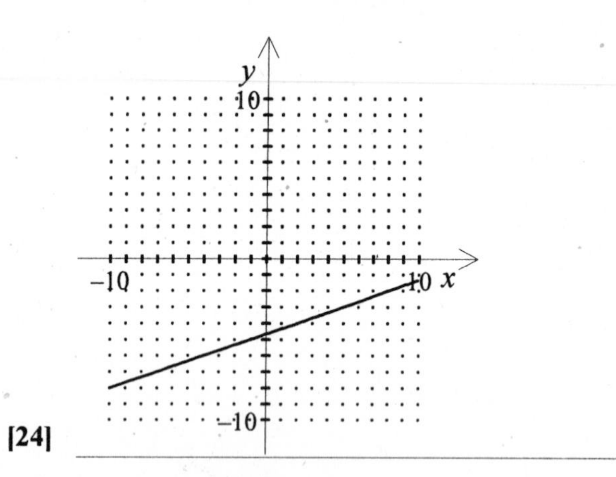

[24]

Objective 2: Find slopes of lines

[25] (B)

[26] (D)

[27] -3

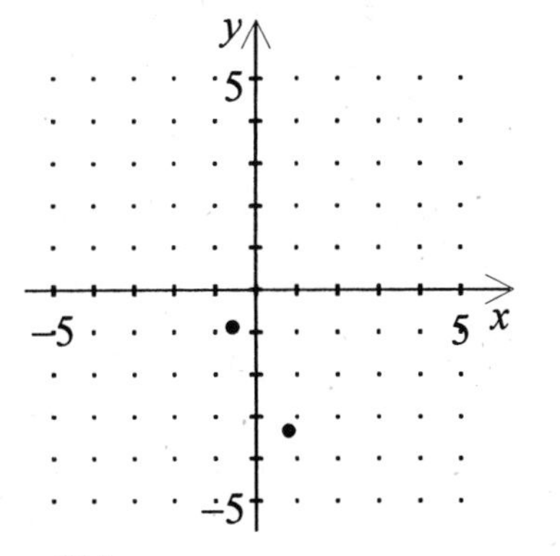

[28] $-\frac{385}{216}$

Objective 3: Write linear equations in two variables

[29] (A)

[30] (C)

[31] $x = 2$

[32] $-\frac{x}{2}+\frac{y}{6}=1$

Objective 4: Use slope to identify parallel and perpendicular lines

[33] (C)

[34] (B)

[35] $y=\frac{8}{7}x+\frac{47}{7}$

[36] Parallel

Objective 5: Use linear equations in two variables to model and solve real-life problems

[37] (A)

[38] (B)

[39] (a) $y = 4x + 9$
(b) The y - intercept represents the average weight of a kitten at birth. The slope indicates the average weekly weight gain for these kittens during their first 5 weeks.

[40] $y = -\frac{1}{4}x + 13$

Section 1.3: Functions

Objective 1: Determine whether relations between two variables are functions

[41] (D)

[42] (C)

[43] Yes, there is only one y value for each x value.

[44] Yes

Objective 2: Use function notation and evaluate functions

[45] (A)

[46] (B)

[47] (a) –27 (b) –3 (c) 29 (d) –25.4

[48] $f(-28) = -17$
$f(21) = 4$

Objective 3: Find domains of functions

[49] (C)

[50] (A)

[51] $x \geq -6,\ x \neq 1,\ 3$

[52] All real numbers

Objective 4: Use functions to model and solve real-life problems

[53] (C)

[54] (A)

[55] 358,000

[56] $H(d) = 1.2d + 11$

Section 1.4: Analyzing Graphs of Functions

Objective 1: Use the Vertical Line Test for functions

[57] (B)

[58] (D)

[59] Function

[60] Function

Objective 2: Find the zeros of functions

[61] (A)

[62] (A)

[63] 64

[64] 0, 3, 7

Objective 3: Determine intervals on which functions are increasing or decreasing

[65] (A)

[66] (D)

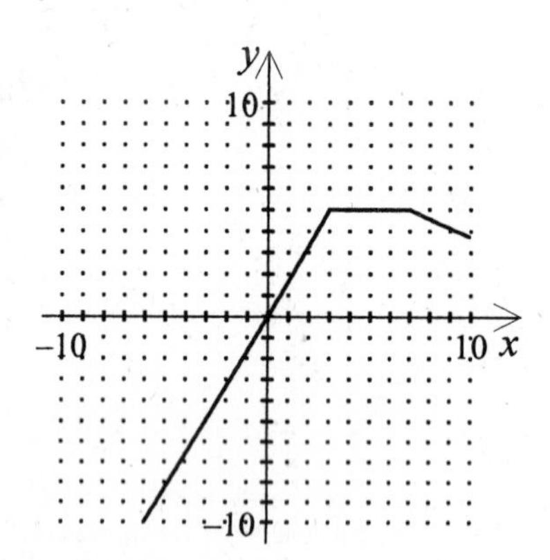

Increasing on $(-\infty, 3]$

Constant on $(3, 7]$

[67] Decreasing on $(7, \infty)$

(a)

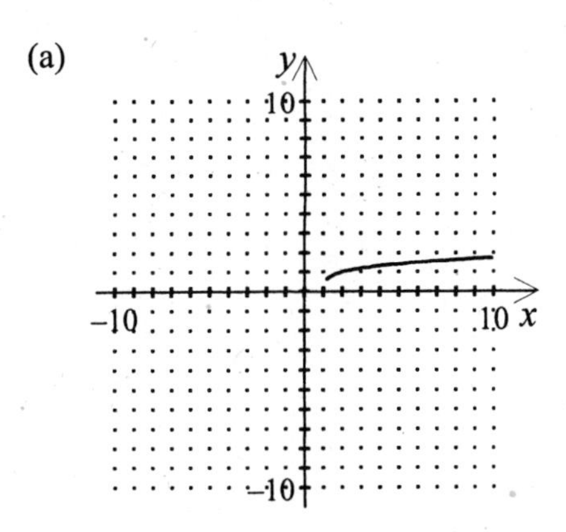

Increasing on $(1, \infty)$

[68] (b) Sample:

x	2	3	4	5
$f(x)$	1	1.19	1.32	1.41

Objective 4: Identify even and odd functions

[69] (C)

[70] (B)

[71] Odd

[72] Neither even nor odd

Section 1.5: A Library of Functions

Objective 1: Identify and graph linear and squaring functions

[73] (D)

[74] (C)

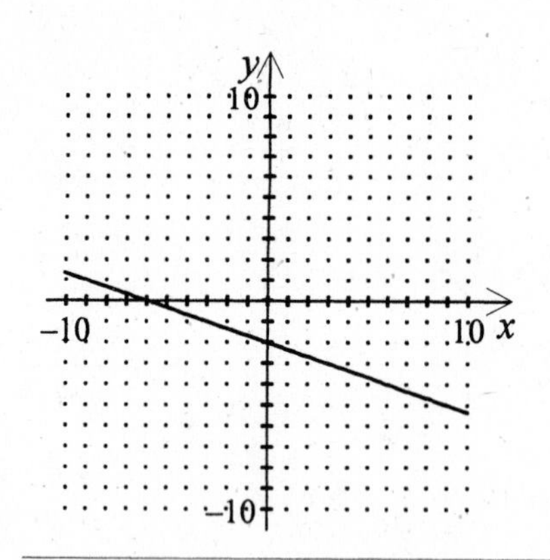

[75]

[76] $f(x) = 2x - 3$

Objective 2: Identify and graph cubic, square root, and reciprocal functions

[77] (A)

[78] (B)

[79] (0, 0)

[80] $(-\infty, 0) \cup (0, \infty)$

Objective 3: Identify and graph step and other piecewise-defined functions

[81] (B)

[82] (A)

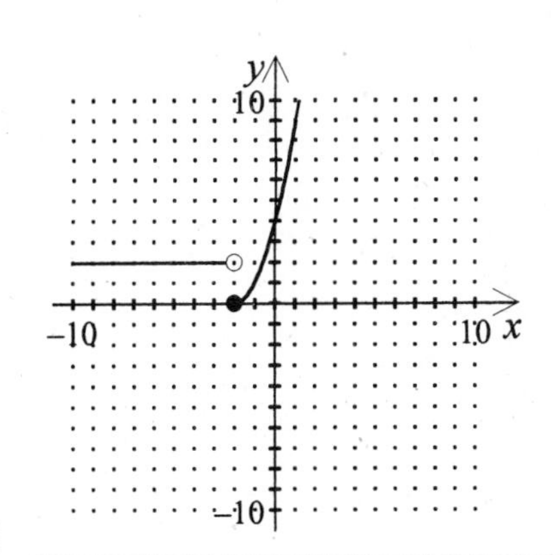

[83] ______

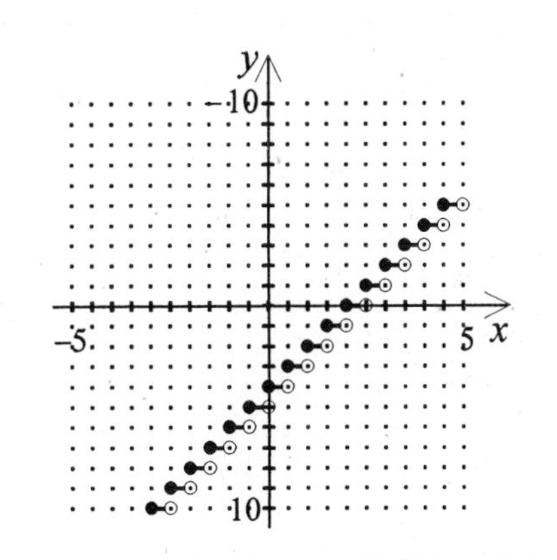

[84] ______

Objective 4: Recognize graphs of common functions

[85] (C)

[86] (D)

[87] Identity Function

[88] $f(x) = \sqrt{x}$

Section 1.6: Shifting, Reflecting, and Stretching Graphs

Objective 1: Use vertical and horizontal shifts to sketch graphs of functions

[89] (A)

[90] (D)

$f(x) = (x+3)^2 - 2$

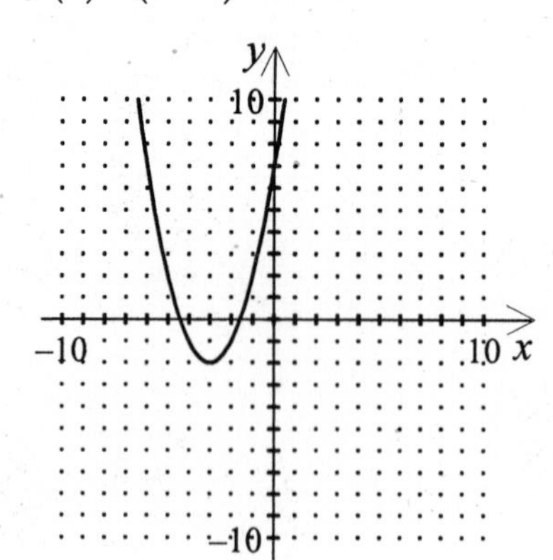

[91]

(a)

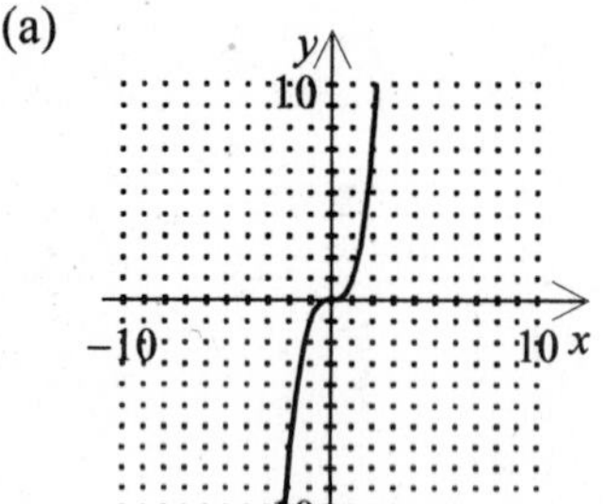

(b)

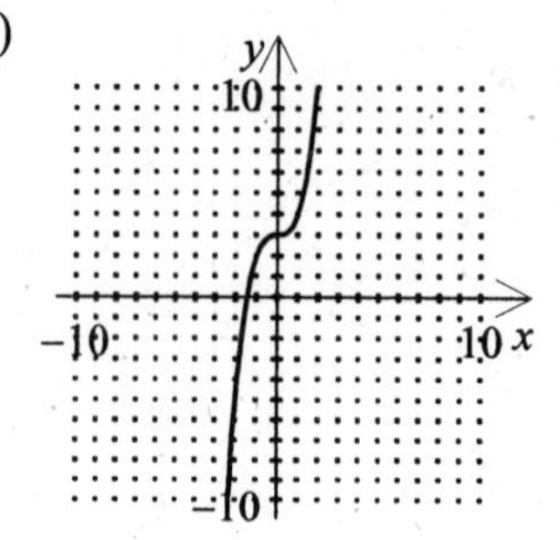

(c)

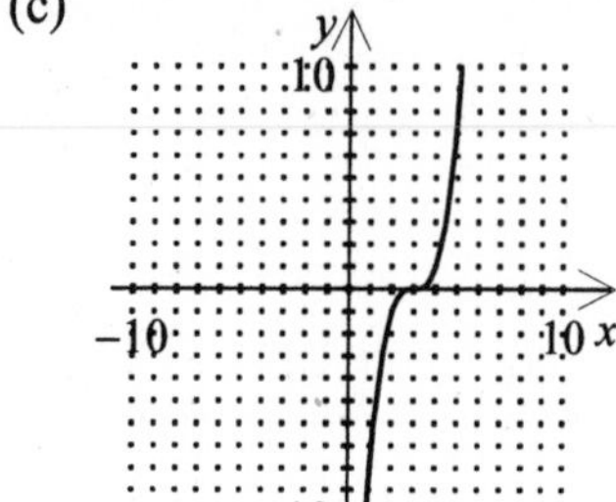

[92]

Objective 2: Use reflections to sketch graphs of functions

[93] (A)

[94] (D)

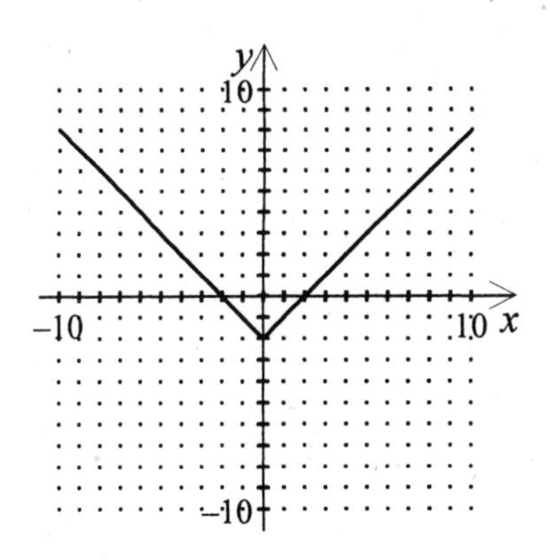

[95] Reflection in the y-axis

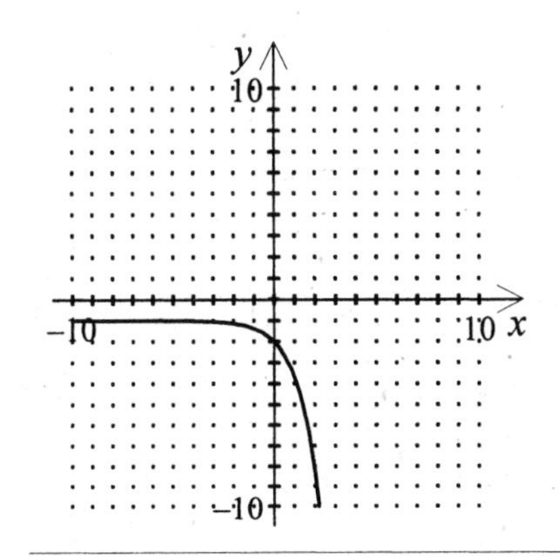

[96] ______

Objective 3: Use nonrigid transformations to sketch graphs of functions

[97] (A)

[98] (B)

[99] $f(x) = 3x^2$

[100] $f(x) = 2x$

Section 1.7: Combinations of Functions

Objective 1: Add, subtract, multiply, and divide functions

[101] (B)

[102] (C)

[103] $5x+4$

[104] $(f-g)(x)=3\sqrt{x}-8,\ x\geq 0$

Objective 2: Find the compositions of one function with another function

[105] (C)

[106] (A)

[107] $\sqrt{x-3}+6$

[108]
(a) $36x^2-60x+22$
(b) $-6x^2+23$

Objective 3: Use combinations of functions to model and solve real-life problems

[109] (B)

[110] (C)

[111] $(C\circ P)(h)=100h+12{,}000$; the cost of operating h hours.

[112]
(a) $r(x)=\dfrac{x}{6}$
(b) $V(r)=\dfrac{4}{3}\pi r^3$
(c) $(V\circ r)(x)=\dfrac{x^3}{162}\pi$
(d) $(V\circ r)(5)=2.424$. The volume of one sphere when the length of each side of the square is 5.

Section 1.8: Inverse Functions

Objective 1: Find inverse functions informally and verify that two functions are inverse functions of each other

[113] (D)

[114] (C)

[115] $f^{-1}(x) = \frac{1}{8}x$

[116] $f^{-1}(x) = x^5$

Objective 2: Use graphs of functions to determine whether functions have inverse functions

[117] (B)

[118] (C)

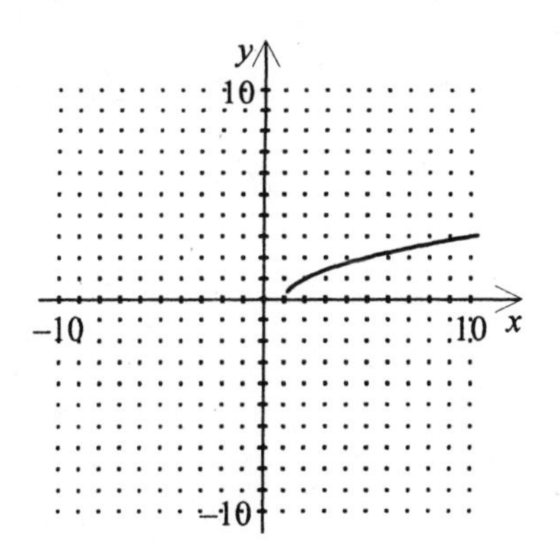

[119] Yes

[120] 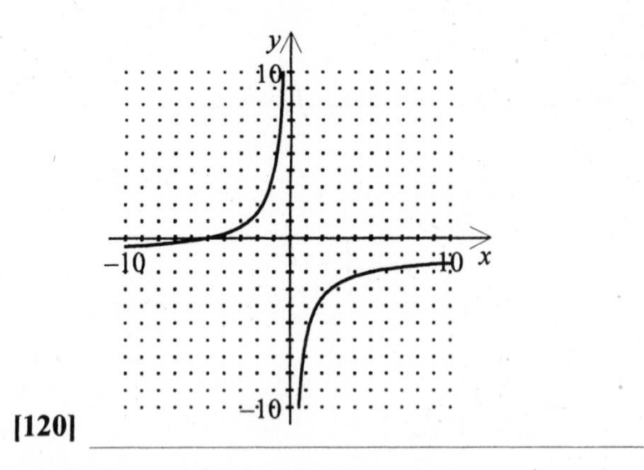

Objective 3: Use the Horizontal Line Test to determine if functions are one-to-one

[121] (D)

[122] (A)

[123] Yes

[124] i and iv only

Objective 4: Find inverse functions algebraically

[125] (B)

[126] (A)

[127] $f^{-1}(x) = 6x + 9$

[128] $f^{-1}(x) = \dfrac{x^2 + 2x + 19}{36},\ x \geq 0$

Section 1.9: Mathematical Modeling

Objective 1: Use mathematical models to approximate sets of data points

[129] (D)

[130] (B)

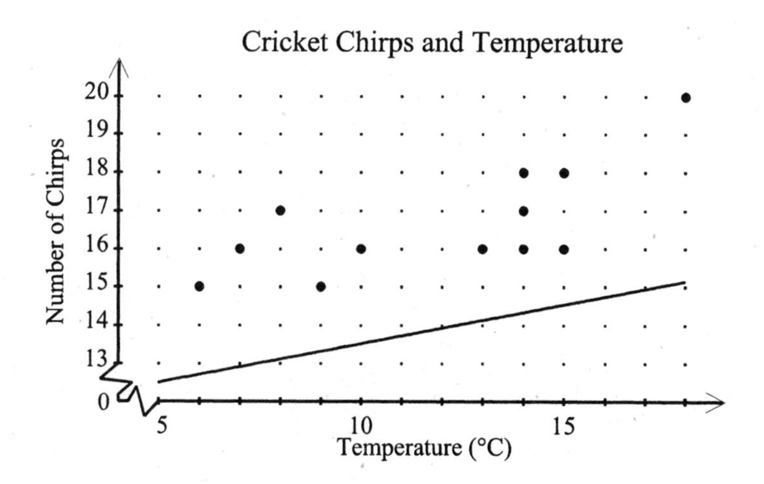

[131] The model is a "bad fit" for the actual data.

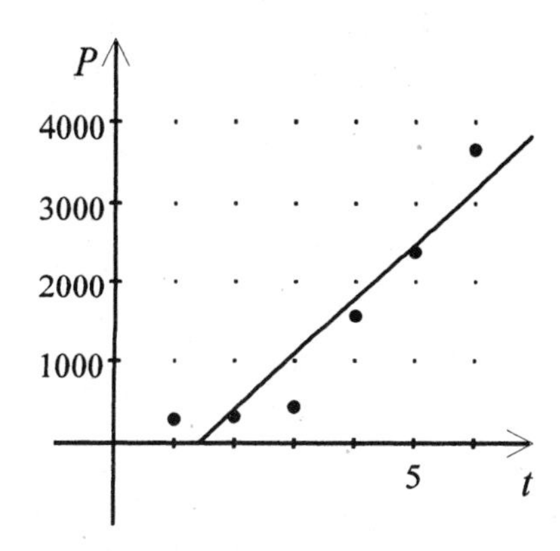

[132] The model is a "good fit" for the actual data.

Objective 2: Write mathematical models for direct variation

[133] (D)

[134] (A)

[135] $y = 7.5x$; 525 units

[136] (a) -4, (b) $w = -4x$

Objective 3: Write mathematical models for direct variation as an nth power

[137] (C)

[138] (B)

[139] No

[140] y varies directly as the cube of x.

Objective 4: Write mathematical models for inverse variation

[141] (A)

[142] (C)

[143] $y = \dfrac{4}{3x}$; $y(4) = \dfrac{1}{3}$

[144] $17.45

Objective 5: Write mathematical models for joint variation

[145] (B)

[146] (C)

[147] $z = \dfrac{5}{12}xy^2$

[148] 45

Objective 6: Use the regression feature of a graphing utility to find the equation of a least squares regression line

[149] (A)

[150] (D)

(a)

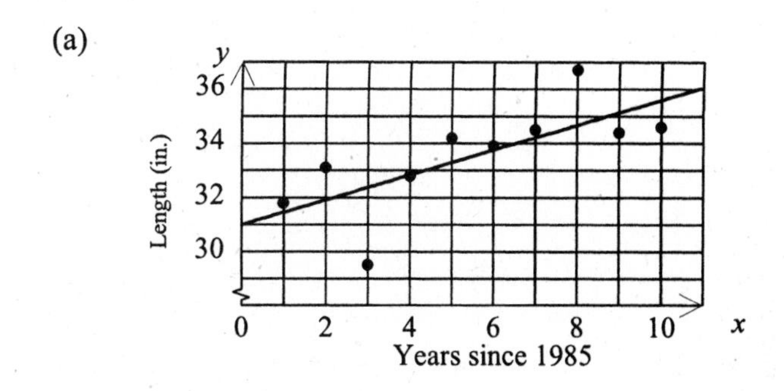

(b) $y = 0.46x + 31.0$

[151] (c) 37.9 in.

(a)

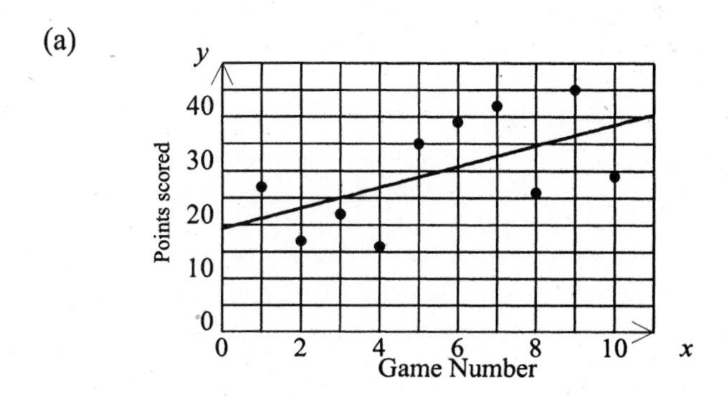

[152] (b) $y = 1.92x + 19.3$

Chapter 2 Polynomial and Rational Functions

Section 2.1: Quadratic Functions

Objective 1: Analyze graphs of quadratic functions

1. Identify the graph of the quadratic function.
   $f(x) = -x^2 - 2$

(A)

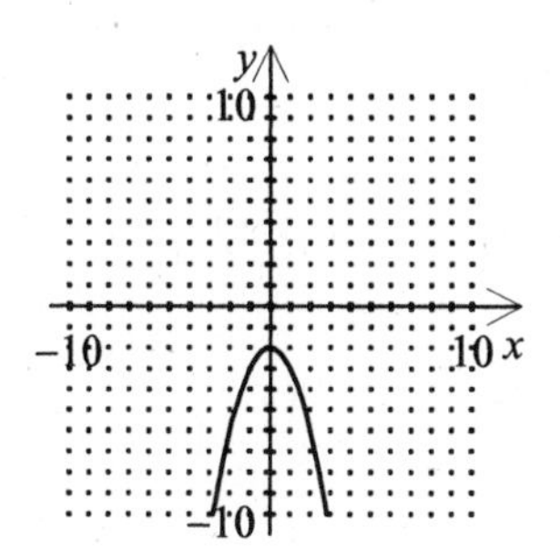

(B)

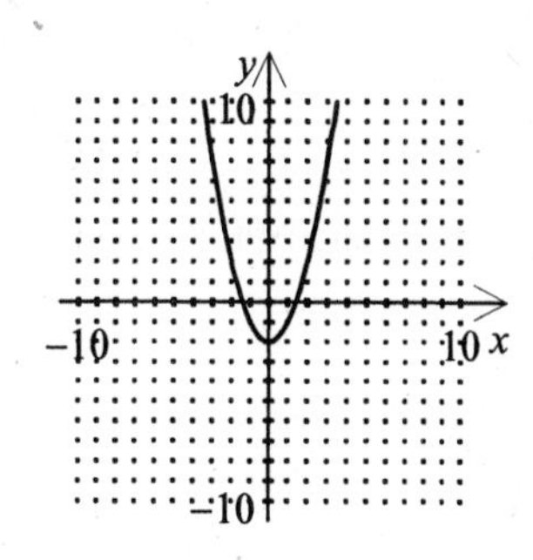

(C)

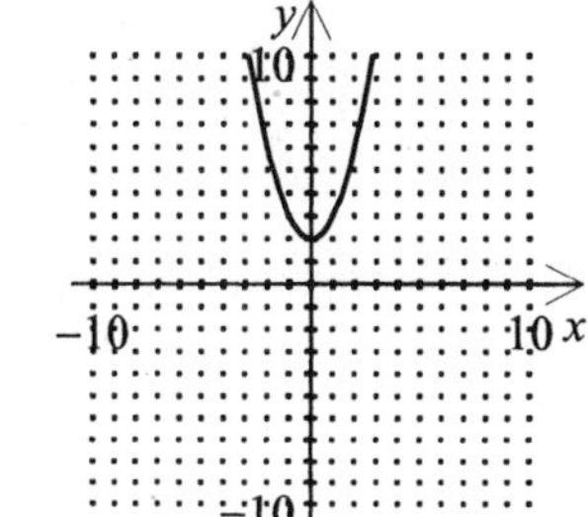

(D)

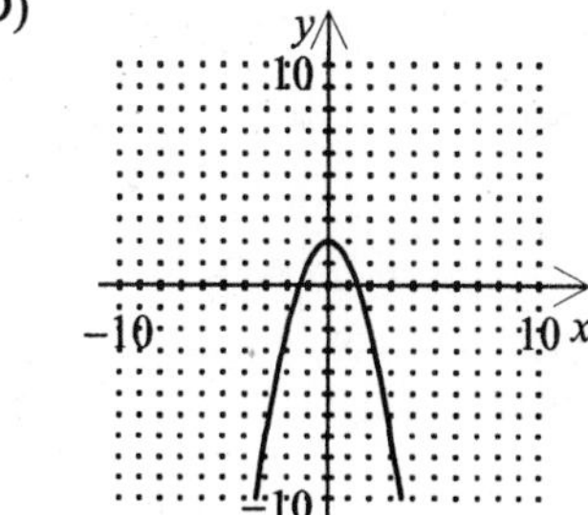

2. For the graph of the quadratic function, identify the direction of the opening and the coordinates of the vertex.
   $f(x) = (x+4)^2 + 3$

(A) Upward; $(-4,\ 3)$

(B) Upward; $(4,\ -3)$

(C) Downward; $(4,\ -3)$

(D) Downward; $(-4,\ 3)$

3. Sketch the graph of $f(x) = \frac{1}{5}x^2$. Identify the vertex.

4. Sketch the graph of the function. Identify the vertex.
   $f(x) = -(x-4)^2 - 1$

Objective 2: Write quadratic functions in standard form and use the results to sketch graphs of functions

5. Identify the equation of a quadratic function with the vertex $\left(0, \frac{1}{3}\right)$, that passes through the point $\left(3, \frac{55}{3}\right)$ and opens upward.

(A) $f(x) = -2x^2 - \frac{1}{3}$ (B) $f(x) = 2x^2 - \frac{1}{3}$ (C) $f(x) = 2x^2 + \frac{1}{3}$ (D) $f(x) = -2x^2 - \frac{2}{3}$

6. Identify the equation of the quadratic function in standard form and find the vertex of the graph.
$f(x) = 12x - 1 - 2x^2$

(A) $f(x) = -2(x-3)^2 + 17$
Vertex: $(3, 17)$

(B) $f(x) = -2(x+3)^2 + 17$
Vertex: $(-17, 3)$

(C) $f(x) = -2(x-3)^2 + 17$
Vertex: $(17, 3)$

(D) $f(x) = -2(x+3)^2 + 17$
Vertex: $(-3, 17)$

7. Use a graphing utility to determine the vertex and x-intercepts of the graph of the quadratic function. Then write the equation of the parabola in standard form.
$f(x) = 3x^2 - 12x + 3$

8. Write the standard form of the equation of the parabola.

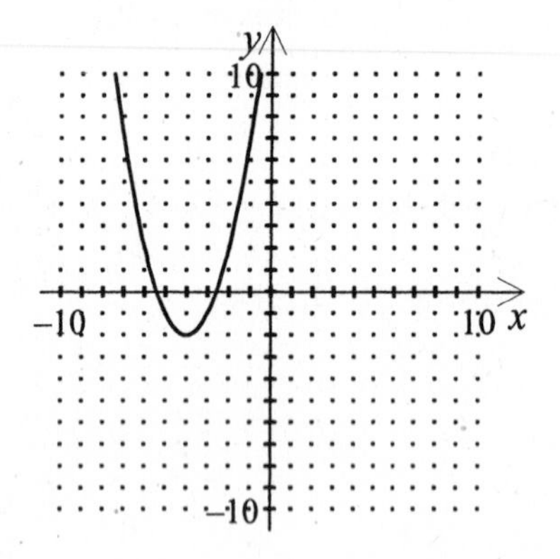

Objective 3: Use quadratic functions to model and solve real-life problems

9. The manager of a new company predicted that the company would lose \$650 its first month, \$550 its second month, and so on. She further predicted that this trend would continue and that the business would continue to improve by \$100 each month.
(a) Identify the quadratic equation in standard form that describes the net financial position as a function of months.
(b) In which month would the net financial position first become positive?

(A) (a) $P(m)=100(m-7)^2-2550$
(b) Month 14

(B) (a) $P(m)=50(m-7)^2-2450$
(b) Month 15

(C) (a) $P(m)=50(m-7)^2-2450$
(b) Month 14

(D) (a) $P(m)=100(m-7)^2-2550$
(b) Month 15

10. Jafco Manufacturing estimates that its profit P in hundreds of dollars is
$P=-4x^2+16x-3$
where x is the number of units produced in thousands. How many units must be produced to obtain the maximum profit?

(A) 2000 units (B) 2 units (C) 16 units (D) 1600 units

11. A farmer has 232 meters of fencing available to enclose a rectangular portion of his land. One side of the rectangle being fenced lies along a river, so only three sides require fencing.
(a) Express the area A of the rectangle as a function of x, where x is the length of the side parallel to the river.
(b) For what value of x is the area largest?

12. The height of an arrow shot into the air is
$h(t)=-16t^2+38.4t$
where $h(t)$ is the height in feet of the arrow above the ground t seconds after it is released. Find the maximum height the arrow reaches by graphing the function.

Section 2.2: Polynomial Functions of Higher Degree

Objective 1: Use transformations to sketch graphs of polynomial functions

13. Identify the equation for the function shown on the graph.

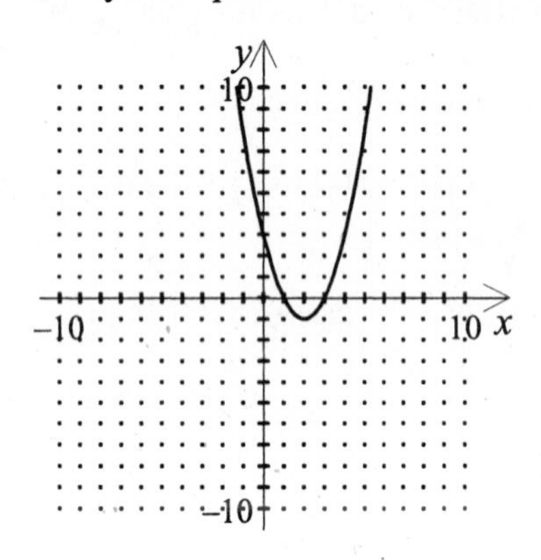

(A) $f(x) = x^3 - 4x^2 + 3x$ (B) $f(x) = x^2 - 4x + 3$

(C) $f(x) = -x^2 - 4x + 3$ (D) $f(x) = -x^3 - 4x^2 + 3x$

14. Identify the graph of the function.
$f(x) = -(x+2)^4 - 2$

(A)

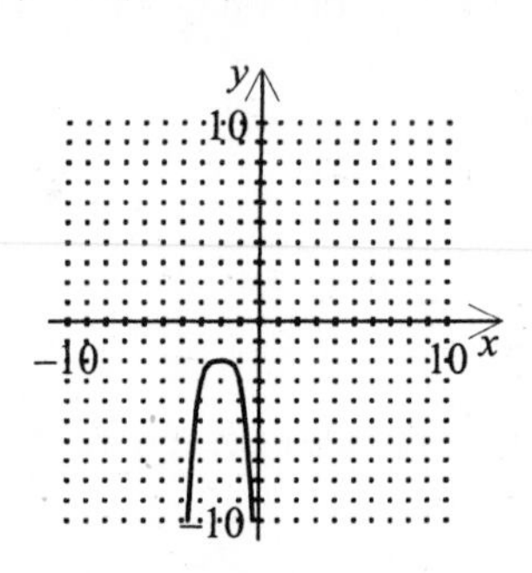

(B)

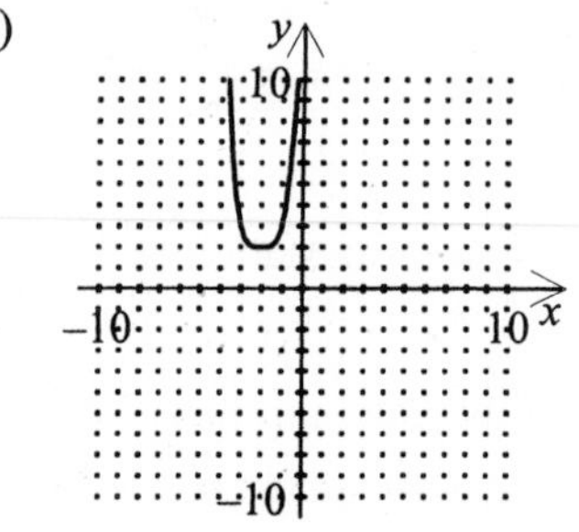

(C)

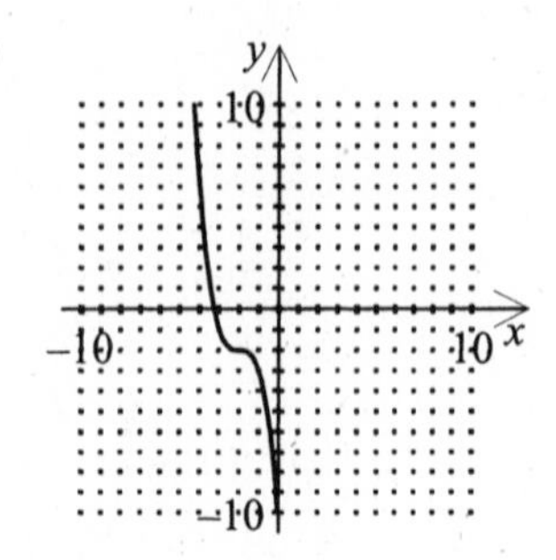

(D)

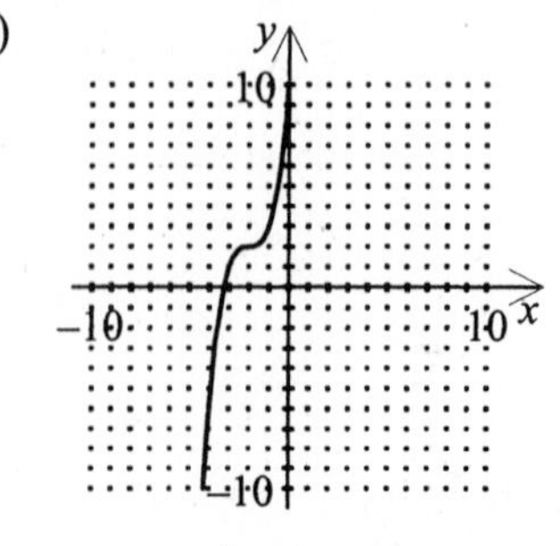

15. Sketch the graph of the function.
$f(x) = 0.5x^3$

16. In the same viewing window of a graphing utility, graph $f(x) = x^2$, $g(x) = x^2 - 3$, $h(x) = -x^2$, and $j(x) = -x^2 - 3$. Describe how the graphs are the same and how they are different.

Objective 2: Use the Leading Coefficient Test to determine the end behavior of graphs of polynomial functions

17. Identify the right-hand and left-hand behavior of the graph of the polynomial function.
$f(x) = 6x^5 + 3x$

(A) Falls to the left.
Rises to the right.

(B) Rises to the left.
Rises to the right.

(C) Falls to the left.
Falls to the right.

(D) Rises to the left.
Falls to the right.

18. Determine which statement is true about the graph of the function.

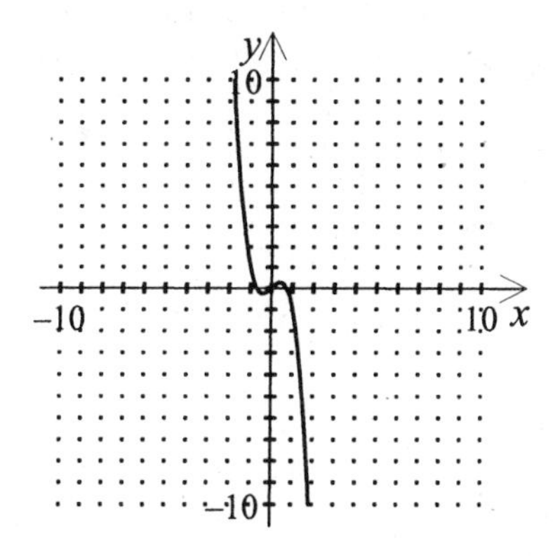

(A) The degree of the function is odd and the leading coefficient is positive.

(B) The degree of the function is even and the leading coefficient is negative.

(C) The degree of the function is odd and the leading coefficient is negative.

(D) The degree of the function is even and the leading coefficient is positive.

19. Identify the right-hand and left-hand behavior of the graph of the polynomial function.
$f(x) = -5x^6 + 4x^2$

20. Tell whether the degree of the function is even or odd and whether the leading coefficient is positive or negative.

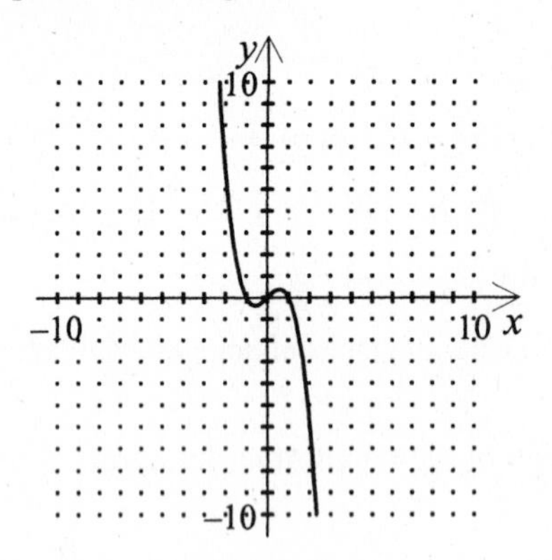

Objective 3: Use zeros of polynomial functions as sketching aids

21. Identify the polynomial function that has zeros at –2, 1, and –1 and matches the graph below.

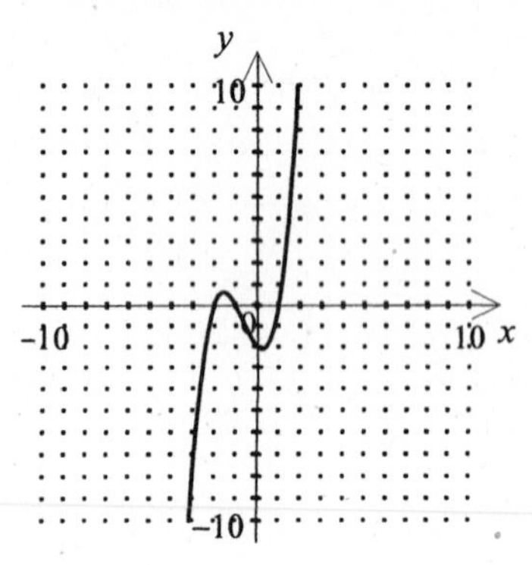

(A) $f(x) = -x^3 - 2x^2 + x - 2$

(B) $f(x) = 2x^2 - x + 1$

(C) $f(x) = x^2 + x - 1$

(D) $f(x) = x^3 + 2x^2 - x - 2$

22. Find all real zeros of the polynomial function.

$f(x) = -5x^4 + 80x^2$

(A) $x = 0,\ x = \pm 4$ (B) $x = 0,\ x = 16$ (C) $x = 0,\ x = 4$ (D) $x = 0,\ x = \pm 16$

23. Estimate the x-intercepts for the graph.

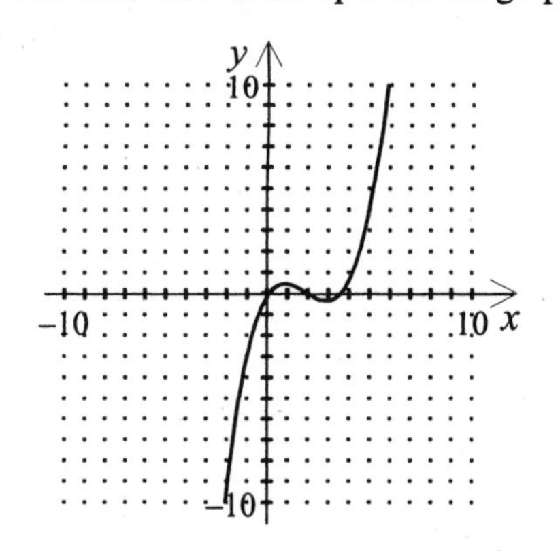

24. Use a graphing utility to graph the function. Use the graph to find any x-intercepts of the graph.
$f(x) = x^3 - x^2 - 9x + 9$

Objective 4: Use the Intermediate Value Theorem to help locate zeros of polynomial functions

25. Determine which of the following correctly uses the Intermediate Value Theorem to show that the graph of the function has a zero in the given interval. Find the value of the zero to two decimal places.
$f(x) = x^3 + 2x^2 - 4x - 5$; $[-3, -2]$

(A) $f(-2) = 3 > f(-3) = -2$
$(-2.79, 0)$

(B) $f(-3) = -2 < 0$ and $f(-2) = 3 > 0$
$(-2.79, 0)$

(C) $f(-3) = -2 < 0$ and $f(-2) = 3 > 0$
$(-5.00, 0)$

(D) $f(-2) = 3 > f(-3) = -2$
$(-5.00, 0)$

26. Use the Intermediate Value Theorem to determine which interval does *not* contain a real zero of the function.
$f(x) = x^3 + 2x^2 - 16x - 31$

(A) $[3, 4]$ (B) $[-2, -1]$ (C) $[4, 5]$ (D) $[-5, -4]$

27. Use the Intermediate Value Theorem to determine an interval bounded by integers one unit in length in which the polynomial function is guaranteed to have a zero.
$f(x) = 7x^5 + 3x^2 - x - 8$

28. Use the Intermediate Value Theorem to show that the graph of the function has a zero in the given interval. Round the zero to two decimal places.
$f(x) = x^3 + 3x^2 - 2x - 5;\ [-4, -3]$

Section 2.3: Polynomial and Synthetic Division

Objective 1: Use long division to divide polynomials by other polynomials

Use long division to divide.

29. $(x^3 - 2x + 8) \div (x + 2)$

(A) $x^2 - 2x - 6 + \dfrac{18}{x+2}$

(B) $x^2 - 4x - 8 + \dfrac{24}{x+2}$

(C) $x^2 - 2x + 2 + \dfrac{4}{x+2}$

(D) $x^2 - 4x + 16 - \dfrac{32}{x+2}$

30. $(-6x^2 - 2x^3 - 1 + 2x) \div (2x + 2)$

(A) $-x^2 - 2x + 4 - \dfrac{9}{2x+2}$

(B) $-x^2 + 2x - 3 - \dfrac{7}{2x+2}$

(C) $-x^2 - 2x + 3 - \dfrac{7}{2x+2}$

(D) $-x^2 + 2x - 4 - \dfrac{9}{2x+2}$

31. $(-x^4 + x^3 + 3x^2 - 5x - 2) \div (x^2 - 2x + 1)$

32. $\dfrac{x^6 - 1}{x - 1}$

Objective 2: Use synthetic division to divide polynomials by binomials of the form (x-k)

33. Use synthetic division to determine how many of the given factors divide the polynomial evenly.
Polynomial: $3x^4 - 16x^3 - 33x^2 + 166x + 120$
Factors: $x + \frac{2}{3},\ x - 5,\ x + \frac{8}{3},\ x - \frac{8}{3},\ x - \frac{2}{3},\ x + 4,\ x + 5$

(A) 3

(B) 2

(C) 1

(D) 4

34. Use synthetic division to determine which one of the following polynomials is *not* a factor of $x^3 - 3x^2 - 25x + 75$.

(A) $x+3$ (B) $x+5$ (C) $x-3$ (D) $x-5$

Use synthetic division to divide.

35. $\left(2x^4 - 6x^3 - 24x - 27\right) \div (x-4)$

36. $\dfrac{x^4 - 33x^2 + 28x + 60}{x+6}$

Objective 3: Use the Remainder Theorem and the Factor Theorem

37. Factor $2x^3 + 3x^2 - 23x - 12$ given that $x+4$ is one of its factors.

(A) $(x+4)(x-3)(2x+1)$ (B) $(x+4)(x+3)(2x-1)$

(C) $(x+4)(x+3)(2x+1)$ (D) $(x+4)(x-3)(2x-1)$

38. Use synthetic division to find $P\left(-\frac{1}{6}\right)$ if $P(x) = 4x^4 - 5x^2 + 5$.

(A) $\frac{77}{36}$ (B) $\frac{3155}{1296}$ (C) $\frac{37}{36}$ (D) $\frac{394}{81}$

39. One of the zeros of $f(x) = x^3 + 12x^2 + 41x + 42$ is –7. Find the other two zeros.

40. Use synthetic division to evaluate $f(x) = x^6 + 2x^5 + 2x^3 - 7x^2 + 33$ for $x = -4$.

Section 2.4: Complex Numbers

Objective 1: Use the imaginary unit i to write complex numbers

41. Express $\sqrt{-29}$ in the form bi where b is a real number.

(A) $-29i$ (B) $\sqrt{-29}i$ (C) $\sqrt{29}i$ (D) $-\sqrt{29}i$

42. Which is the complex number in standard form?
$i+11i^2$

(A) $-11+i$ (B) $-11-i$ (C) $11+i$ (D) $-10i$

43. What real numbers a and b make the equation true?
$(a-1)+(b-6)i = 8+7i$

44. Write the expression $19-\sqrt{-144}$ as a complex number.

Objective 2: Add, subtract, and multiply complex numbers

Perform the indicated operation and write the result in standard form.

45. $(-7-4i)(8-6i)$

(A) $-32+10i$ (B) $-80-10i$ (C) $-80+10i$ (D) $-32+74i$

46. $(4-9i)-(6+7i)$ (A) $-2-16i$ (B) $10+2i$ (C) $87-26i$ (D) $10-2i$

47. $(7+8i)^2$

48. $\sqrt{-192}+\sqrt{-48}$

Objective 3: Use complex conjugates to write the quotient of two complex numbers in standard form

Divide and write the result in standard form.

49. $\dfrac{-5-4i}{4i}$ (A) $\dfrac{16-5i}{4}$ (B) $\dfrac{-4+5i}{4}$ (C) $\dfrac{-4+5i}{16}$ (D) $\dfrac{16-5i}{16}$

50. $\dfrac{-9-i}{-3-5i}$ (A) $\dfrac{16-21i}{16}$ (B) $\dfrac{22+21i}{16}$ (C) $\dfrac{22+21i}{17}$ (D) $\dfrac{16-21i}{17}$

51. $\dfrac{1-5i}{1+7i}$

Divide and write the result in standard form.

52. $\dfrac{-i}{(3+i)^2}$

Objective 4: Find complex solutions of quadratic equations

Use the Quadratic Formula to solve.

53. $2x^2+\frac{1}{2}x+5=0$

(A) $-\frac{2}{3}\pm\frac{17}{3}\sqrt{\frac{4}{5}}\,i$ (B) $-\frac{1}{8}\pm\frac{1}{8}\sqrt{159}\,i$ (C) $\frac{2}{3}\pm\frac{17}{3}\sqrt{\frac{4}{5}}\,i$ (D) $\frac{1}{8}\pm\frac{1}{8}\sqrt{159}\,i$

54. $3x^2+32=0$

(A) $x=\dfrac{\pm\sqrt{6}i}{8}$ (B) $x=\pm4\sqrt{2}i$ (C) $x=\dfrac{\pm4\sqrt{6}i}{3}$ (D) $x=\dfrac{\pm4\sqrt{2}i}{3}$

55. $x^2-8x+32=0$

56. $-9x+6+5x^2=0$

Section 2.5: Zeros of Polynomial Functions

Objective 1: Use the Fundamental Theorem of Algebra to determine the number of zeros of polynomial functions

Determine the maximum number of zeros of the polynomial function.

57. $f(x)=-3x^8-7x^7+9x-3$ (A) 15 (B) –3 (C) 8 (D) 7

58. $f(x)=9x+3$ (A) 2 (B) 1 (C) 4 (D) 3

59. $f(x)=x^3(x+8)^3$

Determine the maximum number of zeros of the polynomial function.

60. $f(x) = 5x^4(2x^4 - 6x^3 + x - 6)$

Objective 2: Find rational zeros of polynomial functions

Find all the real zeros of the function.

61. $f(x) = x^3 - x^2 - 14x + 24$ (A) 3, 2, –4 (B) 4, 3, 2 (C) –2, –3, 4 (D) –4, –2, 3

62. $f(x) = 6x^4 - 27x^3 - 54x^2 + 213x - 90$

(A) $-2,\ -\frac{1}{2},\ 3,\ -5$ (B) $-6,\ 5,\ -\frac{1}{5}$ (C) $-3,\ 1,\ -\frac{1}{3}$ (D) $2,\ \frac{1}{2},\ -3,\ 5$

63. $f(x) = 27x^3 + 27x^2 - 4$

64. Use the Rational Zero Test to determine all possible rational zeros of *f*. Do not find the actual zeros.
$f(x) = 2x^5 + 2x^3 - 3x^2 - 10$

Objective 3: Find conjugate pairs of complex zeros

65. A polynomial of degree 5 whose coefficients are real numbers has the zeros 4, $-6i$, and $-6+i$. Identify the remaining zeros.

(A) $-6i,\ 6-i$ (B) $6i,\ -6+i$ (C) $-4,\ 6i,\ -6-i$ (D) $6i,\ -6-i$

66. Given that one zero of $P(x) = x^3 + 2x^2 + 9x + 68$ is $1-4i$, which of the following is also a zero of $P(x)$?

(A) $-4+i$ (B) $-1-4i$ (C) $4+i$ (D) $1+4i$

67. Find a quadratic function with solutions $3+2i$ and $3-2i$ and a leading coefficient of 1.

68. Given that one zero of $P(x) = x^3 + 9x^2 + 43x + 75$ is $-3-4i$, find all zeros of $P(x)$.

Objective 4: Find zeros of polynomials by factoring

69. Identify the polynomial written in completely factored form.
$6x^3 - 7x^2 - 11x + 12$

(A) $(2x-3)(x-1)(3x+4)$ (B) $(2x-3)(x-1)(3x-4)$

(C) $(2x+3)(x-1)(3x+4)$ (D) $(2x-3)(x+1)(3x+4)$

70. Find the four real zeros of the polynomial.
$f(x) = x^4 + 3x^3 - 12x^2 - 20x + 48$

(A) –2, 2, –4, 3 (B) –2, 2, –4, –3 (C) 2, 2, 4, –3 (D) 2, 2, –4, –3

71. Find all the zeros of the function.
$f(x) = x^4 + 2x^3 - 2x^2 - 8x - 8$

72. Find the complete factorization of $2x^4 + 17x^3 + 49x^2 + 58x + 24$ if $(2x+3)$ is one of the factors.

Objective 5: Use Descartes's Rule of Signs and the Upper and Lower Bound Rules to find zeros of polynomials

73. Use synthetic division to find upper and lower bounds of the real zeros of f.
$f(x) = 6x^3 - 11x^2 - 149x + 154$

(A) Upper: $x = 1$
Lower: $x = -4$

(B) Upper: $x = 5$
Lower: $x = 1$

(C) Upper: $x = 6$
Lower: $x = -5$

(D) Upper: $x = 6$
Lower: $x = -4$

Use Descartes's Rule of Signs to determine the possible number of positive and negative zeros of the function.

74. $f(x) = x^6 - 4x^5 + 3x^4 + 4x^3 - 4x^2 - 4x + 5$

(A) Four, two, or no positive zeros
Three or one negative zeros

(B) Four, two, or no positive zeros
Two or no negative zeros

(C) Five, three, or one positive zeros
Two or no negative zeros

(D) Two or no positive zeros
Four, two, or no negative zeros

75. $f(x) = x^6 + x^5 + 2x^4 + x^3 + x^2 - 4x - 3$

76. Use synthetic division to find upper and lower bounds of the real zeros of *f*.

$$f(x) = 2x^4 + 9x^3 - 3x^2 - 34x - 24$$

Section 2.6: Rational Functions

Objective 1: Find the domains of rational functions

Determine the domain of the function.

77. $f(x) = \dfrac{9x}{x(x-4)}$

(A) All real numbers $x \neq \pm 4, x \neq 0$

(B) All real numbers $x \neq 2$

(C) All real numbers $x \neq \pm 2$

(D) All real numbers $x \neq 4,\ x \neq 0$

78. $f(x) = \dfrac{8x}{x(x^2 - 36)}$

(A) All real numbers $x \neq -6,\ x \neq 6$

(B) All real numbers $x \neq 36,\ x \neq 0$

(C) All real numbers $x \neq 6$

(D) All real numbers $x \neq -6,\ x \neq 6,\ x \neq 0$

79. Determine the domain of *f* and *g*. Complete the table and explain how the functions differ.

$$f(x) = \frac{2x-2}{x^2 - 7x + 6},\quad g(x) = \frac{2}{x-6}$$

x	0	1	2	3	4	5	6
$f(x)$							
$g(x)$							

80. Find the domain of the function.

$$f(x) = \frac{x^2 - x - 20}{x^2 - 11x + 18}$$

Objective 2: Find the horizontal and vertical asymptotes of graphs of rational functions

81. Find the horizontal asymptote of the graph of $f(x)=\frac{7}{x-5}$.

(A) $y=7$ (B) $x=5$ (C) $x=0$ (D) $y=0$

82. Find the vertical asymptote(s), if any, for $f(x)=\frac{5x-3}{x^2+2x-3}$.

(A) $x=1,\ x=-3,\ x=-4$ (B) $x=1,\ x=-3$

(C) $x=-4,\ x=1$ (D) No vertical asymptotes

Identify any horizontal and vertical asymptotes for the graph of the function.

83. $f(x)=\frac{8x+2}{4x+3}$.

84. $f(x)=\frac{3x^2}{x^2-16}$

Objective 3: Analyze and sketch graphs of rational functions

Identify the graph of the rational function. Find any vertical and horizontal asymptotes.

85. $f(x)=\frac{x+3}{x+2}$

(A)

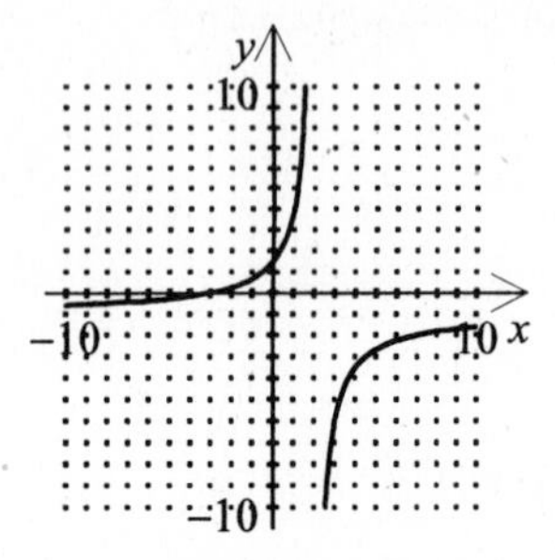

Asymptotes: $y = -1$, $x = 2$

(B)

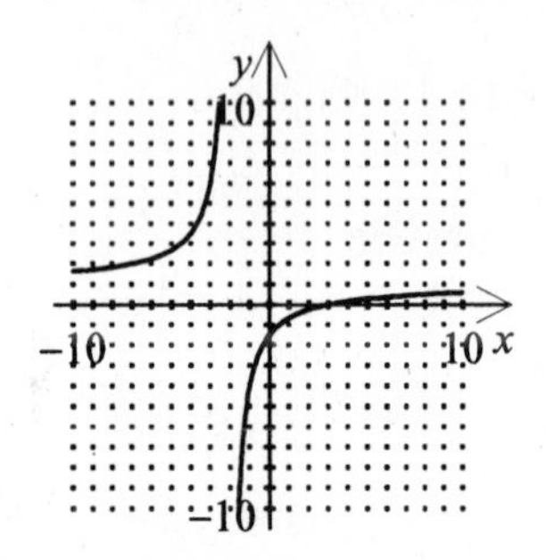

Asymptotes: $y = 1$, $x = -2$

(C)

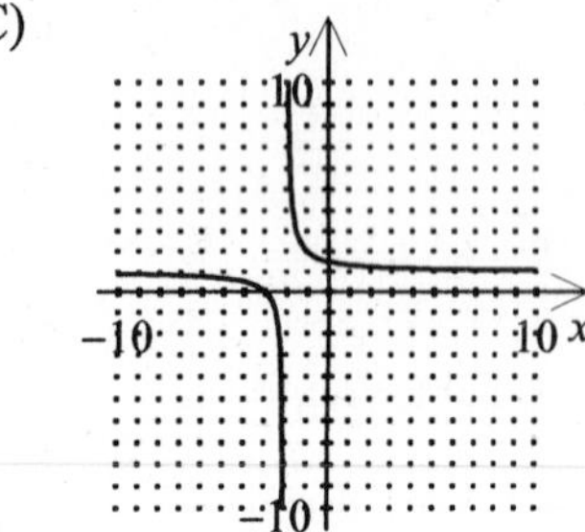

Asymptotes: $y = 1$, $x = -2$

(D)

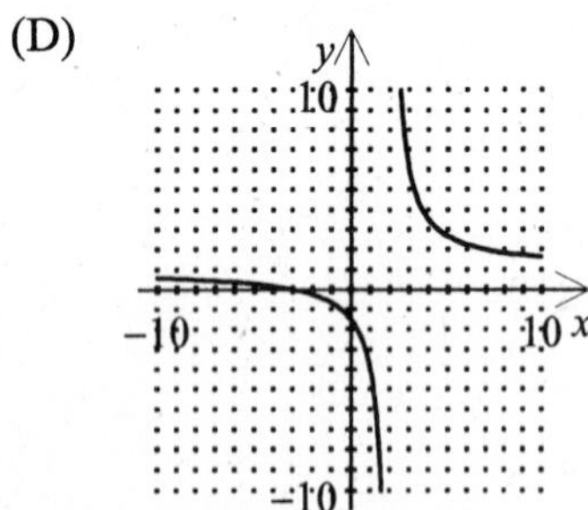

Asymptotes: $y = 1$, $x = 2$

(85.)

Identify the graph of the rational function. Find any vertical and horizontal asymptotes.

86. $f(x)=\dfrac{2x^2-5x-3}{x^2-4}$

(A)

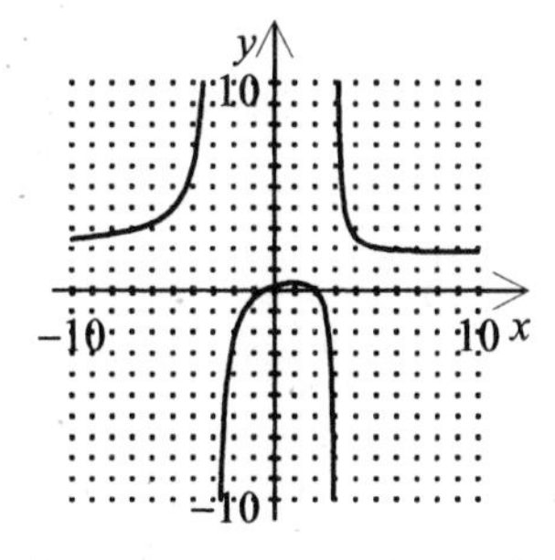

Asymptotes: $x=-3, x=3, y=2$

(B)

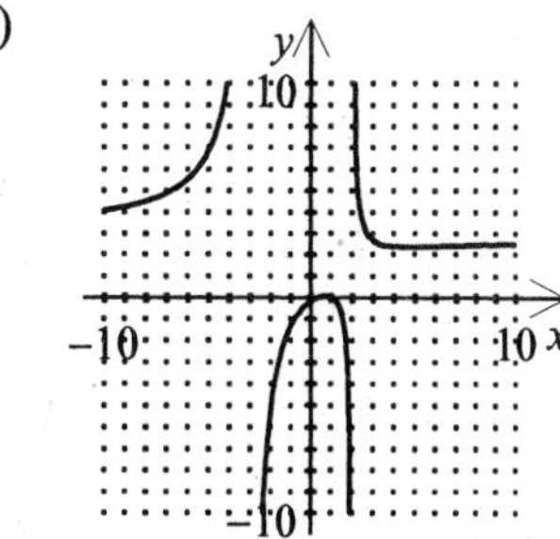

Asymptotes: $x=-3, x=2, y=3$

(C)

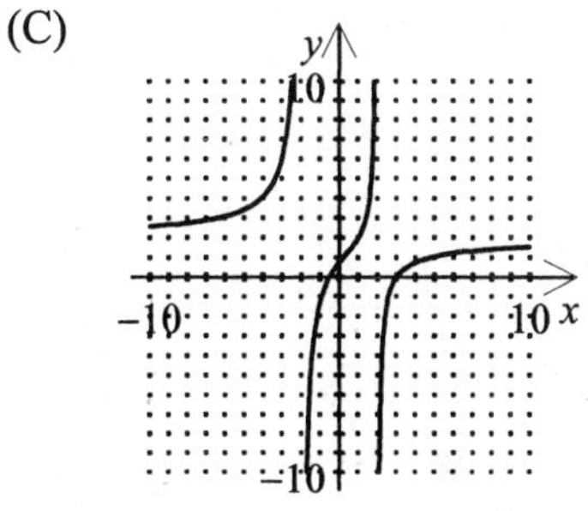

Asymptotes: $x=-2, x=2, y=2$

(D)

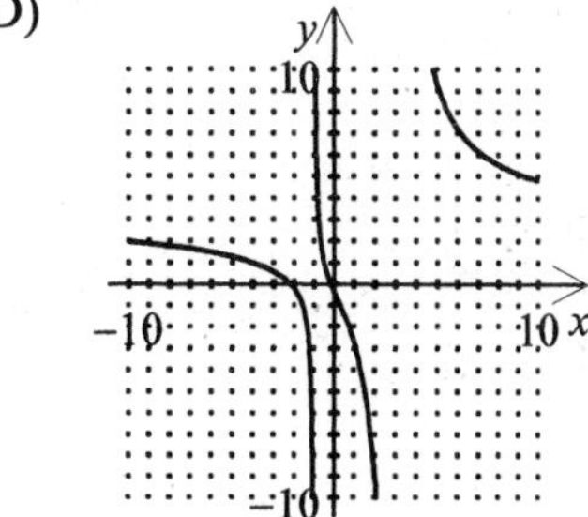

Asymptotes: $x=-1, x=3, y=3$

Sketch the graph of the rational function. Find any vertical and horizontal asymptotes.

87. $f(x)=-\dfrac{3}{x^2-4}$

88. $f(x)=\dfrac{4}{x-2}-1$

Objective 4: Sketch graphs of rational functions that have slant asymptotes

89. Identify the graph of the rational function and find the equation of the slant asymptote.

$$f(x) = \frac{-2x^2 - 3x + 2}{x + 1}$$

(A)

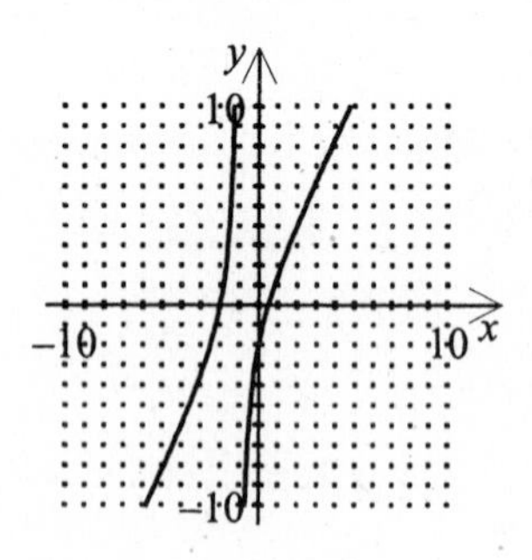

Slant asymptote: $y = -2x - 1$

(B)

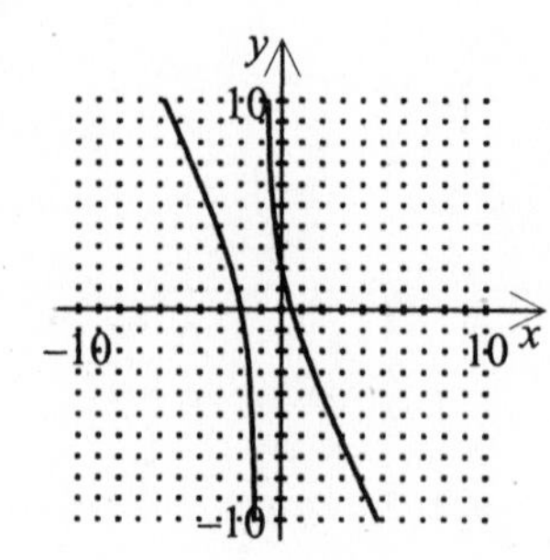

Slant asymptote: $y = -2x - 1$

(C)

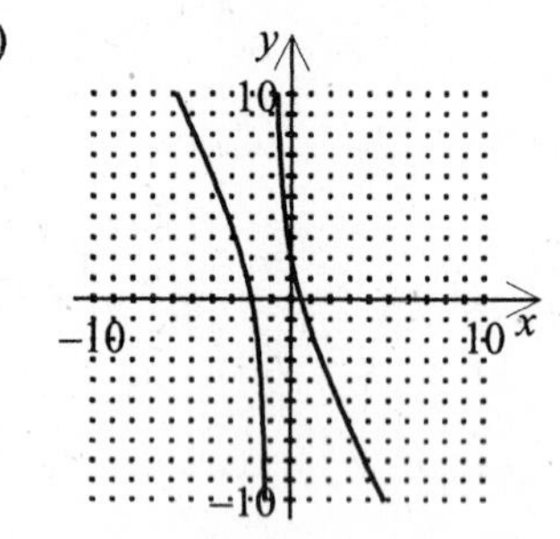

Slant asymptote: $y = 2x + 1$

(D)

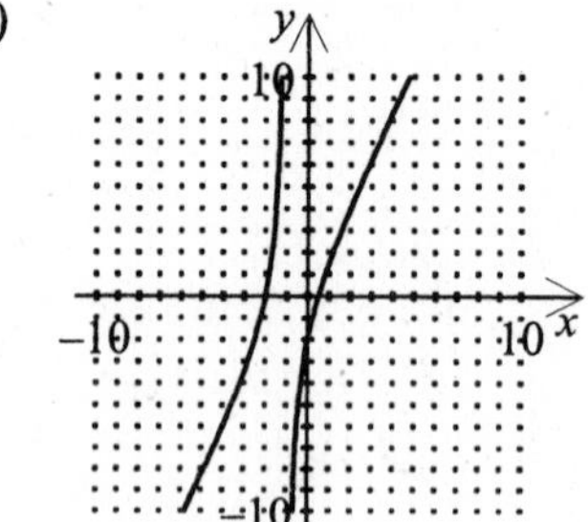

Slant asymptote: $y = 2x + 1$

Find the equation of the slant asymptote of the graph of the rational function.

90. $f(x) = \dfrac{2x^3 - x^2 - 22x - 28}{x^2 - 3x - 4}$

(A) $y = 2x - 5$ (B) $y = -2x + 5$ (C) $y = 2x$ (D) $y = 2x + 5$

91. $f(x) = \dfrac{2x^3 + 4x^2 - x - 10}{x^2 + 3x + 2}$

92. Graph the rational function and find the equation of the slant asymptote.

$$f(x) = \frac{x^2}{x + 3}$$

Objective 5: Use rational functions to model and solve real-life problems

93. Calypso Coffee mixes 100 pounds of a standard coffee containing 25% Columbian coffee beans with x pounds of a premium coffee containing 75% Columbian coffee beans. The concentration of Columbian coffee beans in the final mix is given by

$$C = \frac{3x+100}{4(x+100)}.$$

Identify the graph of the concentration function and find the concentration of Columbian coffee beans the graph approaches as x increases.

(A)

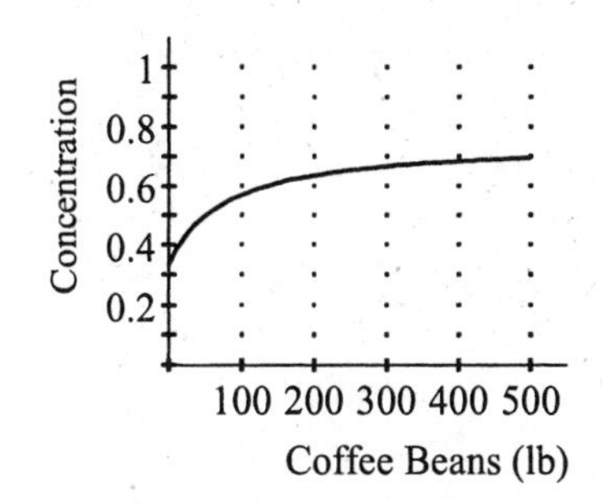

The concentration approaches 75%.

(B)

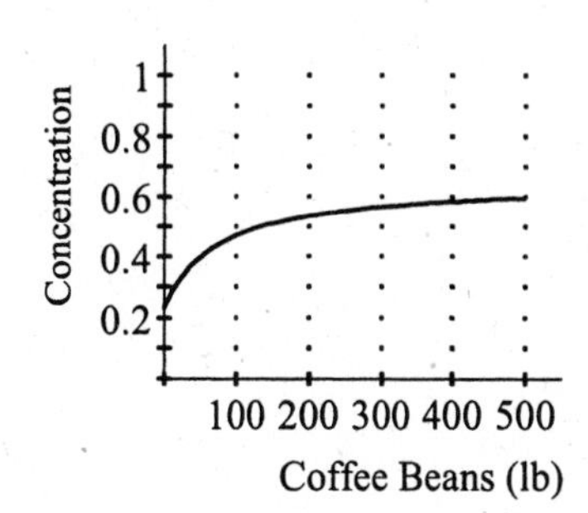

The concentration approaches 65%.

(C)

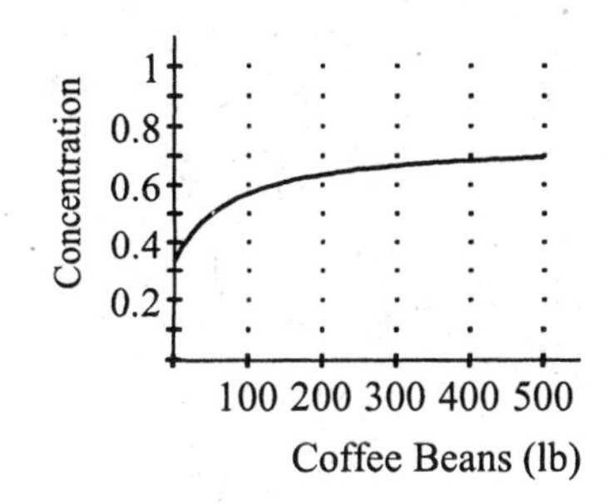

The concentration approaches 50%.

(D)

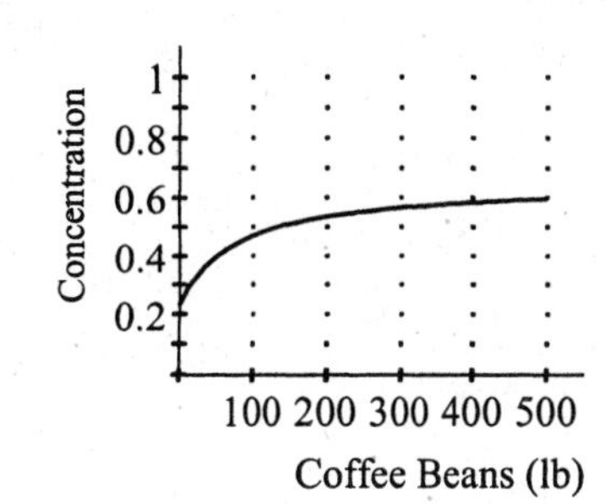

The concentration approaches 75%.

94. After an accident in which organic waste fell into a pond, the decomposition process included oxidation whereby oxygen that was dissolved in the pond water combined with the decomposing waste. The oxygen level of the pond is

$$O = \frac{t^2 - t + 1}{t^2 + 1}$$

where $O = 1$ represents the normal oxygen level of the pond and t represents the number of weeks after the accident. Identify the graph of this model and find the concentration of oxygen after 6 weeks.

(A)

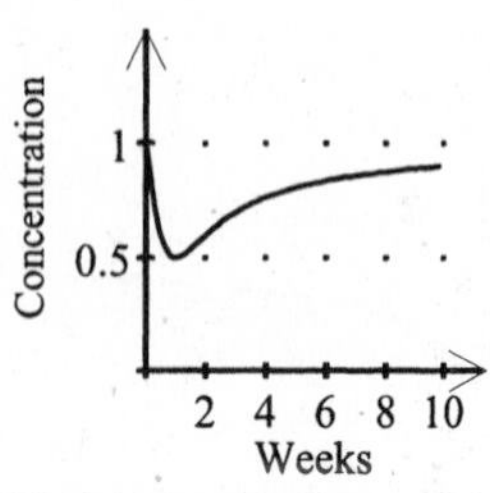

The concentration of O is 0.84.

(B)

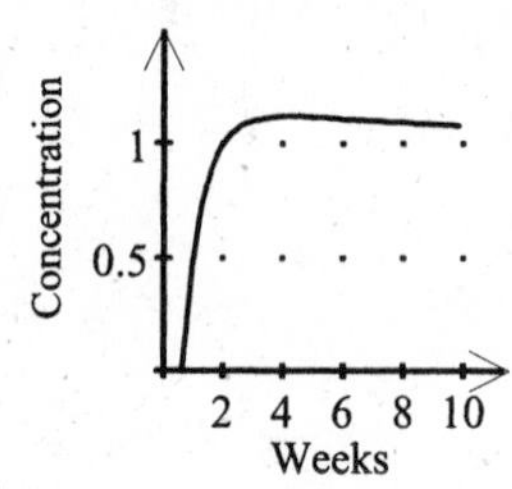

The concentration of O is 1.19.

(C)

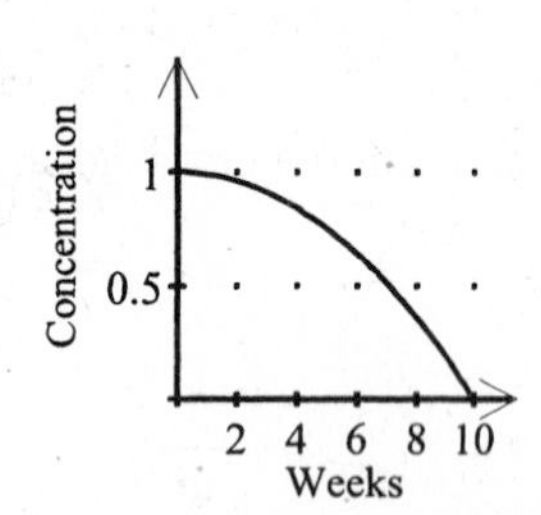

The concentration of O is 0.42.

(D)

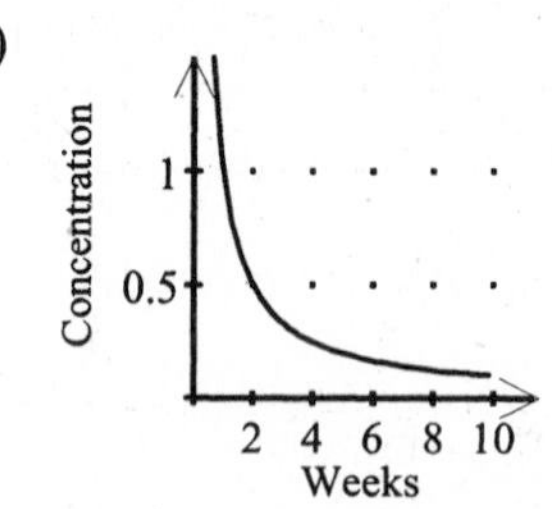

The concentration of O is 0.17.

95. The total revenue R from the sale of a popular new board game is

$$R = \frac{140x^2}{x^2 + 2}$$

where x is the number of years since the game has been released and the revenue R is in millions of dollars. Use a graphing utility to graph the function and find
(a) the revenue generated by the end of the second year.
(b) the revenue generated by the end of the third year.
(c) the maximum revenue that can be expected as the years increase.

96. You are beginning a small business, Barely Bears, that will manufacture plush toy bears. The cost of buying the initial sewing equipment is \$350. Additionally, the materials for each plush bear cost \$2.50. The average cost per bear is

$$\overline{C} = \frac{2.50x + 350}{x}$$

where x represents the number of bears the business has produced. Use a graphing utility to sketch a graph of the average cost equation and then find the cost to manufacture 55 toy bears. What happens to the average cost per bear if you produce 550 bears? What does the limiting cost per bear appear to be?

Section 2.7: Partial Fractions

Objective 1: Recognize partial fraction decompositions of rational expressions

97. Identify the form of the partial fraction decomposition. Do not solve for the constants.

$$\frac{3x^2 + 8x + 1}{x^3 + 6x}$$

(A) $\frac{A}{x} + \frac{Bx}{x^2} + \frac{C}{6}$ (B) $\frac{A}{x} + \frac{B}{x+6} + \frac{C}{x+6}$ (C) $\frac{Ax^2 + Bx + C}{x^3 + 6x}$ (D) $\frac{A}{x} + \frac{Bx + C}{x^2 + 6}$

98. Which is the partial fraction decomposition for the rational expression?

$$\frac{10x^2 - 18}{x\left(x^2 - 9\right)}$$

(A) $\frac{3}{x} + \frac{4}{x^2} + \frac{2}{x-1}$ (B) $\frac{2}{x} + \frac{4x+1}{x^2+3}$ (C) $\frac{2}{x} + \frac{4}{x+3} + \frac{4}{x-3}$ (D) $\frac{6}{x} + \frac{2}{x-3}$

Write the form of the partial fraction decomposition for the rational expression. Do not solve for the constants.

99. $\frac{5x + 4}{x\left(x^2 + 5\right)^2}$

100. $\frac{9x^2 + 5x + 8}{x^3 + 4x^2 + 4x}$

Objective 2: Find partial fraction decompositions of rational expressions

Find the partial fraction decomposition for the rational expression.

101. $\dfrac{-3x^2-4x-13}{\left(x^2+5\right)^2}$

(A) $\dfrac{3}{x^2+5}+\dfrac{4x-2}{\left(x^2+5\right)^2}$

(B) $\dfrac{3}{x^2+5}-\dfrac{4x-2}{\left(x^2+5\right)^2}$

(C) $-\dfrac{3}{x^2+5}+\dfrac{4x-2}{\left(x^2+5\right)^2}$

(D) $-\dfrac{3}{x^2+5}-\dfrac{4x-2}{\left(x^2+5\right)^2}$

102. $\dfrac{5x^2+88x+308}{(x+7)(x+4)(x+10)}$

(A) $-\dfrac{7}{x+7}+\dfrac{2}{x+4}-\dfrac{4}{x+10}$

(B) $\dfrac{7}{x+7}+\dfrac{2}{x+4}+\dfrac{4}{x+10}$

(C) $\dfrac{7}{x+7}+\dfrac{2}{x+4}-\dfrac{4}{x+10}$

(D) $-\dfrac{7}{x+7}-\dfrac{2}{x+4}+\dfrac{4}{x+10}$

103. $\dfrac{-14}{(x+5)(x-2)}$

104. $\dfrac{8x^2+4x+4}{x\left(x^2+2\right)}$

Answer Key for Chapter 2 Polynomial and Rational Functions

Section 2.1: Quadratic Functions

Objective 1: Analyze graphs of quadratic functions

[1] (A)

[2] (A)

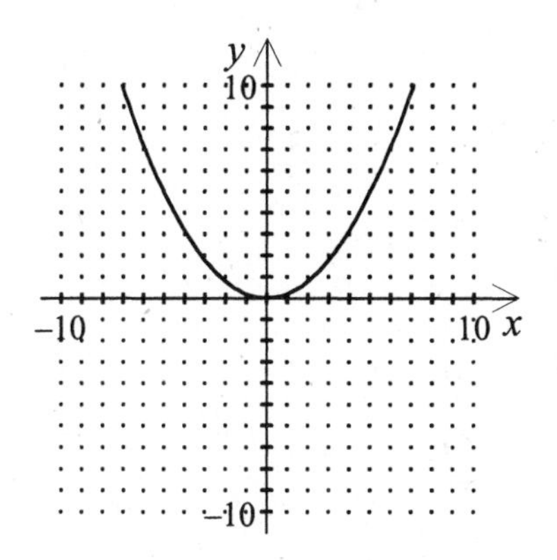

[3] Vertex: $(0, 0)$

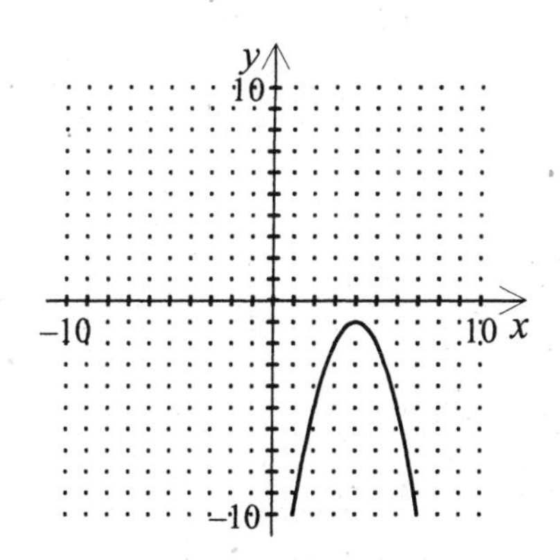

[4] Vertex: $(4, -1)$

Objective 2: Write quadratic functions in standard form and use the results to sketch graphs of functions

[5] (C)

[6] (A)

Vertex: $(2,\ -9)$
x-intercepts: $(3.73,\ 0)$, $(0.27,\ 0)$

[7] $f(x) = 3(x-2)^2 - 9$

[8] $y = (x+4)^2 - 2$

Objective 3: Use quadratic functions to model and solve real-life problems

[9] (B)

[10] (A)

(a) $A(x) = x\left(\dfrac{232 - x}{2}\right) = 116x - \dfrac{1}{2}x^2$

[11] (b) 116 m

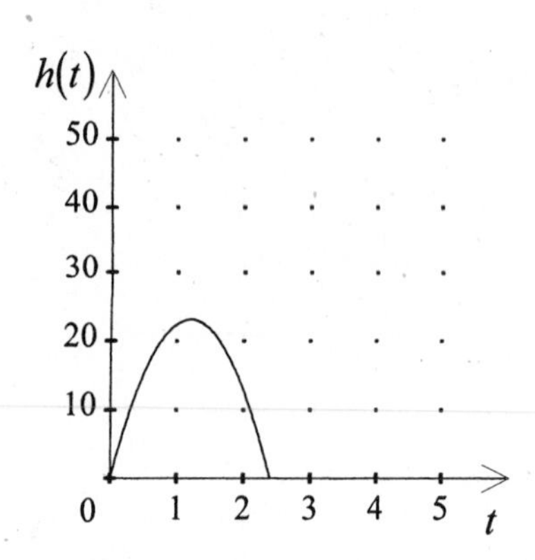

[12] 23.04 ft

Section 2.2: Polynomial Functions of Higher Degree

Objective 1: Use transformations to sketch graphs of polynomial functions

[13] (B)

[14] (A)

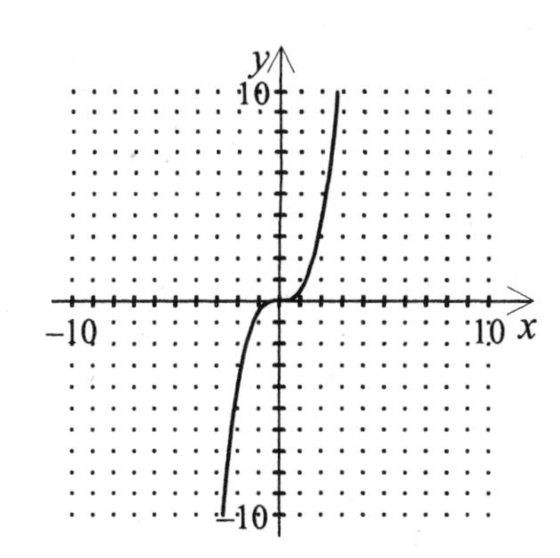

[15]

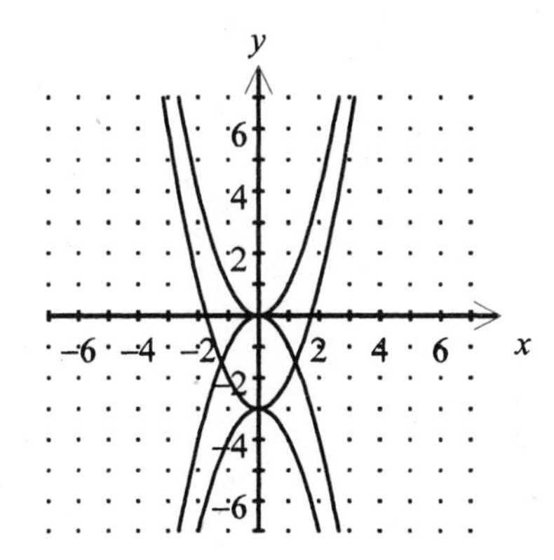

[16] The graphs of $y = x^2$ and $y = x^2 - 3$ both open upward, while the other two graphs open downward. The graphs of $y = x^2$ and $y = -x^2$ pass through the origin, while the graphs of $y = x^2 - 3$ and $y = -x^2 - 3$ both pass through $(0, -3)$.

Objective 2: Use the Leading Coefficient Test to determine the end behavior of graphs of polynomial functions

[17] (A)

[18] (C)

[19] Falls to the left.
Falls to the right.

[20] The degree of the function is odd and the leading coefficient is negative.

Objective 3: Use zeros of polynomial functions as sketching aids

[21] (D)

[22] (A)

[23] $(0, 0), (2, 0), (4, 0)$

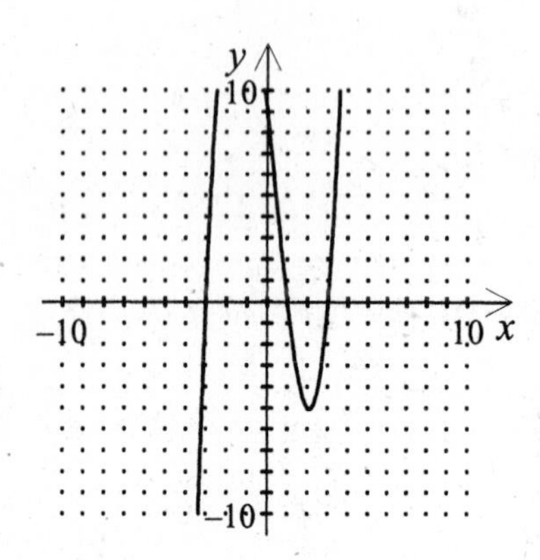

[24] $(3, 0), (1, 0), (-3, 0)$

Objective 4: Use the Intermediate Value Theorem to help locate zeros of polynomial functions

[25] (B)

[26] (C)

[27] $[0, 1]$

$f(-4) = -13 < 0$ and $f(-3) = 1 > 0$

[28] $(-3.13, 0)$

Section 2.3: Polynomial and Synthetic Division

Objective 1: Use long division to divide polynomials by other polynomials

[29] (C)

[30] (C)

[31] $-x^2 - x + 2 - \dfrac{4}{x^2 - 2x + 1}$

[32] $x^5 + x^4 + x^3 + x^2 + x + 1$

Objective 2: Use synthetic division to divide polynomials by binomials of the form (x-k)

[33] (B)

[34] (A)

[35] $2x^3 + 2x^2 + 8x + 8 + \dfrac{5}{x - 4}$

[36] $x^3 - 6x^2 + 3x + 10$

Objective 3: Use the Remainder Theorem and the Factor Theorem

[37] (A)

[38] (D)

[39] –3 and –2

[40] 1841

Section 2.4: Complex Numbers

Objective 1: Use the imaginary unit i to write complex numbers

[41] (C)

[42] (A)

[43] $a = 9,\ b = 13$

[44] $19 - 12i$

Objective 2: Add, subtract, and multiply complex numbers

[45] (C)

[46] (A)

[47] $-15 + 112i$

[48] $12\sqrt{3}i$

Objective 3: Use complex conjugates to write the quotient of two complex numbers in standard form

[49] (B)

[50] (D)

[51] $-\frac{17}{25} - \frac{6}{25}i$

[52] $-\frac{3}{50} - \frac{2}{25}i$

Objective 4: Find complex solutions of quadratic equations

[53] (B)

[54] (C)

[55] $4 \pm 4i$

[56] $\frac{9 \pm \sqrt{39}i}{10}$

Section 2.5: Zeros of Polynomial Functions

Objective 1: Use the Fundamental Theorem of Algebra to determine the number of zeros of polynomial functions

[57] (C)

[58] (B)

[59] 6

[60] 8

Objective 2: Find rational zeros of polynomial functions

[61] (A)

[62] (D)

[63] $-\frac{2}{3}, \frac{1}{3}$

[64] $\pm 1, \pm 2, \pm 5, \pm 10, \pm\frac{1}{2}, \pm\frac{5}{2}$

Objective 3: Find conjugate pairs of complex zeros

[65] (D)

[66] (D)

[67] $f(x) = x^2 - 6x + 13$

[68] $-3 \pm 4i, -3$

Objective 4: Find zeros of polynomials by factoring

[69] (A)

[70] (D)

[71] $\pm 2, \; -1 \pm i$

[72] $(2x+3)(x+2)(x+4)(x+1)$

Objective 5: Use Descartes's Rule of Signs and the Upper and Lower Bound Rules to find zeros of polynomials

[73] (C)

[74] (B)

[75] One positive zero
Five, three, or one negative zeros

[76] Upper: $x = 2$
Lower: $x = -5$

Section 2.6: Rational Functions

Objective 1: Find the domains of rational functions

[77] (D)

[78] (D)

x	0	1	2	3	4	5	6
$f(x)$	$-\frac{1}{3}$	undefined	$-\frac{1}{2}$	$-\frac{2}{3}$	-1	-2	undefined
$g(x)$	$-\frac{1}{3}$	$-\frac{2}{5}$	$-\frac{1}{2}$	$-\frac{2}{3}$	-1	-2	undefined

[79] The functions differ only in where they are undefined.
Domain of f: all real numbers $x \neq 1, \; 6$;
Domain of g: all real numbers $x \neq 6$

[80] All real numbers $x \neq 2, \; x \neq 9$

Objective 2: Find the horizontal and vertical asymptotes of graphs of rational functions

[81] (D)

[82] (B)

[83] Horizontal asymptote: $y = 2$
Vertical asymptote: $x = -\frac{3}{4}$

[84] Horizontal asymptote: $y = 3$
Vertical asymptotes: $x = -4, \quad x = 4$

Objective 3: Analyze and sketch graphs of rational functions

[85] (C)

[86] (C)

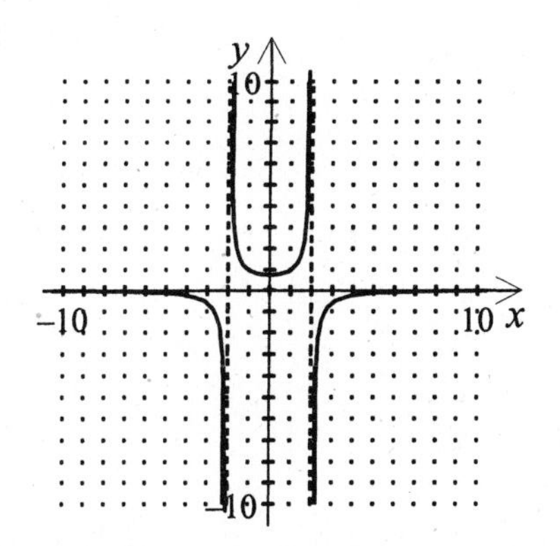

[87] Asymptotes: $y = 0$; $x = 2$; $x = -2$

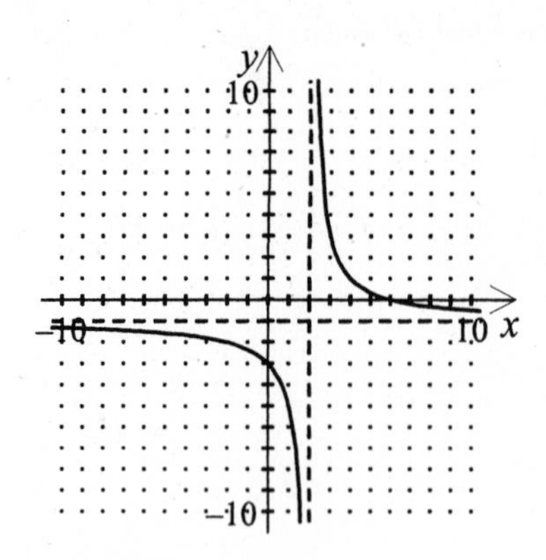

[88] Asymptotes: $x = 2$; $y = -1$

Objective 4: Sketch graphs of rational functions that have slant asymptotes

[89] (B)

[90] (D)

[91] $y = 2x - 2$

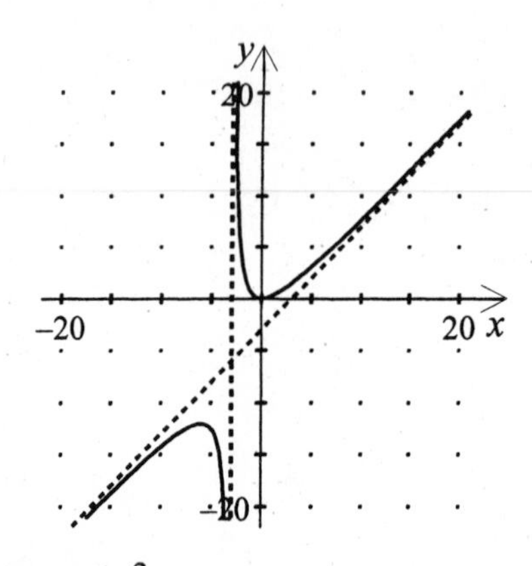

[92] $y = x - 3$

Objective 5: Use rational functions to model and solve real-life problems

[93] (A)

[94] (A)

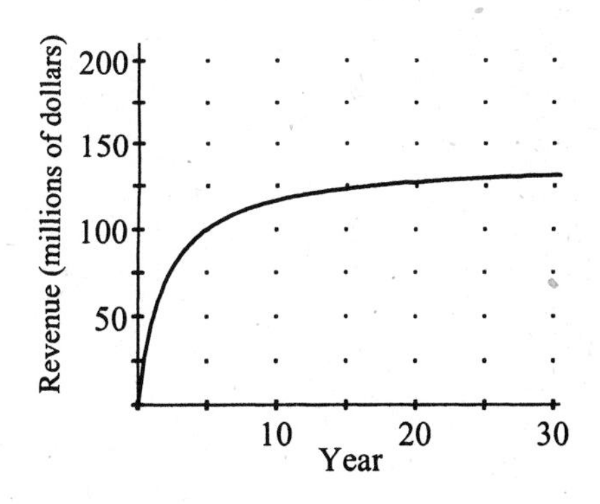

(a) \$93.3 million
(b) \$114.5 million
[95] (c) \$140 million

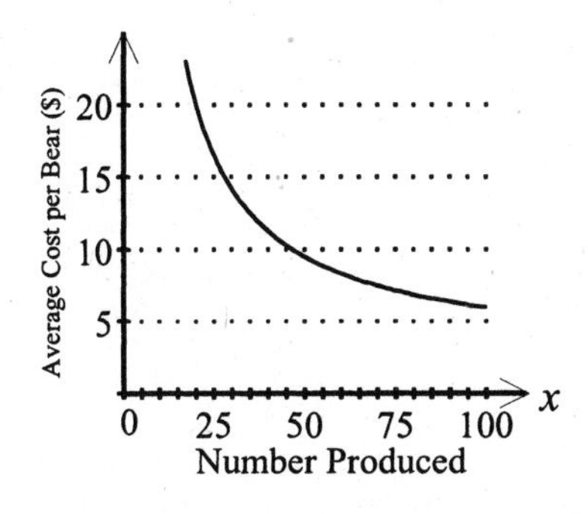

\$8.86 per bear for 55 bears
\$3.14 per bear for 550 bears
[96] The limiting cost per bear is \$2.50.

Section 2.7: Partial Fractions

Objective 1: Recognize partial fraction decompositions of rational expressions

[97] (D)

[98] (C)

[99] $\dfrac{A}{x}+\dfrac{Bx+C}{x^2+5}+\dfrac{Dx+E}{\left(x^2+5\right)^2}$

[100] $\dfrac{A}{x}+\dfrac{B}{x+2}+\dfrac{C}{(x+2)^2}$

Objective 2: Find partial fraction decompositions of rational expressions

[101] (D)

[102] (C)

[103] $\frac{2}{x+5}-\frac{2}{x-2}$

[104] $\frac{2}{x}+\frac{6x+4}{x^2+2}$

Chapter 3 Exponential and Logarithmic Functions

Section 3.1: Exponential Functions and Their Graphs

Objective 1: Recognize and evaluate exponential functions with base a

Evaluate the expression.

1. $500\left(3^{-1.8}\right)$ (A) 69.207 (B) –2916 (C) –2700 (D) –54.243

2. $2^{-\pi}$ (A) –1.142 (B) –6.283 (C) 0.113 (D) 8.825

Evaluate the expression. Round the result to three decimal places.

3. $5^{2.6}$

4. $3.9^{-2/7}$

Objective 2: Graph exponential functions

Identify the graph of the function.

5. $f(x) = 2^{-2-x}$

(A)

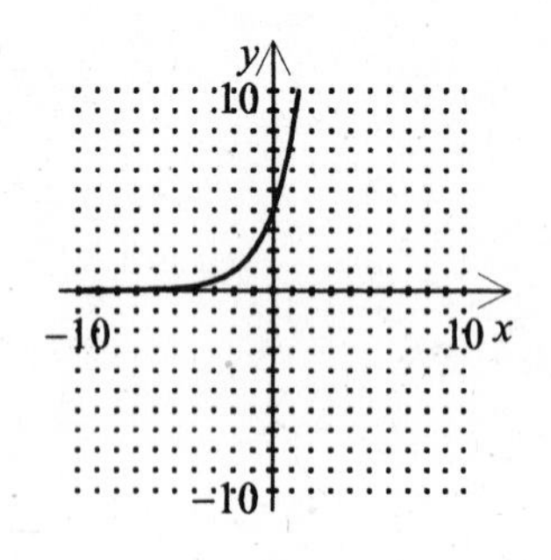

(B)

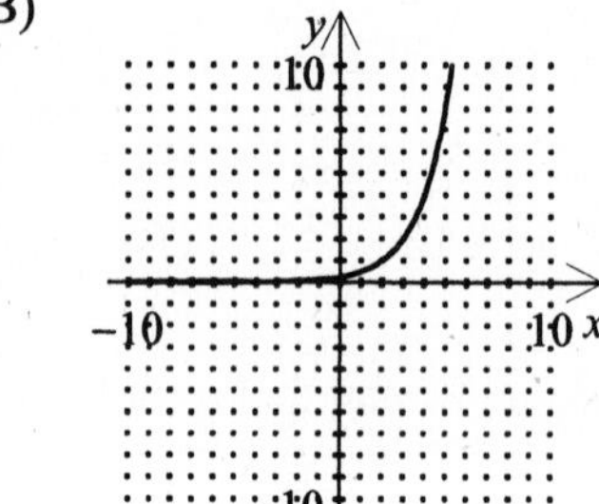

(C)

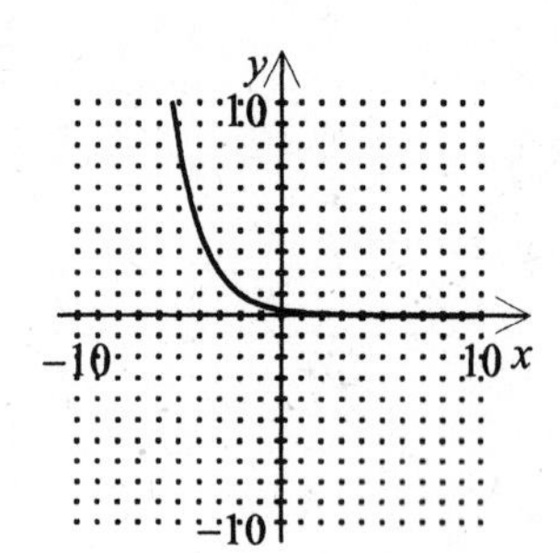

(D)

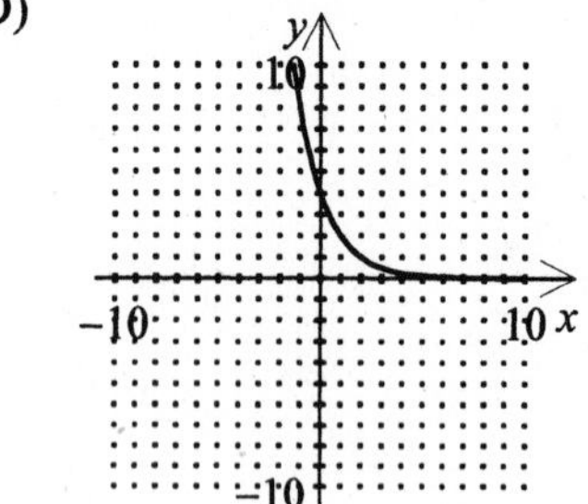

Identify the graph of the function.

6. 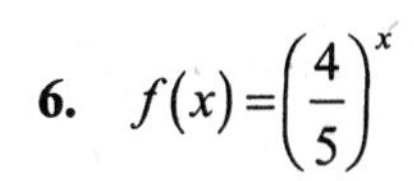
$f(x) = \left(\frac{4}{5}\right)^x$

(A)

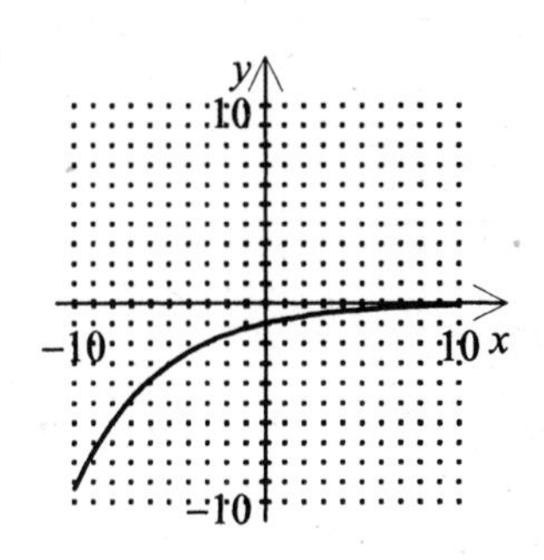

(B)

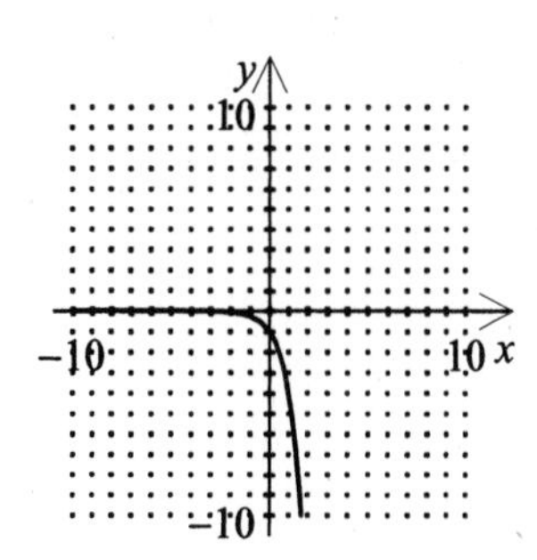

(C)

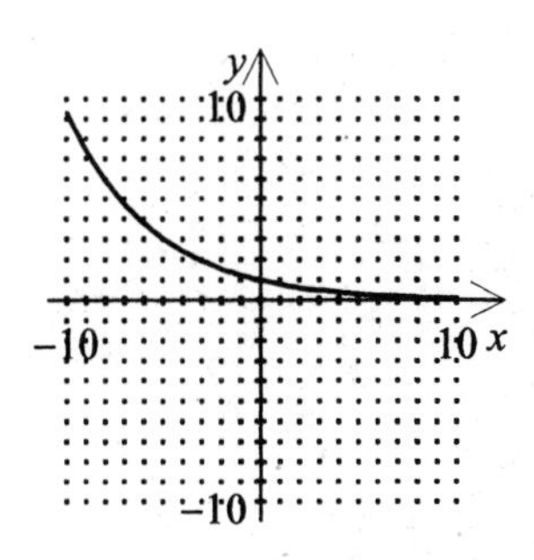

(D)

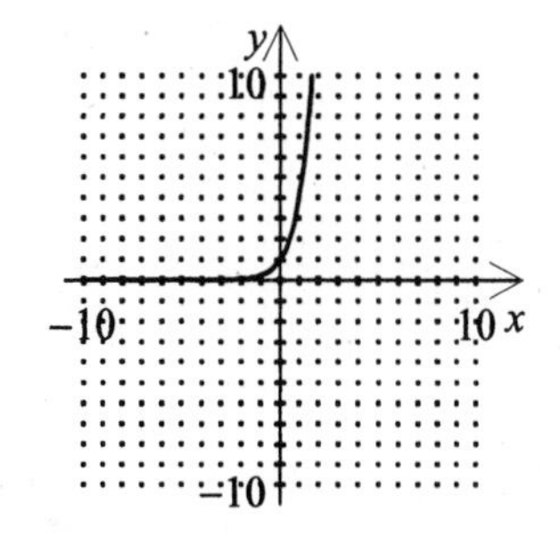

Sketch the graph of the function.

7. $f(x) = \left(\frac{3}{4}\right)^x$

8. $f(x) = \left(\frac{1}{2}\right)^{x+1} + 4$

Objective 3: Recognize and evaluate exponential functions with base e

9. Evaluate the expression. (A) 0.018 (B) 10.873 (C) 54.598 (D) 53.364

e^4

10. Identify the graph of the function.
$f(x) = 2 - e^x$

(A)
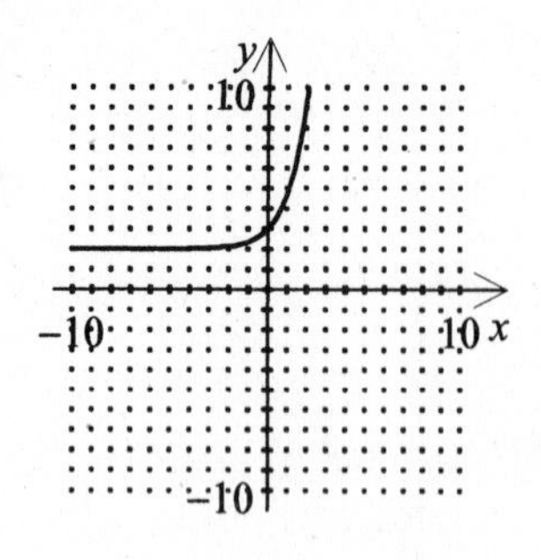

(B)
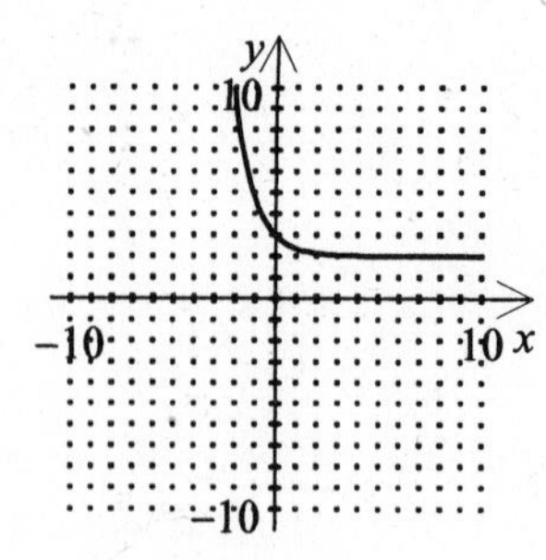

(C)
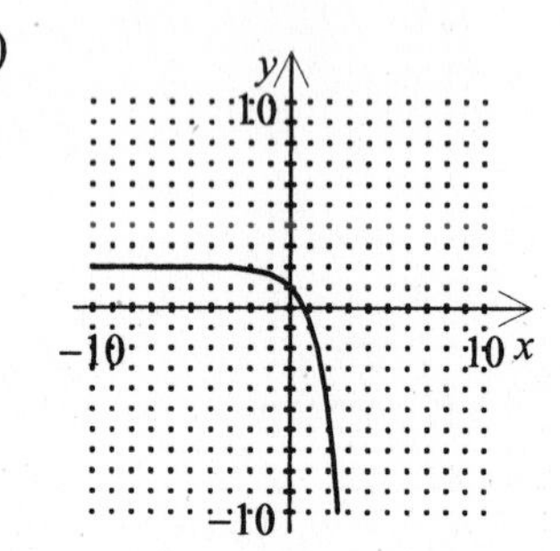

(D)
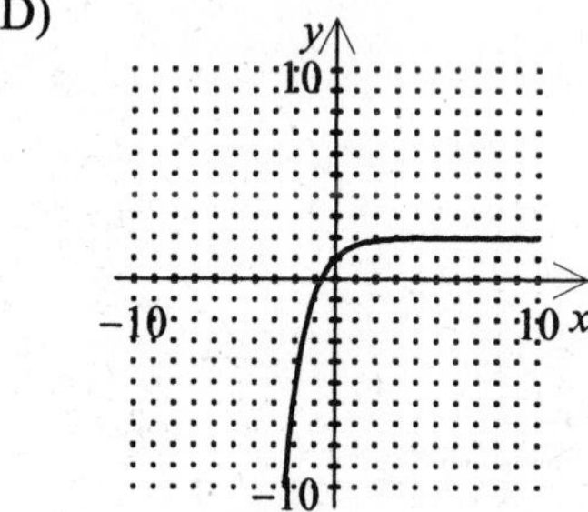

11. Sketch the graph of the function.
$f(x) = -3 + e^x$

12. Evaluate the expression. Round the answer to three decimal places.
$3e^{0.5}$

Objective 4: Use exponential functions to model and solve real-life applications

13. If $6500 is invested in a long-term trust fund with an interest rate of 9% compounded continuously, what is the amount of money in the account after 15 years?

(A) $27,434.52 (B) $23,676.14 (C) $25,073.27 (D) $48,028.86

14. In 1999, the population of a country was estimated at 4 million. For any subsequent year the population $P(t)$ in millions is

$$P(t) = \frac{240}{5 + 54.99e^{-0.0208t}}$$

where t is the number of years since 1999. Use a graphing calculator to estimate the population in 2016.

(A) 5,403,000 (B) 5,503,000 (C) 5,605,000 (D) 5,554,000

15. The half-life of a radioactive substance is 28.8 years. The quantity of the substance present after t years is

$Q = A_0 e^{-t \ln 2/28.8}$

where A_0 is the original amount in the sample. How much of a 350-gram sample of the substance will remain after 51.0 years? Round the answer to the nearest tenth.

16. The average atmospheric pressure P in pounds per square inch is

$P = 14.7e^{-0.21x}$

where x is the altitude in miles above sea level. Find the average atmospheric pressure for an altitude of 2.9 miles. Round the answer to the nearest tenth.

Section 3.2: Logarithmic Functions and Their Graphs

Objective 1: Recognize and evaluate logarithmic functions with base a

17. Evaluate the expression without using a calculator.

$\log_{25} 5$

(A) $-\frac{1}{2}$ (B) 2 (C) -2 (D) $\frac{1}{2}$

18. Identify the logarithmic equation written in exponential form.

$\log_{16} 8 = \frac{3}{4}$

(A) $\left(\frac{3}{4}\right)^{16} = 8$ (B) $16^{3/4} = 8$ (C) $8^{3/4} = 16$ (D) $\left(\frac{3}{4}\right)^{8} = 16$

19. Write the exponential equation in logarithmic form.

$7^6 = 117{,}649$

20. Evaluate the expression without using a calculator.

$\log_{10} \frac{1}{100}$

Objective 2: Graph logarithmic functions

21. Identify the graph of the logarithmic function.
$f(x) = \log_2(x+3)$

(A)

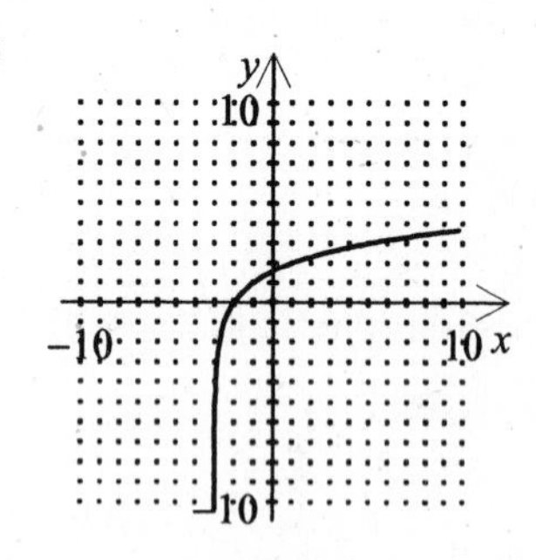

(B)

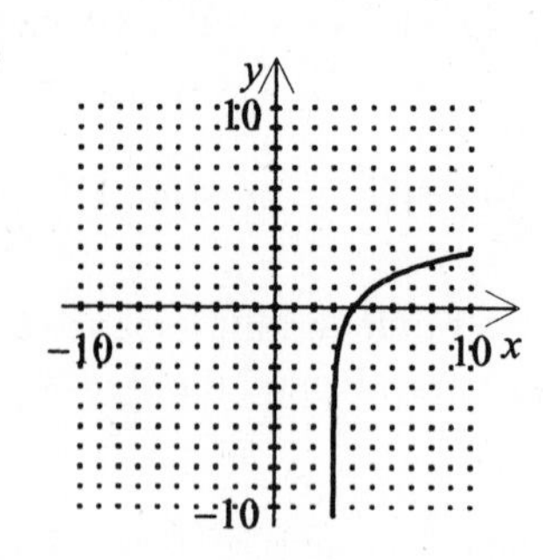

(C)

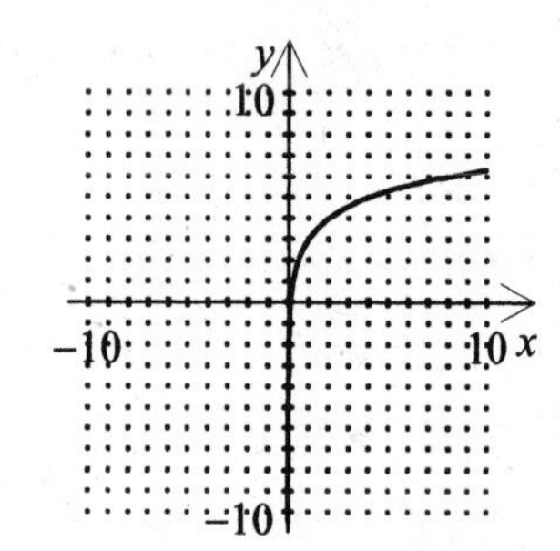

(D)

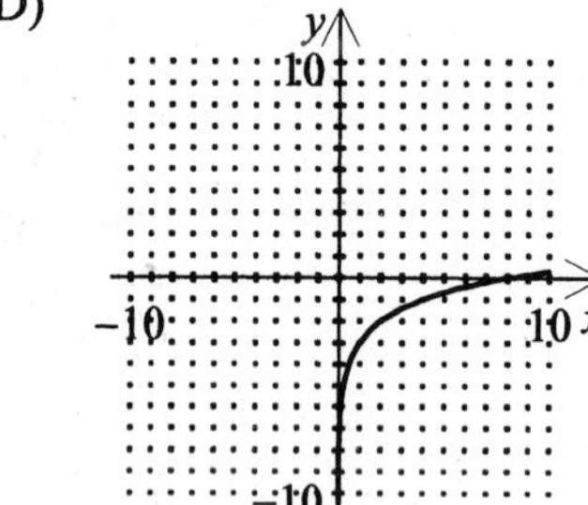

22. Identify the equation represented by the graph.

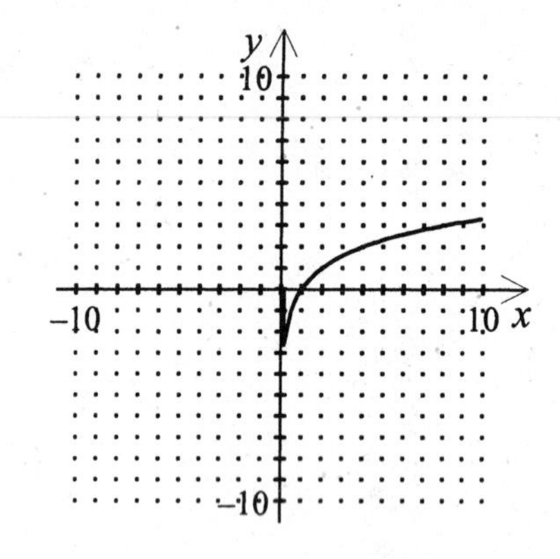

(A) $f(x) = \log_x 2$ (B) $f(x) = \log_x \frac{1}{2}$ (C) $f(x) = \log_2 x$ (D) $f(x) = \log_2 2$

Sketch the graph of the function.

23. $f(x) = \log_5 x$

Sketch the graph of the function.

24. $y = -\log_3 x$

Objective 3: Recognize and evaluate natural logarithmic functions

25. Identify the graph of the logarithmic function.
$f(x) = 2 + \ln(x+4)$

(A)

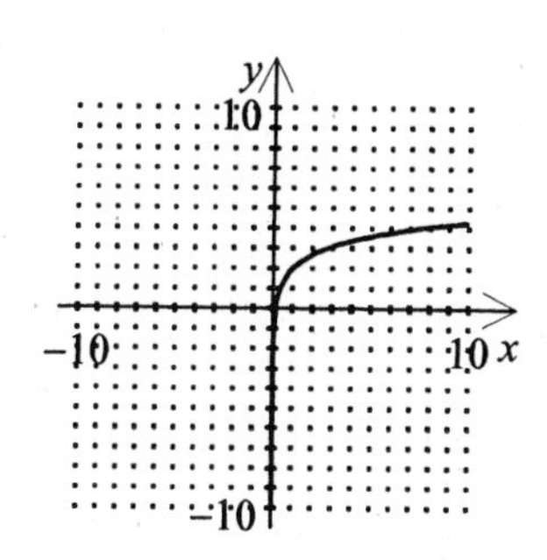

(B)

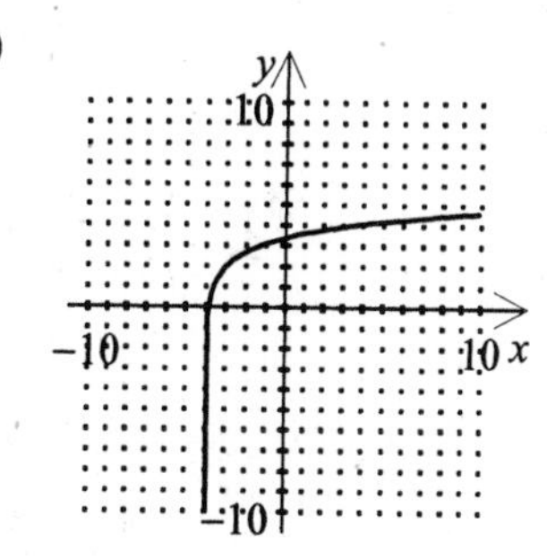

(C)

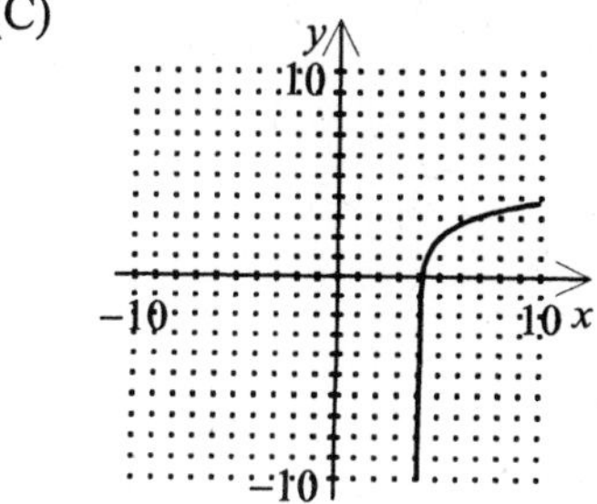

(D)

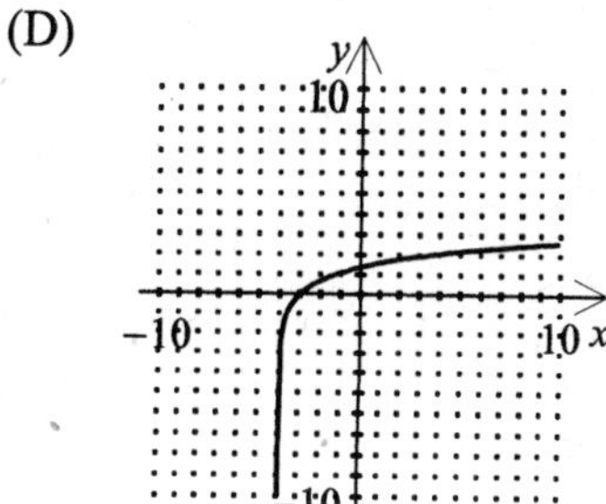

Use a calculator to evaluate the logarithm. Round to three decimal places.

26. ln 66 (A) 8.124 (B) 4.140 (C) 1.820 (D) 4.190

27. $\ln\left(5+\sqrt{3}\right)$

28. Sketch the graph of the function.
$f(x) = \ln(x+3)$

Objective 4: Use logarithmic functions to model and solve real-life applications

29. The time required to grow a certain bacteria in a culture beginning with 10 bacteria is

$$t = \frac{\ln B - \ln 10}{1.504}$$

where B is the number of bacteria and t is the time in hours. How much time is required to grow a culture of 4100 bacteria?

(A) 4.0 hours (B) 14.7 hours (C) 6.0 hours (D) 20.6 hours

30. The number of units sold for a certain product is

$$N = 1700 \ln(9t + 7)$$

where t is the number of years after the product is introduced. What are the expected sales 9 years after the product is introduced?

(A) 7478 units (B) 7611 units (C) 7501 units (D) 33,625 units

31. A company with loud machines must cut its sound intensity to 82% of its original level. If the loudness of a sound β measured in decibels is

$$\beta = 10 \log_{10} \frac{100}{I_0}$$

where I_0 is the percent of the original level to which the sound must be reduced, by how many decibels must the loudness be reduced?

32. The magnitude of an earthquake is

$$M = \frac{2}{3} \log_{10} \frac{E}{10^{11.8}}$$

where E is the energy released. If an earthquake released twice as much energy as an earlier quake that released 10^{19} ergs of energy, find the magnitude of the newer quake to the nearest tenth.

Section 3.3: Properties of Logarithms

Objective 1: Use the change-of-base formula to rewrite and evaluate logarithmic expressions

33. Evaluate the logarithm using the change-of-base formula.

$\log_9 546$

(A) 2.868 (B) 56.724 (C) 0.700 (D) 1.638

34. Which is the logarithm rewritten as a ratio of natural logarithms?
$\log_{7/2} x$

(A) $\dfrac{\ln x}{\ln \frac{7}{2}}$ (B) $\dfrac{\ln \frac{7}{2}}{\ln x}$ (C) $\ln x - \ln \frac{7}{2}$ (D) $\ln\left(\dfrac{\frac{7}{2}}{x}\right)$

35. Evaluate the logarithm using the change-of-base formula. Find the value to three decimal places.
$\log_{2/3} 13$

36. Use the change-of-base formula to rewrite the logarithm as a ratio of logarithms. Then use a graphing utility to sketch the graph.
$f(x) = \log_4 x$

Objective 2: Use properties of logarithms to evaluate or rewrite logarithmic expressions

Find the value of the expression without using a calculator.

37. $2\ln e^8$ (A) 4 (B) $16e$ (C) 16 (D) $\frac{1}{4}$

38. $\log_9 729$ (A) 2 (B) 3 (C) $\frac{2}{9}$ (D) $\frac{1}{9}$

39. Indicate whether the following statement is true or false. If it is false, correct the statement.
$\log_{10} \frac{8}{7} = \log_{10} 8 + \log_{10} 7$

40. Find the value of the expression without using a calculator.
$\log_6 36 + \log_6 30 - \log_6 5$

Objective 3: Use properties of logarithms to expand or condense logarithmic expressions

41. Use the properties of logarithms to expand the expression. (Assume all variables are positive.)

$\log_3 \dfrac{x}{\sqrt[6]{y}}$

(A) $\log_3 x - 6\log_3 y$ (B) $\log_3 x + 6\log_3 y$

(C) $\log_3 x + \dfrac{1}{6}\log_3 y$ (D) $\log_3 x - \dfrac{1}{6}\log_3 y$

Condense the expression to the logarithm of a single quantity.

42. $4\log_{10} x + 3\log_{10}(x-6)$

(A) $\log_{10} x(x-6)$ (B) $\log_{10} x^4(x-6)^3$ (C) $\log_{10} \dfrac{x^4}{(x-6)^3}$ (D) None of these

43. $\left[2\log_4(x+4) + 3\log_4(x+6)\right] - \dfrac{1}{2}\log_4 x$

44. Use the properties of logarithms to expand the expression. (Assume all variables are positive.)

$\log_a \dfrac{2xy^4}{z^3}$

Objective 4: Use logarithmic functions to model and solve real-life applications

45. The amount of time t in years required for a certain radioactive material to decompose is

$t = \dfrac{\ln R - \ln A}{k}$

where R is the mass of the substance remaining after decomposition, A is the original mass of the substance, and k is a constant related to the particular material. Find the time required for 40 grams of a radioactive substance to decompose so that only 20 grams remain if $k = -0.431$.

(A) 16.1 years (B) 1.6 years (C) 0.6 year (D) 8.6 years

46. The magnitude of an earthquake as measured by the Richter Scale is

$R = 0.67\log_{10}(0.37E) + 1.46$

where R is the magnitude of the earthquake and E is the energy in kilowatt-hours released by the earthquake. Which magnitude corresponds to a release of 2.52×10^8 kwh of energy?

(A) 6.2 (B) 6.8 (C) 6.5 (D) None of these

47. The power gain in decibels of an electronic device is determined by the logarithmic expression $10(\log_{10} P_o - \log_{10} P_i)$ where P_o is the output power in watts and P_i is the input power in watts.
(a) Condense the expression using the properties of logarithms.
(b) Determine the power gain of an amplifier with an output of 22.5 W and an input of 0.667 W.

48. The energy E in kilocalories per mole required to transport a substance from the outside to the inside of a living cell is given by
$E = 1.4(\log C_2 - \log C_1)$
where C_1 is the concentration of the substance outside the cell and C_2 is the concentration inside the cell.
(a) Condense the expression using the properties of logarithms.
(b) The concentration of a particular substance outside is four times the concentration inside the cell. How much energy is required to transport the substance from inside to outside the cell? A negative value for E means that energy is needed to prevent the transport of the substance from the region of higher concentration to the region of lower concentration.

Section 3.4: Exponential and Logarithmic Equations

Objective 1: Solve simple exponential and logarithmic equations

Solve for x.

49. $\frac{1}{9} = 27^{6x-3}$ (A) $\frac{7}{18}$ (B) $\frac{1}{6}$ (C) $\frac{11}{18}$ (D) $\frac{1}{18}$

50. $\left(\frac{1}{3}\right)^x = 27$ (A) -3 (B) 3 (C) $\frac{1}{3}$ (D) $-\frac{1}{3}$

51. $e^x = 5$

52. $\log_{10} x = -2$

Objective 2: Solve more complicated exponential equations

Solve the exponential equation algebraically.

53. $7e^{0.06x} + 55 = 76$ (A) 18.310 (B) -0.390 (C) 7.952 (D) 50.000

Solve the exponential equation algebraically.

54. $\dfrac{700}{1+e^{-x}}=675$ (A) -0.711 (B) -0.675 (C) 3.296 (D) 3.234

Solve the exponential equation algebraically. Approximate the result(s) to three decimal places.

55. $3^{2x}-9\left(3^{x}\right)+18=0$

56. $80=4e^{0.095t}$

Objective 3: Solve more complicated logarithmic equations

Find the value of x.

57. $4\ln(4x)=10$ (A) 10 (B) 0.625 (C) 3.046 (D) 0.229

58. $\log_{6}(5x+9)=4$ (A) 3 (B) $\dfrac{4087}{5}$ (C) 261 (D) $\dfrac{1287}{5}$

59. $\log_{3}x+\log_{3}(x-26)=3$

60. Find the value of x. Round to three decimal places.
$\ln\sqrt{4x-5}=1.9$

Objective 4: Use exponential and logarithmic equations to model and solve real-life applications

61. The number of bacteria present in a culture is
$B=100e^{0.262t}$
where t is the time in minutes. Find the time required, to the nearest half minute, to have 3000 bacteria present.

(A) 12.5 min (B) 13.0 min (C) 13.5 min (D) 14.0 min

62. An automobile manufacturer is introducing a new fuel-efficient model and estimates the demand for the car as

$N = 54{,}000 \ln(4t + 2)$

where N is the estimated number of cars to be sold and t is the number of years after the car is introduced. When will the demand be 220,000 cars?

(A) 15.6 years (B) 13.6 years (C) 14.2 years (D) 15.2 years

63. The amount of power generated by a satellite's power supply is

$P = 50e^{-t/300}$

where P is the power in watts and t is the time in days. For how many days will 27 watts of power be available? Round to the nearest whole day.

64. The mathematical model for learning an assembly-line procedure needed for assembling one component of a manufactured item is

$P = \dfrac{0.91}{1 + e^{-0.1n}}$

where P is the proportion of correctly assembled components after n practice sessions. How many practice sessions are required to have at least 65% of the components correctly assembled within the given time period? Round any non-integer to the next highest whole number.

Section 3.5: Exponential and Logarithmic Models

Objective 1: Recognize the five most common types of models involving exponential and logarithmic functions

65. Identify the model shown in the graph.

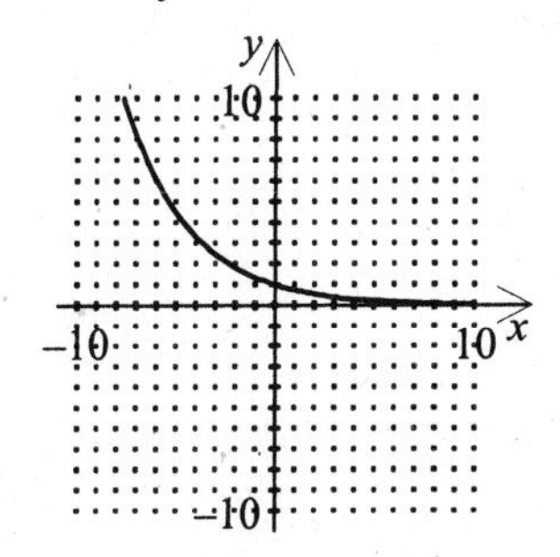

(A) Logarithmic model (B) Exponential decay model

(C) Logistic growth model (D) Exponential growth model

66. Which equation represents the logarithmic model of a function?

(A) $y = 5e^{0.95x}$ (B) $y = 5 + 2\ln x$ (C) $y = 5e^{-(x-1)^2/2}$ (D) $y = 5e^{-0.95x}$

67. Sketch the graph of $y = 3e^{0.32x}$.

68. Identify the model shown in the graph.

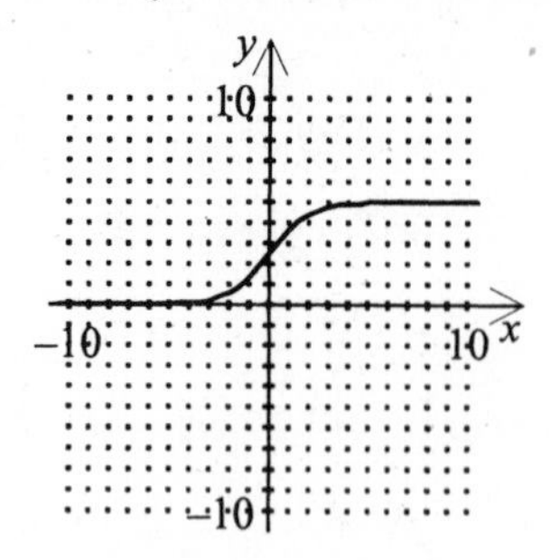

Objective 2: Use exponential growth and decay functions to model and solve real-life problems

69. The number of bison per acre of range in the wild is

$N = 2.5 \times 10^{5-0.003w}$

where N is the number of bison per acre and w is the average weight of the bison in pounds. Find the average weight of a bison in a herd that has an average of five animals per acre of range.

(A) 1616 lb (B) 1410 lb (C) 1723 lb (D) 1566 lb

70. The number of bacteria N in a culture is modeled by

$N = 300e^{kt}$

where t is the time in hours. If $N = 750$ when $t = 6$, what is the time required for the original population to triple in size?

(A) 8.49 hours (B) 7.19 hours (C) 18.00 hours (D) 13.89 hours

71. The population of a bacteria culture with an initial population of 3000 being treated with a new antibiotic can be modeled by

$N = 3000e^{-0.5t}$

where N is the number of bacteria present and t is the time in hours since the treatment began. In how many hours will the culture have a count of 1200? Round the answer to the nearest tenth.

72. A new car with a purchase price of \$23,500 has a value of \$15,300 two years later.

(a) Write the straight-line model $V = mt + b$.

(b) Write the exponential model $V = ae^{kt}$.

(c) Find the book values of the car after three years using each model.

(d) Interpret the meaning of the slope in the straight-line model.

Objective 3: Use Gaussian functions to model and solve real-life problems

73. The life expectancy of a fluorescent tube follows a normal distribution

$y = 0.0007e^{-(x-5500)^2/720{,}000}$

where x is the life expectancy in hours. Use a graphing utility to sketch the function and find the average life of a fluorescent tube.

(A) 7200 hr (B) 4500 hr (C) 6000 hr (D) 5500 hr

74. The ages of the employees of a local company roughly follow a normal distribution

$y = 0.0997e^{-(x-39)^2/32}$

where x is the age of the employee. Use a graphing utility to sketch the function and find the average age of an employee.

(A) 39 (B) 41 (C) 32 (D) 37

75. During boot camp, a drill sergeant measured the weight of the men in his unit. He found the weights roughly followed a normal distribution

$y = 0.0235e^{-(x-154)^2/578}$

where x is the weight of a recruit in pounds. Use a graphing utility to sketch the function and find the average weight of the men.

76. The heights of 500 boys at Madison High School were recorded and found to follow the normal distribution

$y = 0.0997e^{-(x-72)^2/32}$

where x is the height of a male student in inches. Use a graphing utility to sketch the function and find the average height of the boys at Madison High School.

Objective 4: Use logistic growth functions to model and solve real-life problems

77. A virus is accidentally brought to a remote village with a population of 4500 that has never been exposed to the disease. The spread of the virus is modeled by

$$y = \frac{4500}{1+4499e^{-0.3t}}$$

where t is the time in days since the virus was introduced. How many villagers will be infected after 15 days?

(A) 79 (B) 75 (C) 88 (D) 92

78. A rumor begins at a closed-campus high school that the King of Baklava is visiting the school that afternoon. There are 2600 students in the high school and the spread of the rumor is modeled by

$$y = \frac{2600}{1+2599e^{-0.6t}}$$

where y is the number of students who have heard the rumor and t is the time in minutes. How long before 60% of the students have heard the rumor?

(A) 15.2 min (B) 13.8 min (C) 13.1 min (D) 11.7 min

79. Bacteria growing in a petri dish is modeled by

$$y = \frac{500{,}000}{1+50e^{-0.43t}}$$

where t is the time in hours. Find the number of bacteria after 7 hours.

80. A new product was heavily advertised for several months, but then advertising was sharply reduced. The peak sales were 300,000 units but are now decreasing and can be modeled by

$$S = \frac{300{,}000}{1+0.5e^{0.03t}}$$

where t is the time in months. In how many months will the sales be 30% of the peak sales? Round the answer to the nearest tenth.

Objective 5: Use logarithmic functions to model and solve real-life problems

81. Chemists measure the acidity of an aqueous solution using pH. The pH is

$$\text{pH} = -\log_{10}\left[\text{H}^+\right]$$

where pH is a measure of the hydrogen ion concentration $\left[\text{H}^+\right]$ measured in moles of hydrogen per liter of a solution. Find the pH of a solution if $\left[\text{H}^+\right] = 4.0 \times 10^{-3}$.

(A) 4.6 (B) 2.4 (C) 11.6 (D) 9.4

82. The level of sound β in decibels with an intensity of I is

$$\beta = 10 \log_{10} \frac{I}{I_0}$$

where I_0 is 10^{-12} watt/m^2, corresponding roughly to the faintest sound that can be heard by the human ear. Find the level of sound β of a snowmobile with $I = 10^{-2}$ watt/m^2.

(A) 90 decibels (B) 95 decibels (C) 100 decibels (D) 140 decibels

83. Due to the installation of a highway sound barrier, the noise level near a busy highway was reduced from 89 decibels to 79 decibels. Find the percent decrease, to the nearest tenth of one percent, in the intensity level of the noise as a result of the sound barrier.

84. Chemists measure the acidity of an aqueous solution using pH. The pH is

$$\text{pH} = -\log_{10}\left[\text{H}^+\right]$$

where pH is a measure of the hydrogen ion concentration $\left[\text{H}^+\right]$ measured in moles of hydrogen per liter of a solution. Compute $\left[\text{H}^+\right]$ for a solution in which $\text{pH} = 4.60$.

Answer Key for Chapter 3 Exponential and Logarithmic Functions

Section 3.1: Exponential Functions and Their Graphs

Objective 1: Recognize and evaluate exponential functions with base a

[1] (A)

[2] (C)

[3] 65.663

[4] 0.678

Objective 2: Graph exponential functions

[5] (C)

[6] (C)

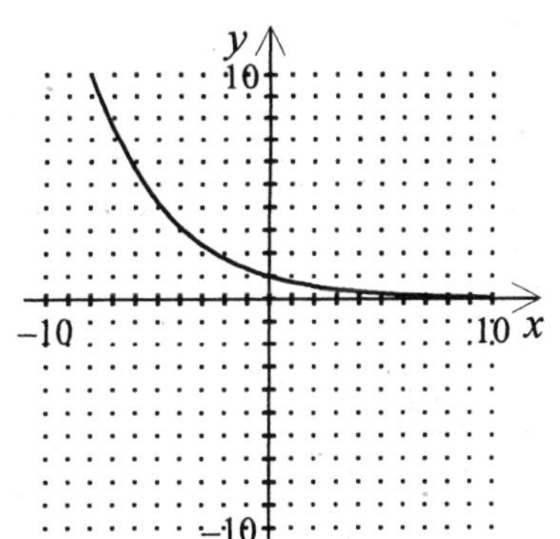

[7]

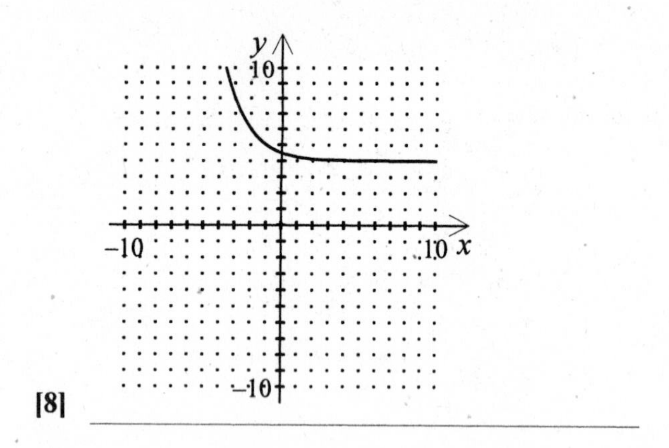

[8] ______

Objective 3: Recognize and evaluate exponential functions with base e

[9] (C)

[10] (C)

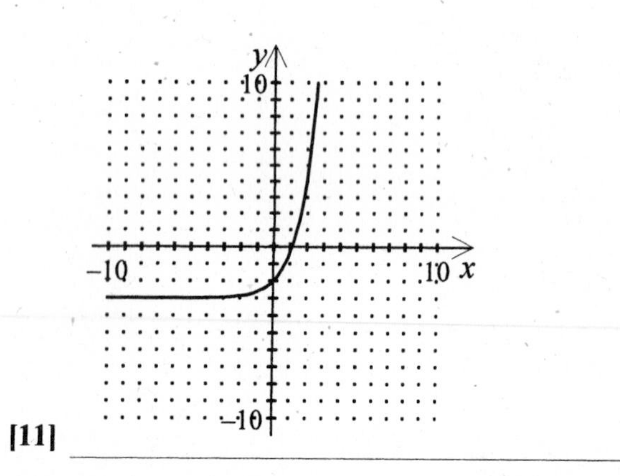

[11] ______

[12] 4.946

Objective 4: Use exponential functions to model and solve real-life applications

[13] (C)

[14] (B)

[15] 102.6 g

[16] 8.0 lb/in^2

Section 3.2: Logarithmic Functions and Their Graphs

Objective 1: Recognize and evaluate logarithmic functions with base a

[17] (D)

[18] (B)

[19] $\log_7 117{,}649 = 6$

[20] –2

Objective 2: Graph logarithmic functions

[21] (A)

[22] (C)

[23]

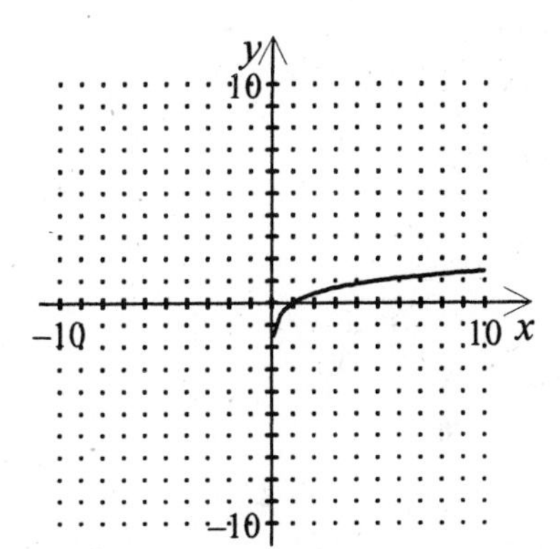

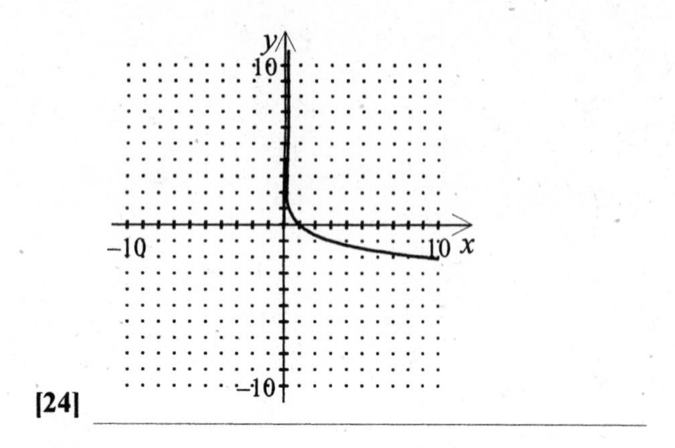

[24]

Objective 3: Recognize and evaluate natural logarithmic functions

[25] (B)

[26] (D)

[27] 1.907

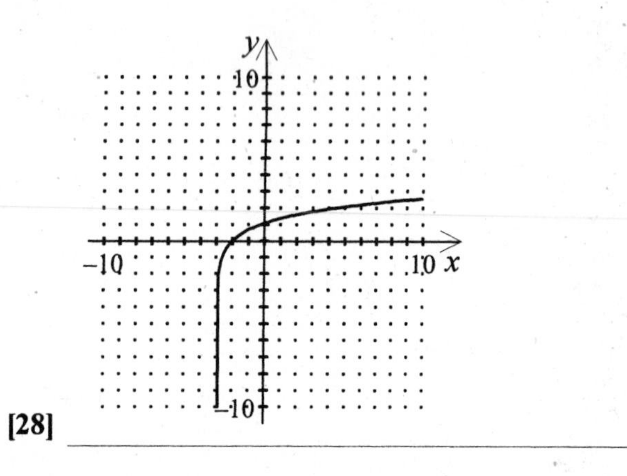

[28]

Objective 4: Use logarithmic functions to model and solve real-life applications

[29] (A)

[30] (B)

[31] 0.862 decibels

[32] Magnitude of 5.0

Section 3.3: Properties of Logarithms

Objective 1: Use the change-of-base formula to rewrite and evaluate logarithmic expressions

[33] (A)

[34] (A)

[35] –6.326

$$f(x) = \frac{\log_{10} x}{\log_{10} 4}$$

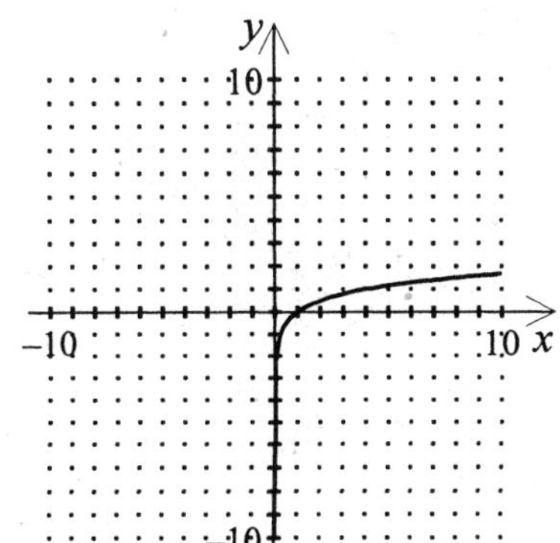

[36]

Objective 2: Use properties of logarithms to evaluate or rewrite logarithmic expressions

[37] (C)

[38] (B)

[39] False. Answers may vary. Sample answer: $\log_{10} \frac{8}{7} = \log_{10} 8 - \log_{10} 7$.

[40] 3

Objective 3: Use properties of logarithms to expand or condense logarithmic expressions

[41] (D)

[42] (B)

[43] $\log_4 \dfrac{(x+4)^2(x+6)^3}{\sqrt{x}}$

[44] $\log_a 2 + \log_a x + 4\log_a y - 3\log_a z$

Objective 4: Use logarithmic functions to model and solve real-life applications

[45] (B)

[46] (B)

[47] (a) $10(\log_{10} P_o - \log_{10} P_i) = \log_{10}\left(\dfrac{P_o}{P_i}\right)^{10}$

(b) 15.281 dB

[48] (a) $E = \log\left(\dfrac{C_2}{C_1}\right)^{1.4}$

(b) -0.843 kcal/mol

Section 3.4: Exponential and Logarithmic Equations

Objective 1: Solve simple exponential and logarithmic equations

[49] (A)

[50] (A)

[51] $\ln 5 \approx 1.609$

[52] 0.01

Objective 2: Solve more complicated exponential equations

[53] (A)

[54] (C)

[55] 1, 1.631

[56] 31.534

Objective 3: Solve more complicated logarithmic equations

[57] (C)

[58] (D)

[59] $x = 27$

[60] 12.425

Objective 4: Use exponential and logarithmic equations to model and solve real-life applications

[61] (B)

[62] (C)

[63] 185 days

[64] 10

Section 3.5: Exponential and Logarithmic Models

Objective 1: Recognize the five most common types of models involving exponential and logarithmic functions

[65] (B)

[66] (B)

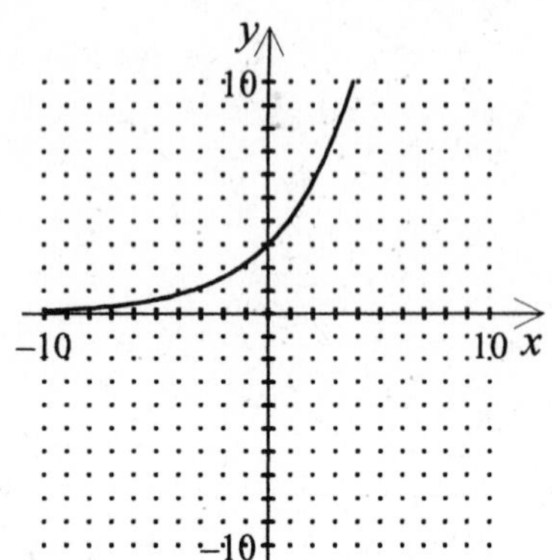

[67]

[68] Logistic growth model

Objective 2: Use exponential growth and decay functions to model and solve real-life problems

[69] (D)

[70] (B)

[71] 1.8 hr

[72]
(a) $V = -4100t + 23{,}500$
(b) $V = 23{,}500e^{-0.215t}$
(c) Straight-line, \$11,200; Exponential, \$12,345
(d) The slope represents the yearly decrease in value.

Objective 3: Use Gaussian functions to model and solve real-life problems

[73] (D)

[74] (A)

[75] 154 lb

[76] 72 in.

Objective 4: Use logistic growth functions to model and solve real-life problems

[77] (C)

[78] (B)

[79] 144,317

[80] 51.4 months

Objective 5: Use logarithmic functions to model and solve real-life problems

[81] (B)

[82] (C)

[83] 90.0%

[84] 2.51×10^{-5} mol/L

Chapter 4 Trigonometry

Section 4.1: Radian and Degree MeasureSection

Objective 1: Describe angles

1. Identify the angle that is in standard position.

(A)

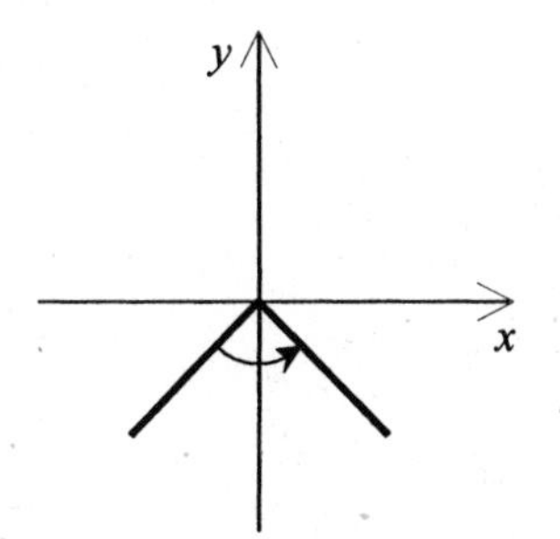

(B)

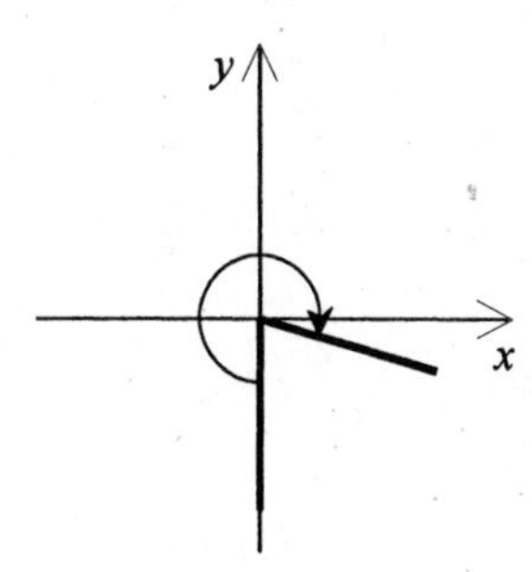

(C)

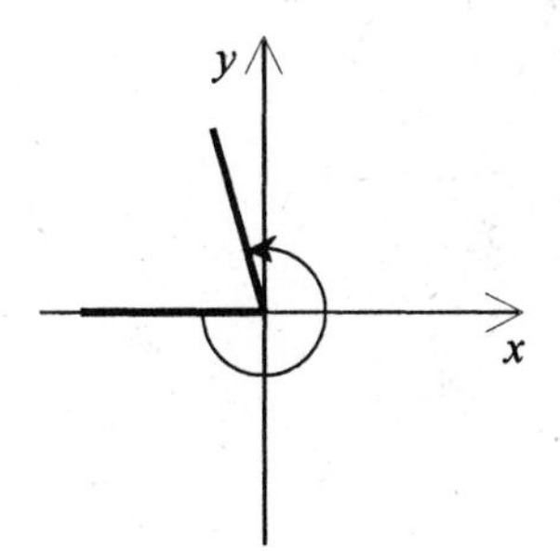

(D)

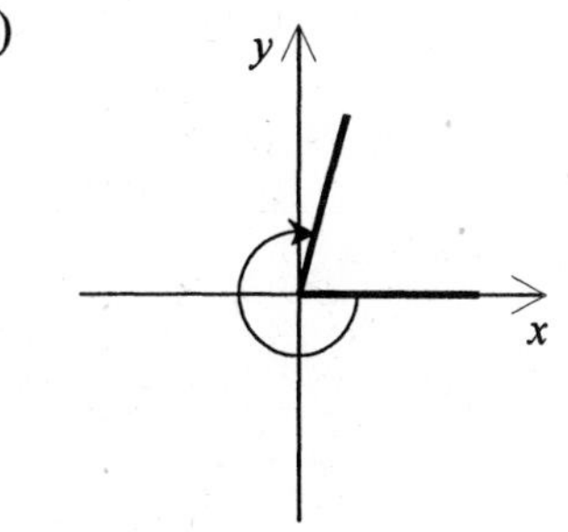

2. Select the correct description of $\overrightarrow{NM}$.

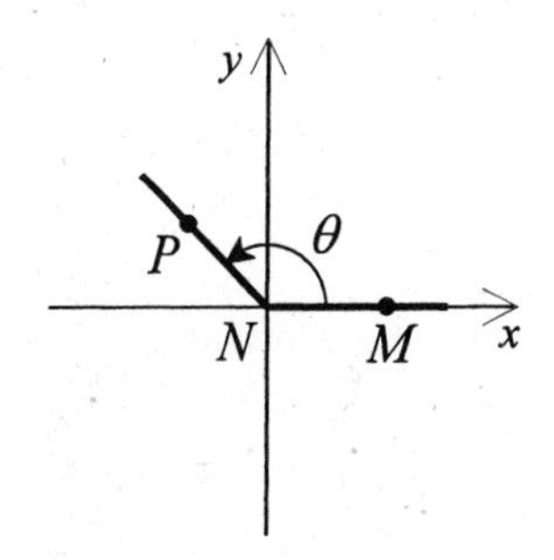

(A) Vertex of angle θ

(B) Initial side of angle θ

(C) Coterminal side of angle θ

(D) Terminal side of angle θ

3. Sketch a positive angle with its terminal side in Quadrant II.

4. Sketch one positive and one negative angle, each of which is coterminal with the angle shown.

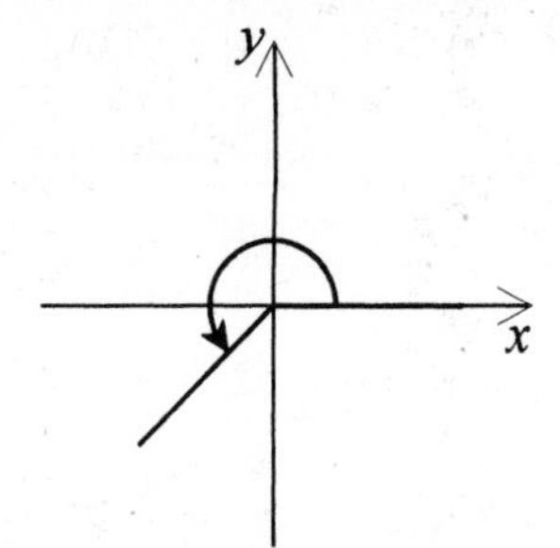

Objective 2: Use radian measure

5. Convert the measure to radians.

$1\frac{1}{6}$ revolutions counterclockwise from the x-axis

(A) $\frac{7\pi}{6}$ (B) -420 (C) $\frac{7}{3}$ (D) $\frac{7\pi}{3}$

6. In which quadrant is the terminal side of the angle θ?

$\theta = -\frac{11\pi}{6}$

(A) Quadrant I (B) Quadrant II (C) Quadrant III (D) Quadrant IV

7. Sketch the angle in standard position.

$-\frac{\pi}{2}$

8. Find one positive angle and one negative angle that are coterminal with the given angle.

$-\frac{5\pi}{3}$

Objective 3: Use degree measure

9. Express the angle in radian measure in terms of π. Do not use a calculator.

120°

(A) $\frac{2\pi}{3}$ (B) $\frac{4\pi}{3}$ (C) $\frac{3\pi}{2}$ (D) $\frac{3\pi}{4}$

10. In which quadrant is the terminal side of the angle θ?

$\theta = -185°$

(A) Quadrant I (B) Quadrant II (C) Quadrant III (D) Quadrant IV

11. Express the angle in degree measure.

$\frac{7\pi}{12}$

12. Convert the measure from degrees to radians. Round to three decimal places.

97°

Objective 4: Use angles to model and solve real-life problems

13. The needle of the scale in the bulk food section of a supermarket is 27 cm long. Find the distance the tip of the needle travels when it rotates 74°.

(A) 34.9 cm (B) 5.6 cm (C) 17.4 cm (D) 199.8 cm

14. A point on the rim of a wheel has a linear speed of 33 cm/s. If the radius of the wheel is 50 cm, what is the angular speed of the wheel in radians per second?

(A) 2.1 rad/s (B) 0.7 rad/s (C) 1.3 rad/s (D) 0.3 rad/s

15. A 85-foot long irrigation sprinkler line rotates around one end as shown. The sprinkler moves through an arc of 250° in 8.85 hours. Find the speed of the moving end of the sprinkler in feet per minute. Round your answer to the nearest tenth.

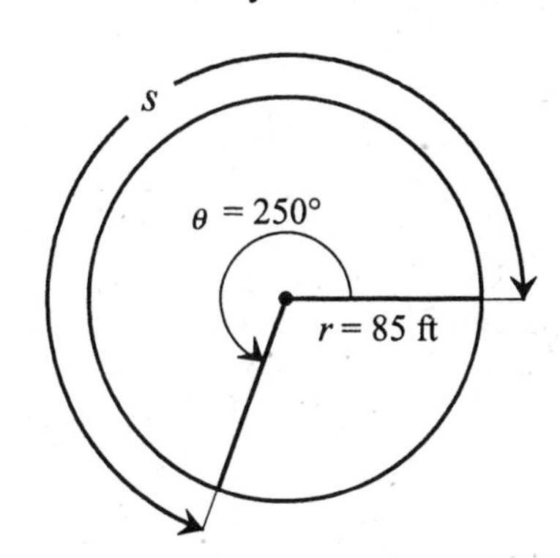

16. One city is 450 miles due north of another city. Assuming that the earth is a sphere of radius 3964 miles, what is the difference in their latitudes?

Section 4.2: Trigonometric Functions: The Unit Circle

Objective 1: Identify a unit circle and its relationship to real numbers

17. Identify the point that lies on the unit circle.

(A) $\left(\frac{67}{35}, -\frac{32}{35}\right)$ (B) $\left(-\frac{47}{48}, -\frac{1}{48}\right)$ (C) $\left(\frac{1}{5}, \frac{4}{5}\right)$ (D) $\left(\frac{9}{41}, \frac{40}{41}\right)$

18. A number line has its 0 fixed at $(1, 0)$ on a unit circle. What central angle, in radians, is formed by the radius to $(1, 0)$ and the radius to the point that corresponds to 2.44 on the number line when it is wrapped around the circle?

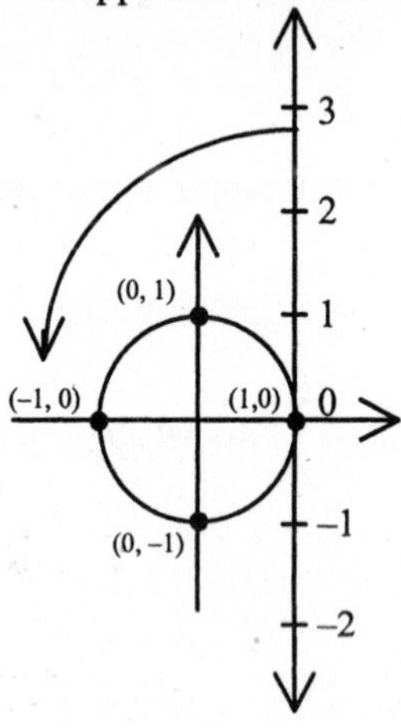

(A) 139.80 π (B) 139.80 (C) 2.44 π (D) 2.44

19. The length of an arc along a wheel which has a radius of 1 foot is found by wrapping a tape measure around the wheel. What is the arc length that corresponds to a central angle of $\frac{3\pi}{4}$?

20. If the point $(0.67, y)$ is on the unit circle in Quadrant I, what is the value of y? Round to three decimal places.

Objective 2: Evaluate trigonometric functions using the unit circle

21. Use the unit circle and a straightedge to approximate the value of the expression.

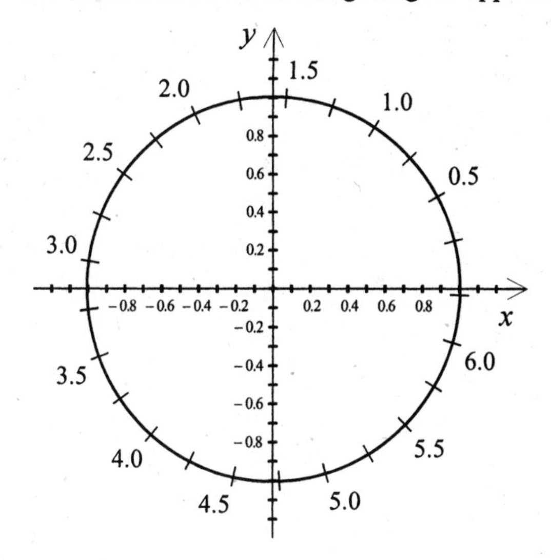

sin 2.25

(A) 0.78 (B) 0.63 (C) –0.78 (D) –0.63

Use the unit circle and symmetry to help you evaluate the function(s).

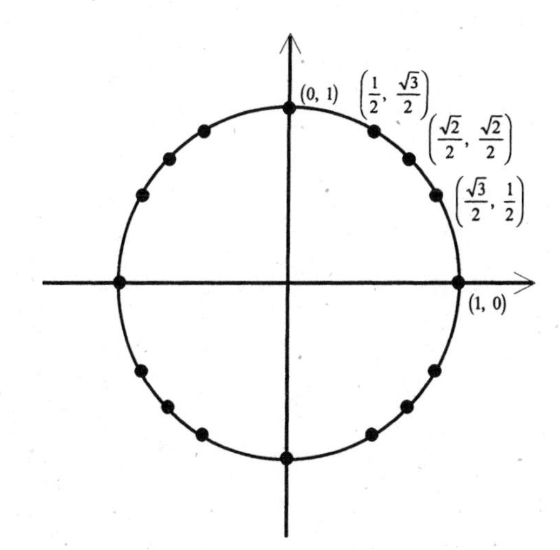

22. $\tan\left(-\frac{7\pi}{4}\right)$ (A) $\frac{\sqrt{3}}{3}$ (B) $\frac{\sqrt{2}}{2}$ (C) 1 (D) $-\frac{\sqrt{2}}{2}$

Use the unit circle and symmetry to help you evaluate the function(s).

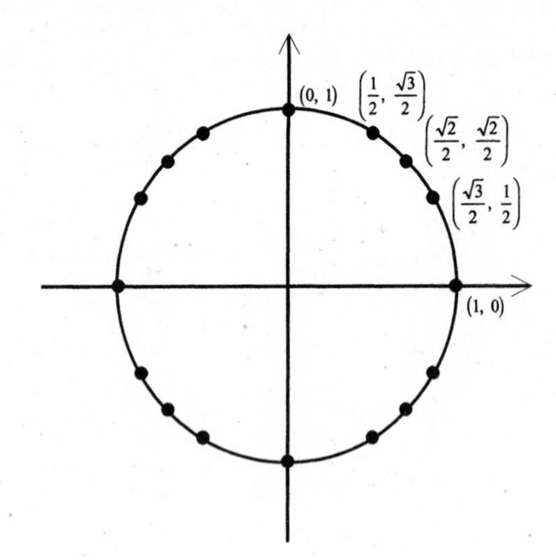

23. All six trigonometric functions of the real number $t = \dfrac{5\pi}{4}$

24. Use the coordinates of a point on the unit circle to write the trigonometric ratio.

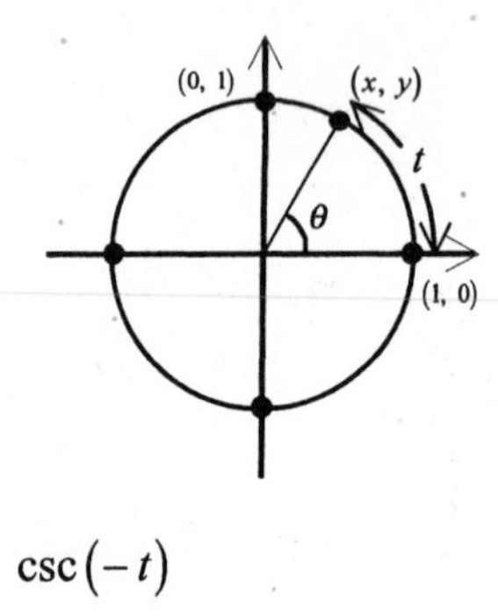

$\csc(-t)$

Objective 3: Use the domain and period to evaluate sine and cosine functions

25. Use the period of the function to select the expression that has the same value as the given expression.

$\cos\left(-\dfrac{47\pi}{8}\right)$

(A) $\cos\dfrac{\pi}{8}$ (B) $\cos\dfrac{\pi}{4}$ (C) $\cos\dfrac{7\pi}{8}$ (D) $\cos\dfrac{\pi}{16}$

26. Use the given trigonometric value to evaluate the indicated expression.
If $\cos t = \frac{13}{16}$, find $\cos(t - 4\pi)$.

(A) $\frac{13}{16}$ (B) $\frac{16}{13}$ (C) $\frac{3}{16}$ (D) $-\frac{13}{16}$

27. Use the period of the function to identify an angle in the interval $[0, \pi]$ that generates the same value for the function. Show your reasoning and give the value of the function.
$\sin\left(-\frac{23\pi}{6}\right)$

28. Identify the trigonometric function as an even function or an odd function.
$y = \csc\theta$

Objective 4: Use a calculator to evaluate trigonometric functions

Use a calculator to evaluate the expression.

29. $\tan(-5.2)$ (A) –10.9882 (B) –0.0910 (C) 1.8856 (D) 0.5303

30. $\sec 8$ (A) –6.8729 (B) –0.1455 (C) –13.7457 (D) 1.0488

31. Use a calculator to evaluate the cotangent, secant, and cosecant of the real number. Round to four decimal places.
$t = -5.8$

32. Use a calculator to evaluate the sine, cosine, and tangent of the real number. Round to four decimal places.
$t = \frac{7\pi}{15}$

Section 4.3: Right Triangle Trigonometry

Objective 1: Evaluate trigonometric functions of acute angles

33. Identify the ratio that defines the trigonometric function of the angle θ.
$\sec\theta$

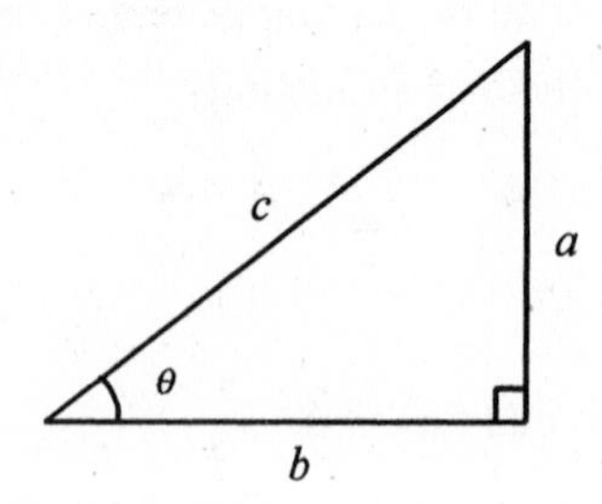

(A) $\frac{a}{c}$ (B) $\frac{b}{a}$ (C) $\frac{c}{b}$ (D) $\frac{b}{c}$

34. Identify the ratio $\frac{a}{c}$ for the indicated angle and find its value.

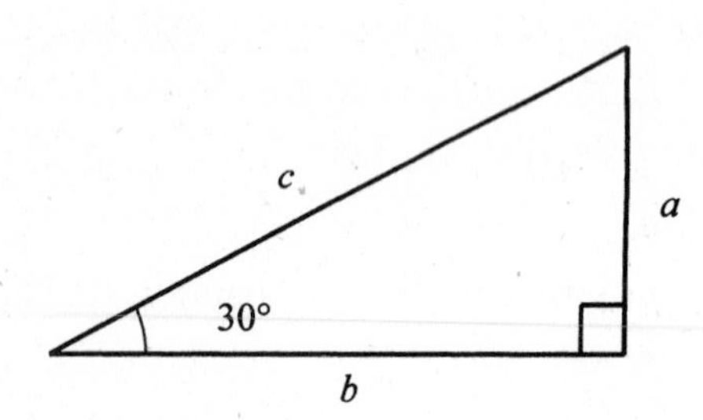

(A) $\cos 30° = \frac{\sqrt{3}}{2}$ (B) $\tan 30° = \frac{\sqrt{3}}{3}$ (C) $\csc 30° = \frac{1}{2}$ (D) $\sin 30° = \frac{1}{2}$

35. Find the exact value of the six trigonometric functions of the angle θ given in the figure. (Use the Pythagorean Theorem to find the third side of the triangle.)

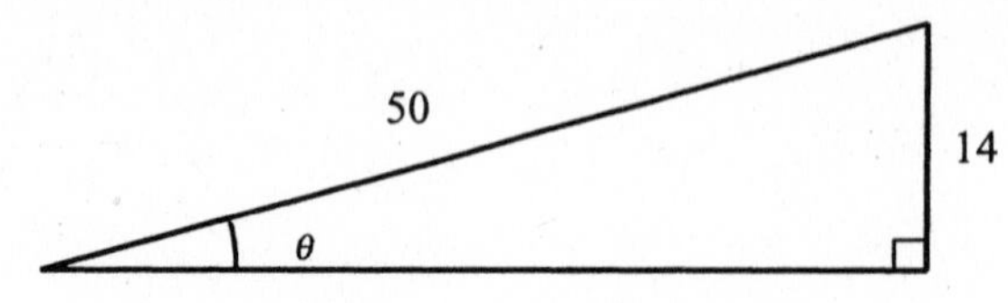

36. Sketch a right triangle corresponding to the trigonometric function of the acute angle θ. Use the Pythagorean Theorem to determine the third side and then find the indicated trigonometric function of θ.

If $\tan\theta = 1$, find $\sin\theta$.

Objective 2: Use the fundamental trigonometric identities

37. Let θ be an acute angle. Use the given function value and trigonometric identities to find the indicated trigonometric function.

If $\cos\theta = \dfrac{11}{61}$, find $\sec\theta$.

(A) $\dfrac{60}{61}$ (B) $\dfrac{11}{60}$ (C) $\dfrac{61}{11}$ (D) $\dfrac{61}{60}$

38. Use the fundamental trigonometric identities to determine the simplified form of the expression.

$$\frac{\sec\beta}{\csc\beta}$$

(A) $\sin\beta$ (B) $\cos\beta$ (C) $\tan\beta$ (D) $\cot\beta$

39. Use the fundamental trigonometric identities to determine a simplified form of the expression.

$$\frac{\cos^2\beta}{1-\sin^2\beta}$$

40. Let θ be an acute angle. Use the given function value and trigonometric identities to find the indicated trigonometric function.

If $\sin\theta = \dfrac{6}{7}$, find $\cos\theta$.

Objective 3: Use a calculator to evaluate trigonometric functions

Use a calculator to evaluate the function. (Be sure the calculator is in the correct angle mode.)

41. $\csc 22°\,36'\,17''$ (A) 0.3844 (B) 2.6017 (C) –1.2846 (D) –1.7367

42. $\sin 24.35°$ (A) 0.4123 (B) -0.7052 (C) -0.9031 (D) 2.4254

Use a calculator to evaluate the function. Round your answer to four decimal places. (Be sure the calculator is in the correct angle mode.)

43. $\csc 82.77°$

44. $\sin 8° 38'$

Objective 4: Use trigonometric functions to model and solve real-life problems

45. The cable supporting a ski lift rises 5 feet for each 8 feet of horizontal length. The top of the cable is fastened 750 feet above the cable's lowest point. Find the lengths b and c, and find the measure of angle θ.

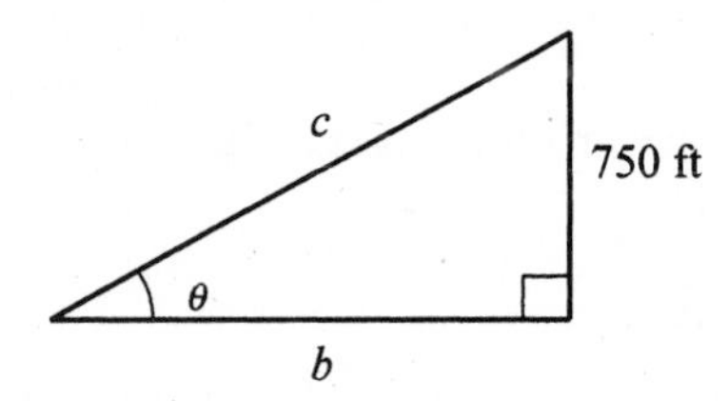

(A) $b = 1415$ ft
$c = 1200$ ft
$\theta = 38.7°$

(B) $b = 469$ ft
$c = 884$ ft
$\theta = 58.0°$

(C) $b = 1200$ ft
$c = 1415$ ft
$\theta = 32.0°$

(D) $b = 884$ ft
$c = 469$ ft
$\theta = 0.6°$

46. A 14-foot ladder makes an angle of 63° with the ground as it leans against a wall. How far up the wall does the ladder reach?

(A) 15.71 ft (B) 6.36 ft (C) 12.47 ft (D) 27.48 ft

47. A photographer points a camera at a window in a nearby building forming an angle of 41° with the camera platform. If the camera is 54 meters from the building, how high above the platform is the window? Round to two decimal places.

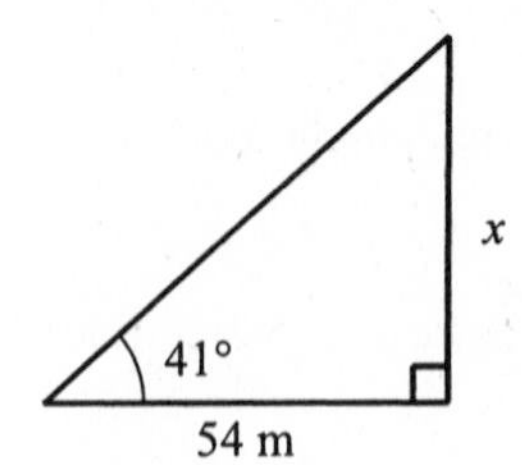

48. The tailgate of a truck is $2\frac{1}{2}$ feet above the ground. The incline of a ramp used for loading the truck is 14°. Find the length of the ramp. Round your answer to one decimal place.

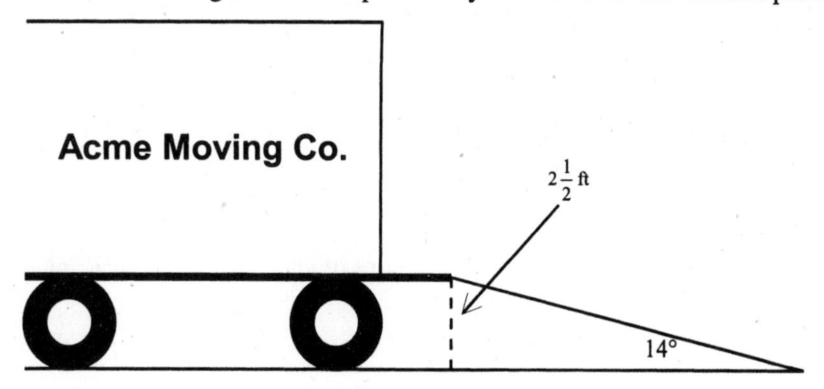

Section 4.4: Trigonometric Functions of Any Angle

Objective 1: Evaluate trigonometric functions of any angle

49. Given $\tan\theta = -\frac{12}{35}$ and $\sin\theta > 0$, find $\cos\theta$.

(A) $\cos\theta = -\frac{35}{37}$ (B) $\cos\theta = -\frac{12}{37}$ (C) $\cos\theta = \frac{12}{37}$ (D) $\cos\theta = \frac{35}{37}$

50. Identify the quadrant in which θ lies.
$\cos > 0$ and $\cot < 0$

(A) Quadrant I (B) Quadrant II (C) Quadrant III (D) Quadrant IV

51. Evaluate the sine, cosine, and tangent functions of the quadrant angle.
$\theta = 0$

52. The point given is on the terminal side of an angle in standard position. Find the cotangent, the secant, and the cosecant of the angle.
$(-9, -40)$

Objective 2: Use reference angles to evaluate trigonometric functions

Find the reference angle θ'.

53. $\theta = \frac{3\pi}{4}$ (A) $-\frac{\pi}{4}$ (B) $-\frac{5\pi}{4}$ (C) $\frac{\pi}{4}$ (D) $\frac{3\pi}{4}$

54. $\theta = 4$ (A) 2.2832 (B) 4 (C) 2.4292 (D) 0.8584

55. Rewrite the indicated trigonometric function in terms of the angle's reference angle. Use the same function.
$\sin(-474°)$

56. Angle θ is shown along with the value of its reference angle. Rewrite the indicated function in terms of the reference angle θ' and give the value of the function. Round your answer to four decimal places.
$\cos\theta$

Objective 3: Evaluate trigonometric functions of real numbers

57. Find the exact value of the function.
$\csc(-150°)$
(A) $\frac{\sqrt{3}}{3}$ (B) $-\sqrt{3}$ (C) –1 (D) –2

58. Use a calculator to approximate two values of $\theta\ (0 \le \theta < 2\pi)$ that satisfy the equation.
$\sin\theta = 0.4794$

(A) 1.285, 2.642 (B) 1.285, 3.427 (C) 0.500, 3.427 (D) 0.500, 2.642

59. Find the exact value of the function.

$\sin \dfrac{4\pi}{3}$

60. Find the indicated trigonometric value in the specified quadrant.

θ is in Quadrant IV and $\tan\theta = -4$. Find $\sec\theta$.

Section 4.5: Graphs of Sine and Cosine Functions

Objective 1: Sketch the graphs of basic sine and cosine functions

61. Give the number of full cycles of the function that are found in the interval.

$y = 3\sin x$ on the interval $[-7\pi,\ 7\pi]$

(A) 7 (B) 14 (C) 8 (D) 6

62. Identify the function shown in the graph.

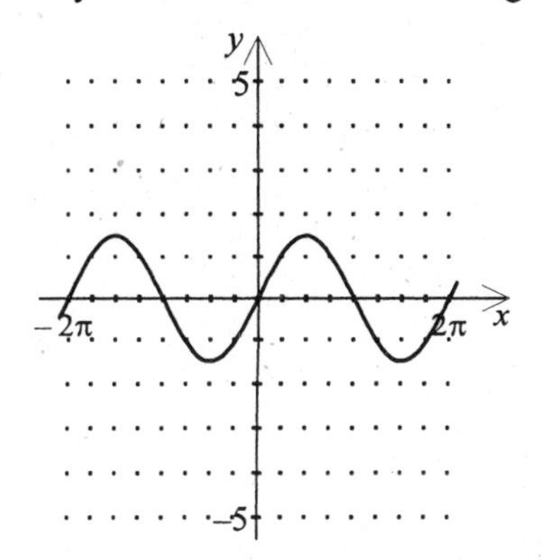

(A) $y = \cos 1.5x$ (B) $y = \sin 1.5x$ (C) $y = 1.5\sin x$ (D) $y = 1.5\cos x$

63. List the key points on the graph of the function.

$y = 2.5\sin x$ on the interval $[0, 2\pi]$

64. Sketch the graph of the function.

$y = 1.5\sin x$ on the interval $[-2\pi, 2\pi]$

Objective 2: Use amplitude and period to help sketch the graphs of sine and cosine functions

65. Find the amplitude and the period of the function.
$y = -3\cos 2x$

(A) Amplitude $= 3$ Period $= \pi$ (B) Amplitude $= -3$ Period $= 2$ (C) Amplitude $= 3$ Period $= 2$ (D) Amplitude $= -3$ Period $= \pi$

66. Identify the function that has the given amplitude and period.
Amplitude $= 7$, period $= 4\pi$

(A) $y = 7\sin\dfrac{\pi x}{4}$ (B) $y = 7\sin\dfrac{x}{2}$ (C) $y = 3.5\sin\dfrac{x}{2}$ (D) $y = 3.5\sin 4\pi x$

67. Sketch a graph of the function that has the given amplitude and period.
Sine function, amplitude $= 2.5$, period $= 3\pi$

68. Find the amplitude and the period of the graphed function.

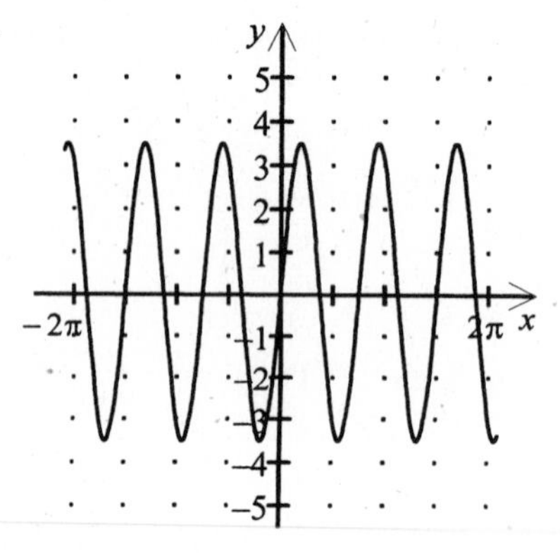

Objective 3: Sketch translations of the graphs of sine and cosine functions

69. Identify the graph of the function.

$$y = -0.5\sin\frac{x}{4} - 2.5$$

(A)

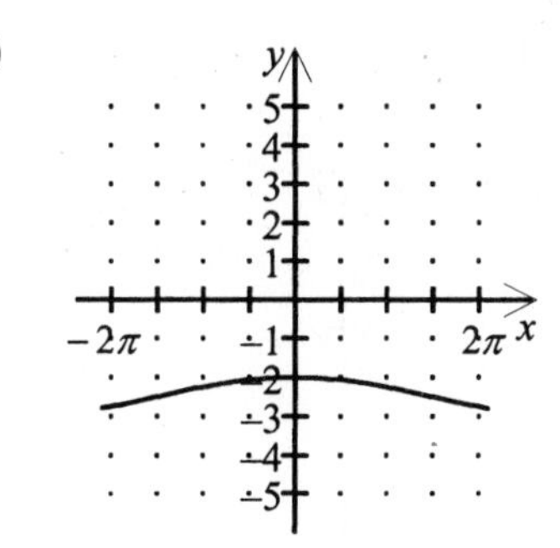

(B)

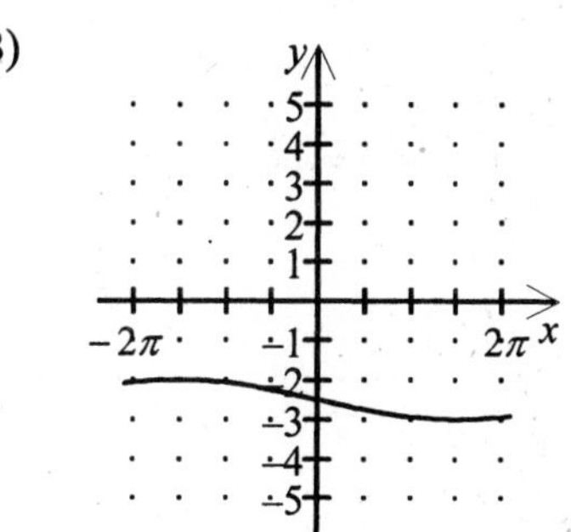

(C)

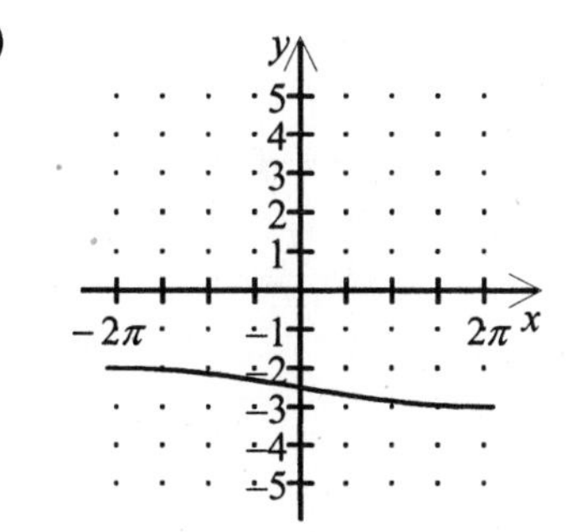

(D)

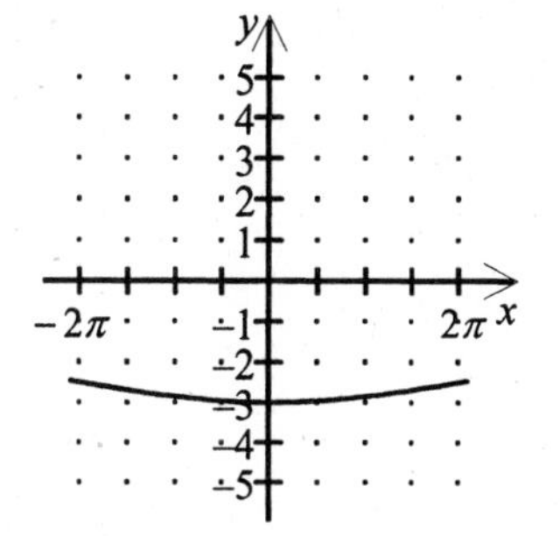

70. Identify the graph of the cosine function that has the required phase shift and vertical translation.

phase shift $= \dfrac{4\pi}{3}$; vertical shift $= -1$

(A)

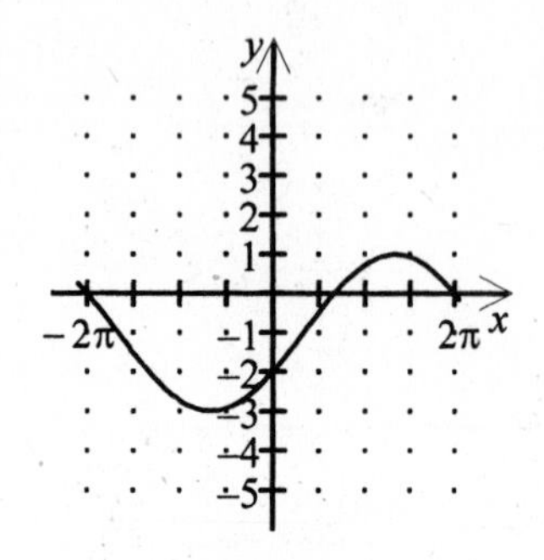

(B)

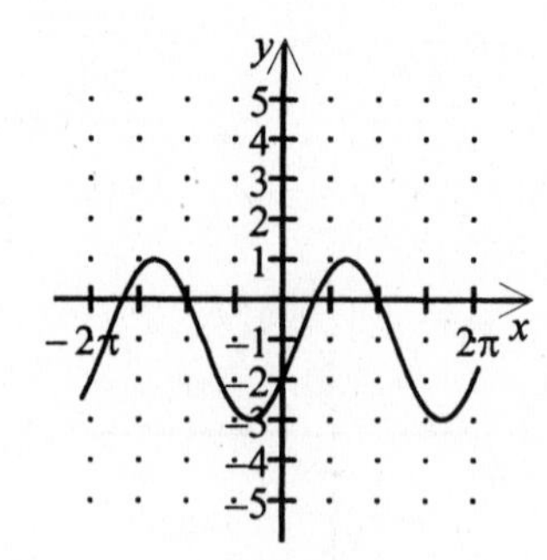

(C)

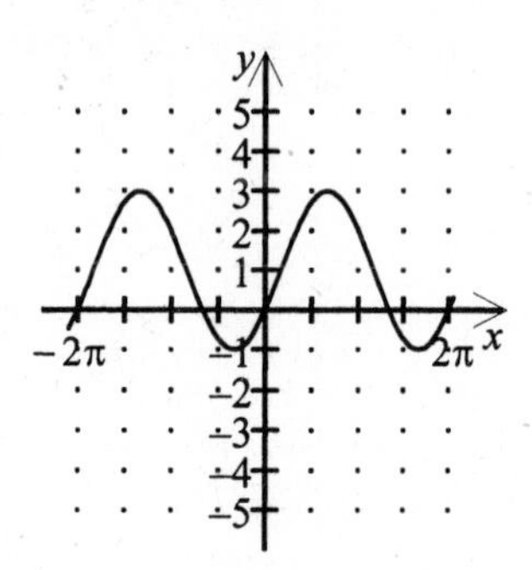

(D)

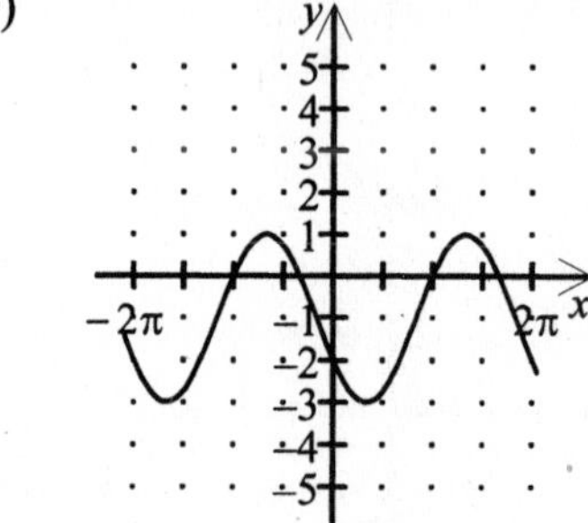

71. List the start and end points of any full cycle on the graph of the function. $y = -\cos(6x + 5\pi)$

72. Sketch the graph of the function.

$y = -1.5\cos\left(3x - \dfrac{\pi}{2}\right) - 2$, on the interval $[-2\pi, 2\pi]$

Objective 4: Use sine and cosine functions to model real-life data

73. The data below represent the average monthly cost of natural gas in an Oregon home.

Month	Aug	Sep	Oct	Nov	Dec	Jan
Cost ($)	22.90	28.79	45.38	66.35	89.32	104.41

Month	Feb	Mar	Apr	May	Jun	Jul
Cost ($)	110.80	104.91	88.83	66.85	44.88	28.79

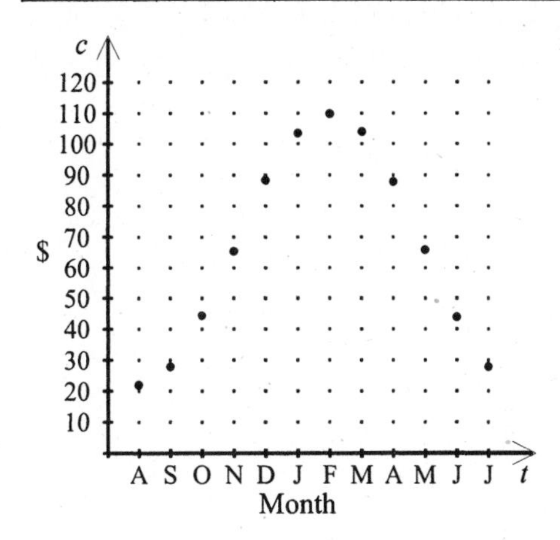

Which sine function best describes the data?

(A) $c(t) = 44.0 \sin\left(\frac{\pi t}{6} - \frac{2\pi}{3}\right) + 66.8$

(B) $c(t) = 44.0 \sin\left(\frac{\pi t}{6} - \frac{\pi}{12}\right) + 66.8$

(C) $c(t) = 44.0 \sin\left(\frac{\pi t}{8} - 12\right) + 22.9$

(D) $c(t) = 44.0 \sin\left(\frac{\pi t}{4} - \frac{2\pi}{3}\right) + 22.9$

74. A weight attached to the end of a spring is pulled down 7 centimeters below its equilibrium point and released. It takes 6 seconds for it to complete one cycle of moving from 7 centimeters below the equilibrium point to 7 centimeters above the equilibrium point and then returning to its low point. Identify the sinusoidal function that best represents the position of the moving weight and the approximate position of the weight at 9 seconds.

(A) $h(t) = 6\cos\frac{2\pi t}{7}$, -5.8 cm

(B) $h(t) = -7\cos\frac{\pi t}{3}$, 7 cm

(C) $h(t) = 7\sin\frac{\pi t}{3}$, 7 cm

(D) $h(t) = -6\sin\frac{2\pi t}{7}$, -1.3 cm

75. Jantje is checking her bicycle on a stand. The bicycle's wheel has a radius of 39 centimeters and its center is 63 centimeters above the ground. Jantje is rotating the wheel at a constant angular speed of $\frac{1\pi t}{2}$ radians per second.

(a) Describe the function $h(t)$ using the sine function where h is the height in feet and t is the time in minutes.

(b) Find $h(t)$ when $t = 7.5$ seconds. Round your answer to the nearest tenth of a centimeter.

76. The high and low tides of a coastal city on May 3, 2000 are shown in the table. Find an equation for $h(t)$, the height of the water as a function of t, the number of hours after midnight. Assume that the water height can be modeled by a cosine function. If needed, round to three decimal places.

Tide	Time	Height
High	3:00 A.M.	107 cm
Low	9:24 A.M.	39 cm

Section 4.6: Graphs of Other Trigonometric Functions

Objective 1: Sketch the graphs of tangent functions

77. Identify the graph of the function.

$$y = -\tan\frac{9x}{4}$$

(A)

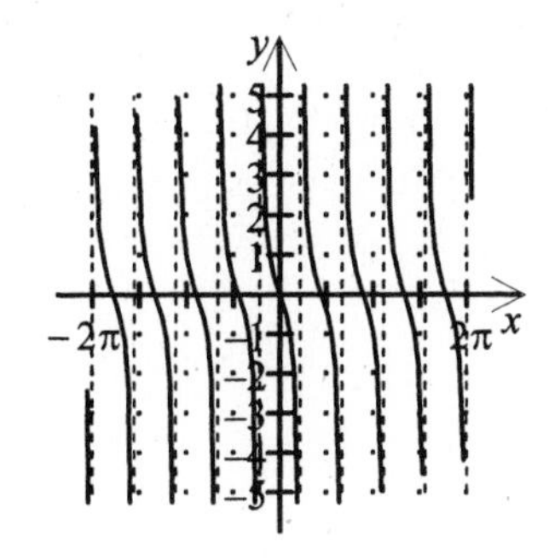

(B)

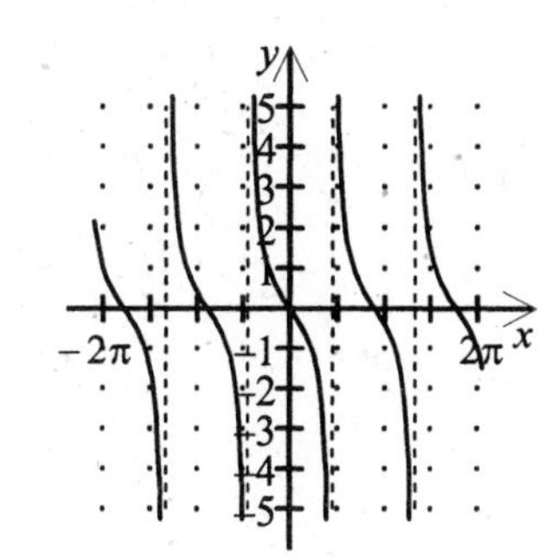

(C)

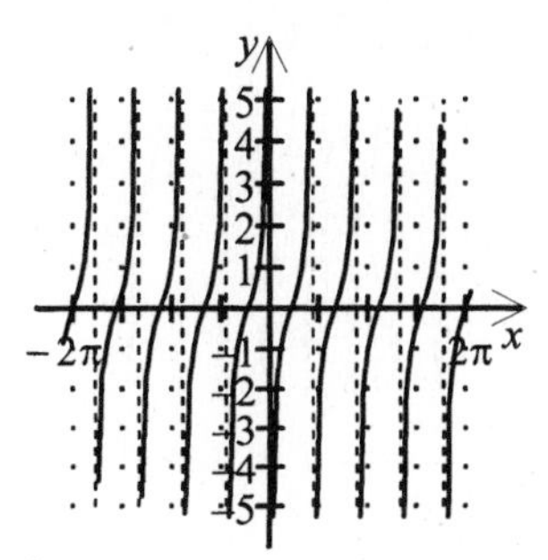

(D)

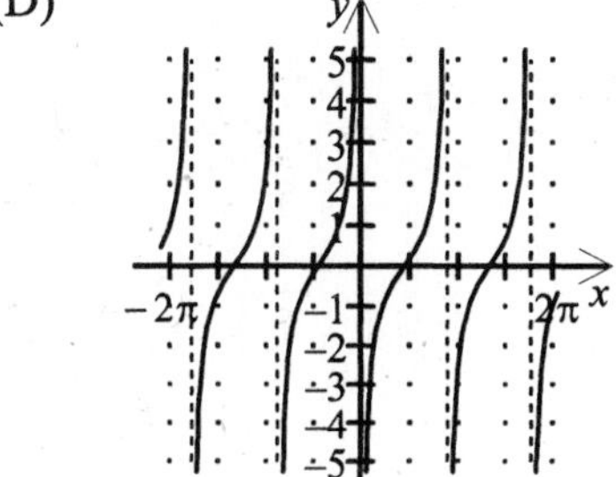

78. Identify the function that is graphed.

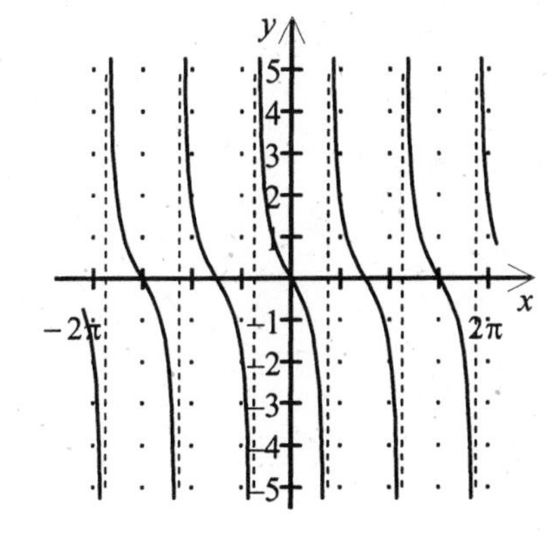

(A) $y = -\tan\frac{3x}{4}$ (B) $y = -\cot\frac{3x}{4}$ (C) $y = -\tan\frac{4x}{3}$ (D) $y = \cot\frac{4x}{3}$

79. Sketch the graph of the function and write equations for 2 consecutive asymptotes.

$$y = -\tan\left(\frac{1x}{2} + \frac{\pi}{8}\right)$$

80. Sketch the graph of the function. Include any asymptotes.

$$y = -2.5\tan\frac{3\pi x}{10}$$

Objective 2: Sketch the graphs of cotangent functions

81. Identify the graph of the function.

$$y = -\cot\frac{9x}{4}$$

(A)

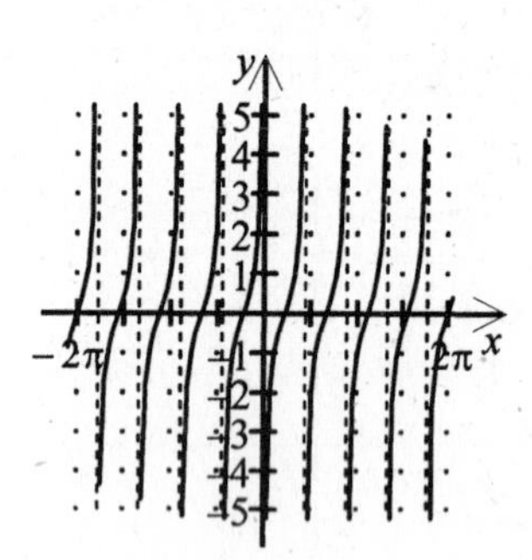

(B)

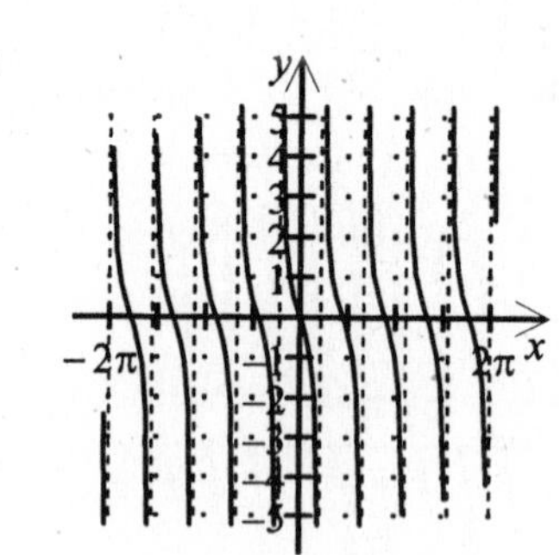

(C)

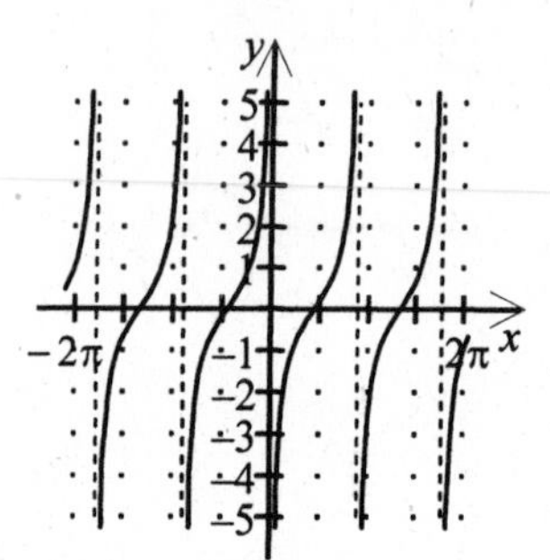

(D)

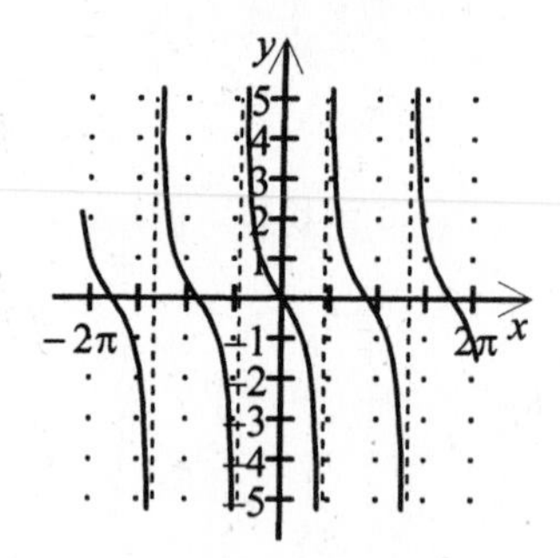

82. Identify the function that is graphed.

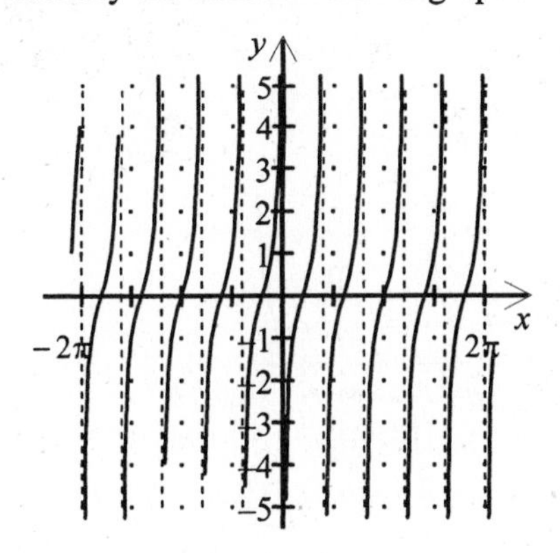

(A) $y = \tan\dfrac{5x}{2}$ (B) $y = -\tan\dfrac{2x}{5}$ (C) $y = -\cot\dfrac{5x}{2}$ (D) $y = -\cot\dfrac{2x}{5}$

83. Sketch the graph of the function and write equations for 2 consecutive asymptotes.

$y = -\cot\left(2x - \dfrac{3\pi}{2}\right)$

84. Sketch the graph of the function. Include any asymptotes.

$y = -3.5\cot \pi x$

Objective 3: Sketch the graphs of secant and cosecant functions

85. Identify the equation that matches the description of the function.

A secant function has a phase shift of $\dfrac{\pi}{4}$ and a period of $\dfrac{8\pi}{7}$.

(A) $y = \sec\left(\dfrac{7x}{4} - \dfrac{7\pi}{16}\right)$ (B) $y = \sec\left(\dfrac{7x}{4} + \dfrac{\pi}{4}\right)$

(C) $y = \sec\left(\dfrac{8x}{7} - \dfrac{7\pi}{16}\right)$ (D) $y = \sec\left(\dfrac{7x}{4} + \dfrac{7\pi}{16}\right)$

86. Identify the graph of the function.

$y = \csc \frac{7x}{4}$

(A)

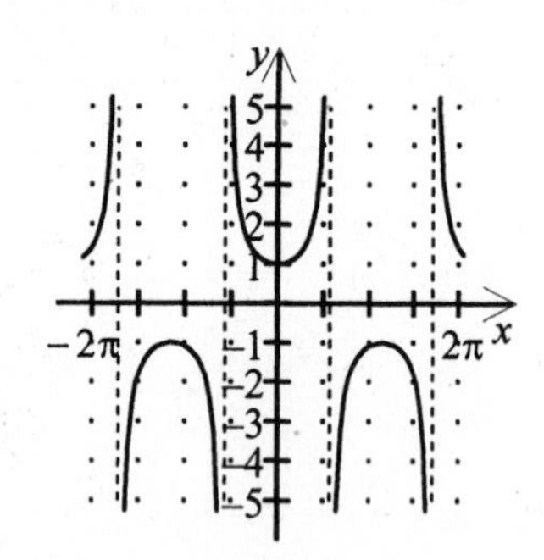

(B)

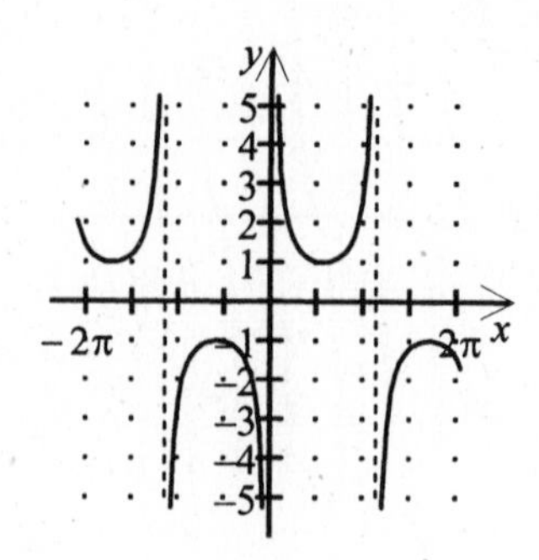

(C)

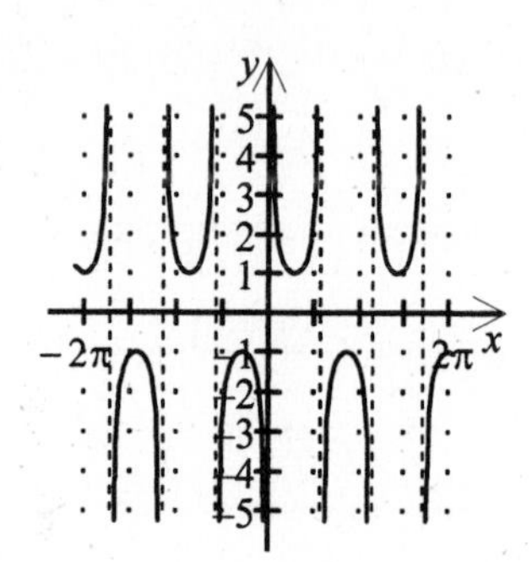

(D)

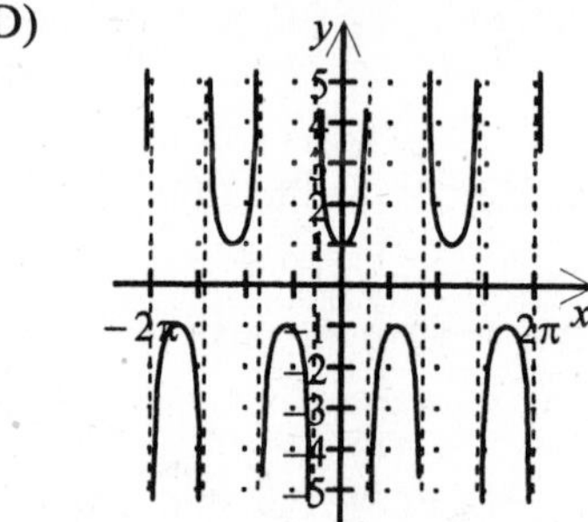

87. Sketch the graph of the function and write equations for 2 consecutive asymptotes.

$y = -\sec\left(\frac{9x}{4} + \frac{9\pi}{16}\right)$

88. Give the period and the phase shift of the function.

$y = -\csc\left(\frac{4x}{5} + \frac{7\pi}{5}\right)$

Objective 4: Sketch the graphs of damped trigonometric functions

89. Identify the graph of the function.

$y = \left|\frac{x}{4}\right| \cos \frac{4x}{3}$

(A)

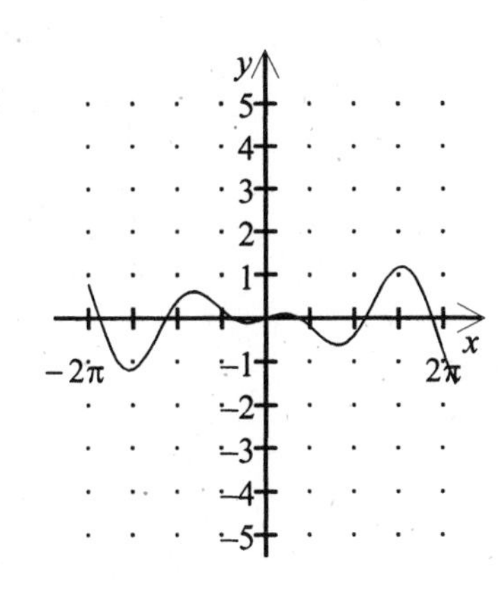

(B)

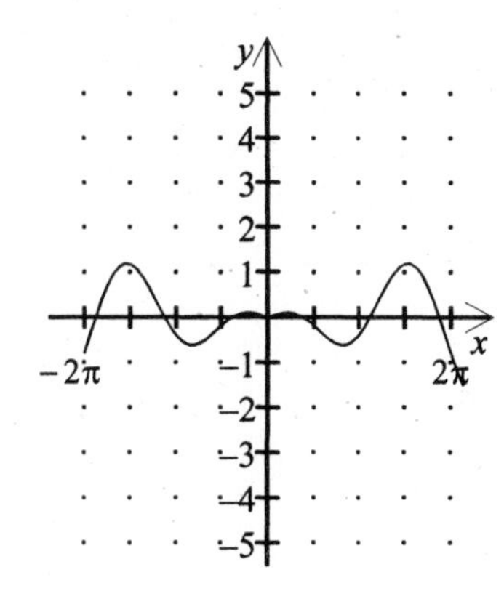

(C)

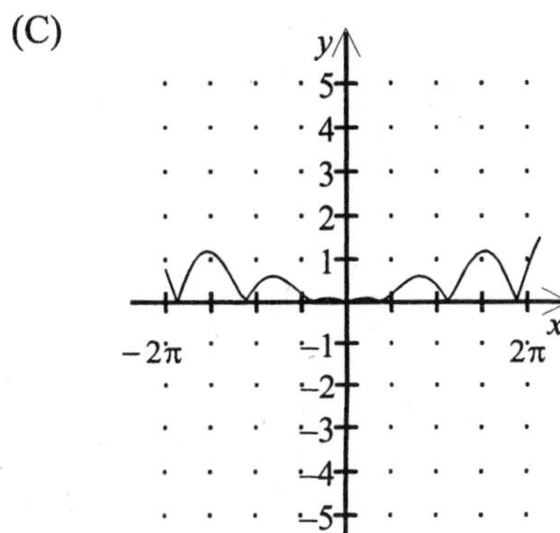

(D)

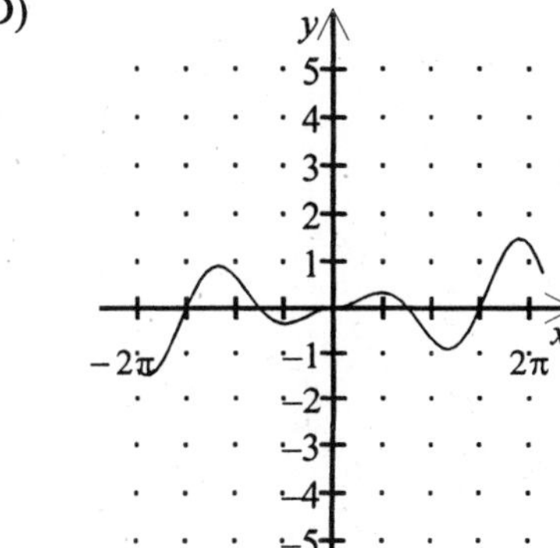

90. Identify the graph that shows the boundaries outlining the maximum and minimum values of the function.

$y = e^{-x/3} \sin 3x$

(A)

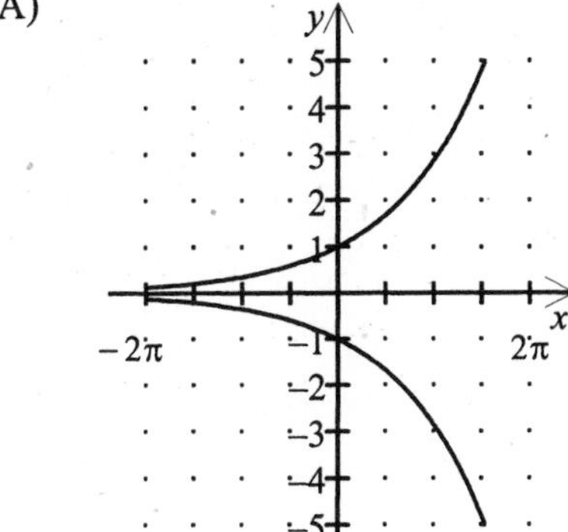

(B)

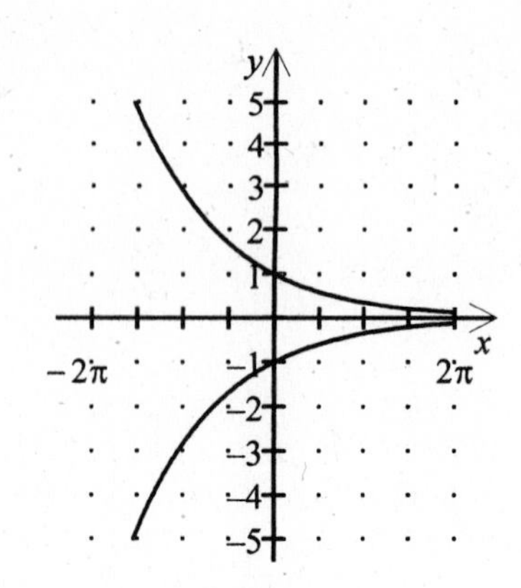

(C)

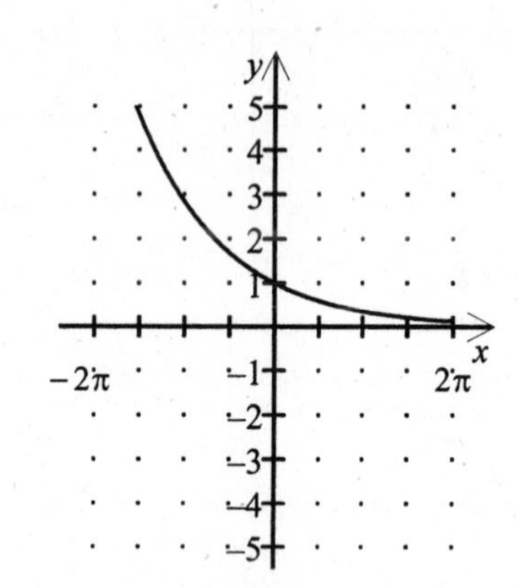

(D)

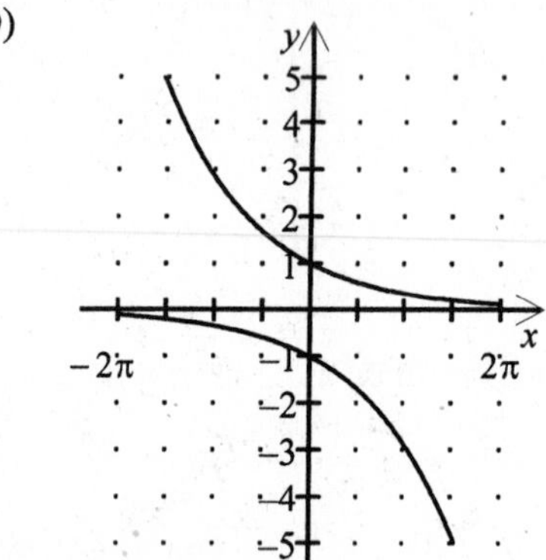

(90.)

91. What is the damping factor in the equation?

$f(x) = 2^{-x}\sin(-4x)$

92. Use a graphing utility to sketch the graph of the function.

$y = \dfrac{\cos 2x}{|x|}$

Section 4.7: Inverse Trigonometric Functions

Objective 1: Evaluate the inverse sine function

93. If possible, evaluate the expression without the aid of a calculator.

$\sin^{-1}\left(-\frac{1}{2}\right)$

(A) 0 (B) $-\frac{\pi}{6}$ (C) $-\frac{\pi}{2}$ (D) Not possible

94. Use a calculator to approximate the expression.

arcsin 0.21

(A) 4.80 (B) 272.84 (C) 0.21 (D) 0.04

95. Determine the exact value of the missing coordinate of the point marked on the graph of the function.

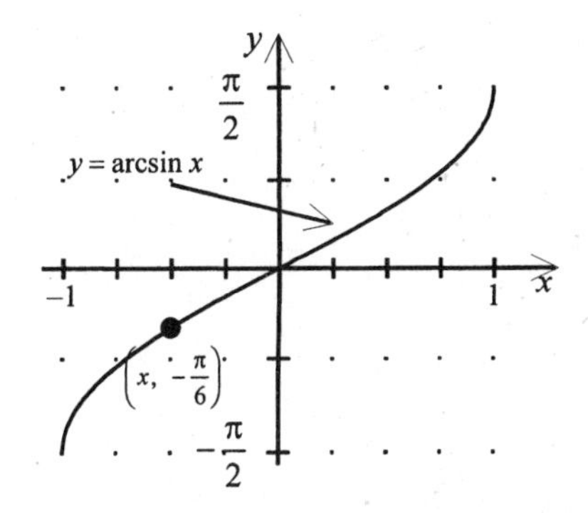

96. Give the exact value in radians.

$\theta = \arcsin\left(-\frac{\sqrt{3}}{2}\right)$

Objective 2: Evaluate the other inverse trigonometric functions

Use a calculator to approximate the expression.

97. arccos 0.56 (A) 1.56 (B) 0.98 (C) 1.00 (D) 1.18

Use a calculator to approximate the expression.

98. $\arctan(-0.24)$ (A) -0.2355 (B) -0.0042 (C) -4.0864 (D) -238.731

99. Give the exact value in radians.

$\theta = \arccos\left(-\frac{1}{2}\right)$

100. Determine the exact value of the missing coordinate of the point indicated on the graph of the function.

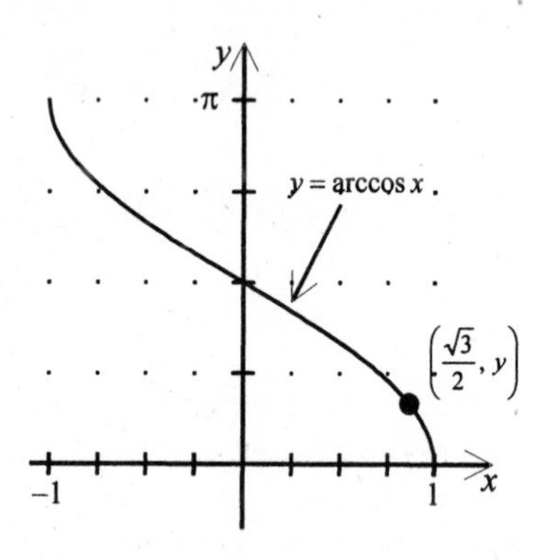

Objective 3: Evaluate the compositions of trigonometric functions

Use the properties of inverse functions to evaluate the expression.

101. $\sin\left(\arctan\frac{2}{3}\right)$ (A) $\frac{3\sqrt{13}}{13}$ (B) $\frac{2}{3}$ (C) $\frac{\sqrt{13}}{2}$ (D) $\frac{2\sqrt{13}}{13}$

102. $\tan\left(\arcsin\frac{2\sqrt{x}}{1+x}\right)$ (A) $\frac{\sqrt{x}(1-x)}{2x}$ (B) $\frac{1-x}{1+x}$ (C) $\frac{\sqrt{x}(1+x)}{2x}$ (D) $\frac{2\sqrt{x}}{1-x}$

103. $\arccos\left(\cos\left(-\frac{35\pi}{12}\right)\right)$

104. $\sin\left(\arcsin\frac{\sqrt{2}}{2}\right)$

Section 4.8: Applications and Models

Objective 1: Solve real-life problems involving right triangles

105. At a distance of 48 feet from the base of a flag pole, the angle of elevation to the top of a flag that is 4.1 feet tall is 32.5°. The angle of elevation to the bottom of the flag is 28.9°. The pole extends 1 foot above the flag. Find the height of the pole.

(A) 27.8 ft (B) 30.6 ft (C) 31.6 ft (D) 26.8 ft

106. An airplane is flying east at a constant altitude of 30,500 meters. When first seen to the east of an observer, the angle of elevation to the airplane is 63°. After 72 seconds, the angle of elevation is 48.4°. Find the speed of the airplane.

(A) 150 m/s (B) 268 m/s (C) 160 m/s (D) 257 m/s

107. A flat roof rises at a 31° angle from the front wall of a storage shed to the back wall. The front wall is 9.5 feet tall and the back wall is 16.7 feet tall. Find the length of the roof line and the depth of the shed from front to back. Round your answers to the nearest tenth of a foot.

108. The pilot of a small private plane can look forward and see the control tower for a small airstrip. Beyond that is a large factory that is 5.3 miles from the airstrip. The angles of depression are 13.1° and 3.8°, respectively. Find the airplane's altitude, to the nearest ten feet.

Objective 2: Solve real-life problems involving directional bearings

109. A ship leaves port at 8 miles per hour, with a heading of N 40° W. There is a warning buoy located 6 miles directly north of the port. What is the bearing of the warning buoy as seen from the ship after 1.5 hours?

(A) N 22.5° E (B) S 67.5° W (C) N 22.5° W (D) S 67.5° E

110. A hiker travels at 3.3 miles per hour at a heading of N 47° E from a ranger station. After 3.5 hours, how far north and how far east is the hiker from the ranger station?

(A) 8.4 miles north and 7.9 miles east (B) 1.4 miles north and 11.5 miles east

(C) 11.5 miles north and 1.4 miles east (D) 7.9 miles north and 8.4 miles east

111. A private seaplane travels in a straight line from a local airstrip to a remote lake that is 142 miles east and 329 miles north of the airstrip. Find the lake's bearing and distance from the airstrip. Round your answers to the nearest tenth.

112. What direction bearing is shown?

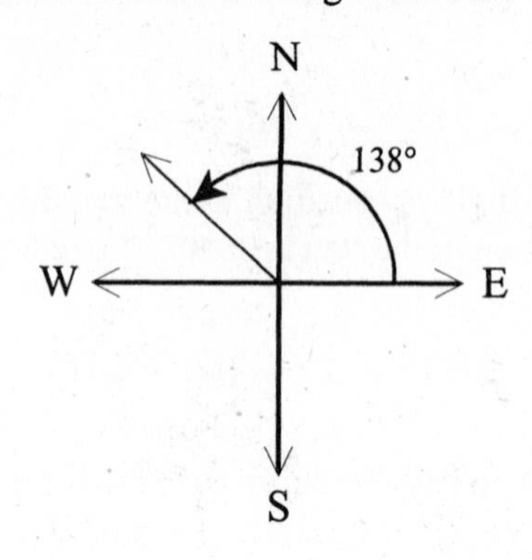

Objective 3: Solve real-life problems involving harmonic motion

113. A rowboat is observed from a dock as it bobs up and down in simple harmonic motion because of wave action. The boat moves from a high point of 4.8 feet below the dock to a low point of 6.6 feet below the dock and back to its high point 6 times every minute. Let t be time, in minutes, and h be the distance below the dock, in feet. Identify the equation that describes the boat's motion.

(A) $h(t) = -5.7 + 0.9\cos\frac{\pi}{3}t$

(B) $h(t) = 5.7 - 0.9\cos 12\pi t$

(C) $h(t) = -1.8\sin\frac{\pi}{3}t$

(D) $h(t) = 1.8\sin 12\pi t + 5.7$

114. A mass attached to a spring vibrates up and down in simple harmonic motion according to the equation

$$h(t) = -2\sin\frac{3\pi}{8}t$$

where h is in centimeters and t is in seconds. Identify the amplitude and the period of the vibrations.

(A) Amplitude = –2 cm

Period $= \frac{3\pi}{8}$ sec

(B) Amplitude = 2 cm

Period $= \frac{3\pi}{8}$ sec

(C) Amplitude = 2 cm

Period $= \frac{16\pi}{3}$ sec

(D) Amplitude = 2 cm

Period $= \frac{16}{3}$ sec

115. If you watch from ground level, a child riding on a merry-go-round will seem to be undergoing simple harmonic motion from side to side. Assume the merry-go-round is 14.4 feet across and the child completes 8 rotations in 64 seconds. Write a sine function that describes d, the child's apparent distance from the center of the merry-go-round, as a function of time t.

116. If a pendulum bob is released from its farthest right position and it swings in small arcs, then the distance x from the equilibrium position can be described by a cosine function. Assume a pendulum bob swings in an arc 5 centimeters wide with a period of 1.2 seconds. Let x be positive if the bob is to the right of the equilibrium position and let t be the time after the bob's release. What is the equation for the bob's horizontal location and where is the bob after 1.5 seconds? Round as needed to two decimal places.

Answer Key for Chapter 4 Trigonometry

Section 4.1: Radian and Degree MeasureSection

Objective 1: Describe angles

[1] (D)

[2] (B)

Answers may vary. Sample answer:

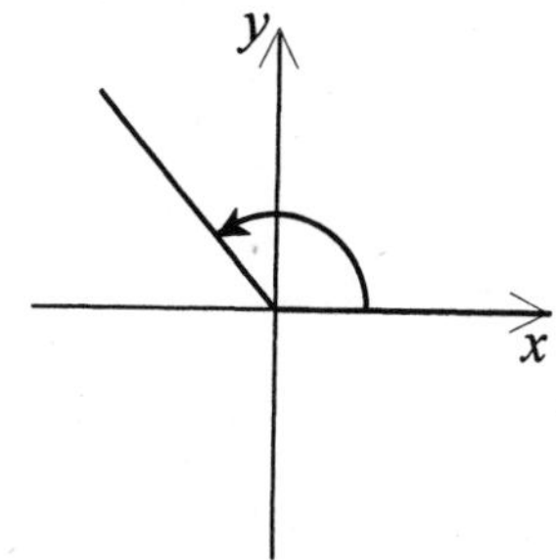

[3]

Answers may vary. Sample answers:

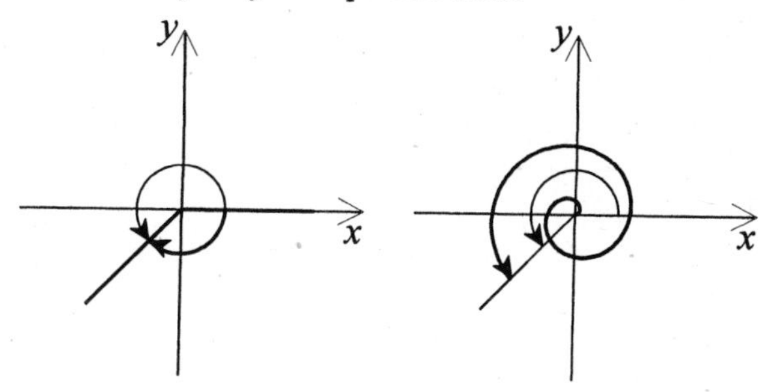

[4]

Objective 2: Use radian measure

[5] (D)

[6] (A)

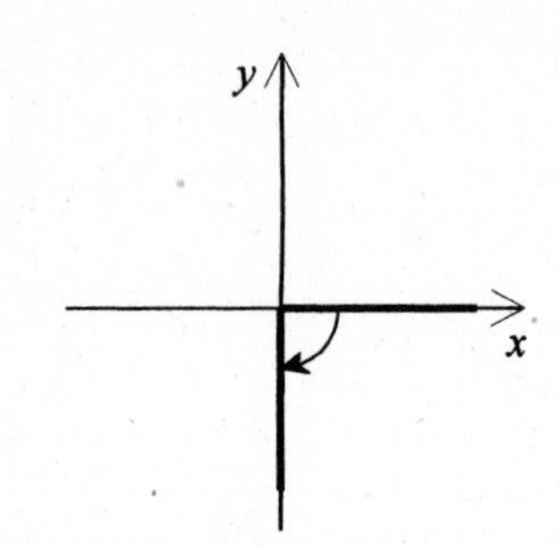

[7]

Positive coterminal angle: $\frac{\pi}{3}$

[8] Negative coterminal angle: $-\frac{11\pi}{3}$

Objective 3: Use degree measure

[9] (A)

[10] (B)

[11] 105°

[12] 1.693

Objective 4: Use angles to model and solve real-life problems

[13] (A)

[14] (B)

[15] 0.7 ft/min

[16] 6° 30′ 16″

Section 4.2: Trigonometric Functions: The Unit Circle

Objective 1: Identify a unit circle and its relationship to real numbers

[17] (D)

[18] (D)

[19] 2.36 ft

[20] 0.742

Objective 2: Evaluate trigonometric functions using the unit circle

[21] (A)

[22] (C)

[23] $\sin t = -\frac{\sqrt{2}}{2}$, $\csc t = -\sqrt{2}$
$\cos t = -\frac{\sqrt{2}}{2}$, $\sec t = -\sqrt{2}$
$\tan t = 1$, $\cot t = 1$

[24] $-\frac{1}{y}$

Objective 3: Use the domain and period to evaluate sine and cosine functions

[25] (A)

[26] (A)

[27] Answers may vary. Sample answer: $\sin\left(-\frac{23\pi}{6}\right) = \sin\left(-4\pi + \frac{\pi}{6}\right) = \sin\frac{\pi}{6} = \frac{1}{2}$.

[28] odd function

Objective 4: Use a calculator to evaluate trigonometric functions

[29] (C)

[30] (A)

[31] $\cot t = 1.9060$; $\sec t = 1.1293$; $\csc t = 2.1524$

[32] $\sin t = 0.9945$; $\cos t = 0.1045$; $\tan t = 9.5144$

Section 4.3: Right Triangle Trigonometry

Objective 1: Evaluate trigonometric functions of acute angles

[33] (C)

[34] (D)

[35] $\sin\theta = \frac{14}{50}$; $\cos\theta = \frac{48}{50}$; $\tan\theta = \frac{14}{48}$; $\cot\theta = \frac{48}{14}$; $\sec\theta = \frac{50}{48}$; $\csc\theta = \frac{50}{14}$

[36] Answers may vary. Sample answer: $\sin\theta = \frac{\sqrt{2}}{2}$.

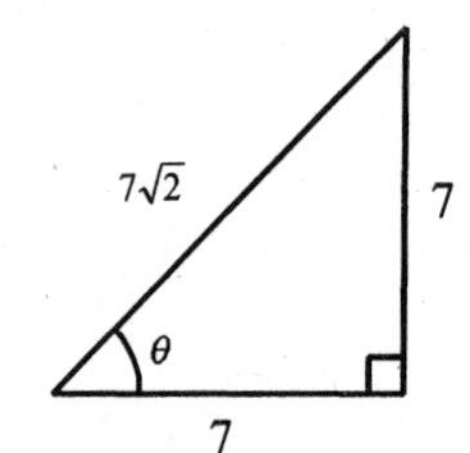

Objective 2: Use the fundamental trigonometric identities

[37] (C)

[38] (C)

[39] Answers may vary. Sample answer: 1

[40] $\frac{\sqrt{13}}{7}$

Objective 3: Use a calculator to evaluate trigonometric functions

[41] (B)

[42] (A)

[43] 1.0080

[44] 0.1501

Objective 4: Use trigonometric functions to model and solve real-life problems

[45] (C)

[46] (C)

[47] 46.94 m

[48] 10.3 ft

Section 4.4: Trigonometric Functions of Any Angle

Objective 1: Evaluate trigonometric functions of any angle

[49] (A)

[50] (D)

[51] $\sin 0 = 0, \ \cos 0 = 1, \ \tan 0 = 0$

[52] $\cot\theta = \frac{9}{40}, \ \sec\theta = -\frac{41}{9}, \ \csc\theta = -\frac{41}{40}$

Objective 2: Use reference angles to evaluate trigonometric functions

[53] (C)

[54] (D)

[55] $-\sin 66°$

[56] $\cos\theta = \cos\theta' = 0.6561$

Objective 3: Evaluate trigonometric functions of real numbers

[57] (D)

[58] (D)

[59] $-\dfrac{\sqrt{3}}{2}$

[60] $\sec\theta = \sqrt{17}$

Section 4.5: Graphs of Sine and Cosine Functions

Objective 1: Sketch the graphs of basic sine and cosine functions

[61] (A)

[62] (C)

[63] $(0,\ 0),\ \left(\dfrac{\pi}{2},\ 2.5\right),\ (\pi,\ 0),\ \left(\dfrac{3\pi}{2},\ -2.5\right)$

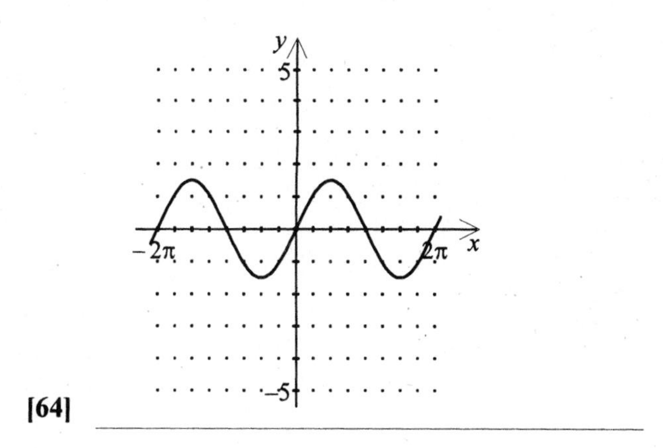

[64]

Objective 2: Use amplitude and period to help sketch the graphs of sine and cosine functions

[65] (A)

[66] (B)

Answers may vary. Sample answer:

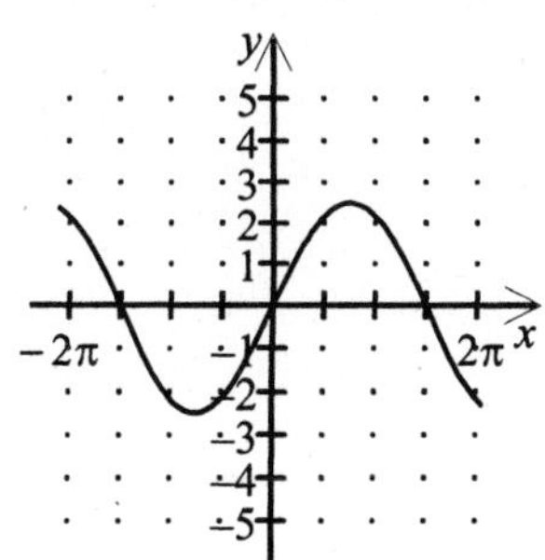

[67]

[68] Amplitude = 3.5 ; period = $\dfrac{3\pi}{4}$

Objective 3: Sketch translations of the graphs of sine and cosine functions

[69] (C)

[70] (D)

[71] Answers may vary. Sample answer: $\left(-\dfrac{5\pi}{6},\ -1\right), \left(-\dfrac{\pi}{2},\ -1\right)$

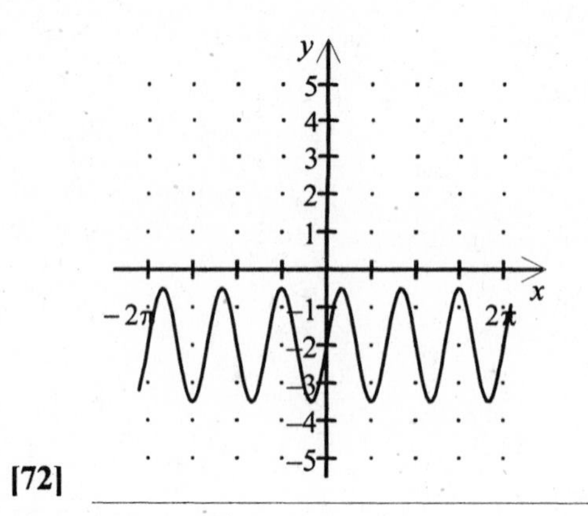

[72]

Objective 4: Use sine and cosine functions to model real-life data

[73] (A)

[74] (B)

(a) $h(t) = 63 + 39\sin\left(\frac{1\pi t}{2}\right)$

[75] (b) 35.4 centimeters

[76] $h(t) = 34\cos(0.491t - 1.473) + 73$

Section 4.6: Graphs of Other Trigonometric Functions

Objective 1: Sketch the graphs of tangent functions

[77] (A)

[78] (C)

Answers may vary. Sample answer: $x = \dfrac{3\pi}{4}$, $x = -\dfrac{5\pi}{4}$

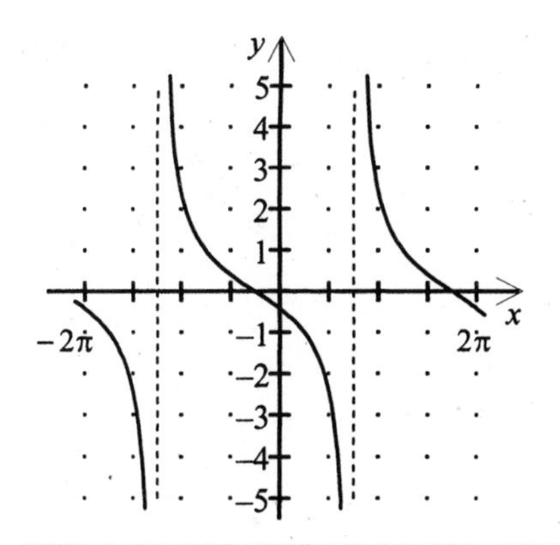

[79]

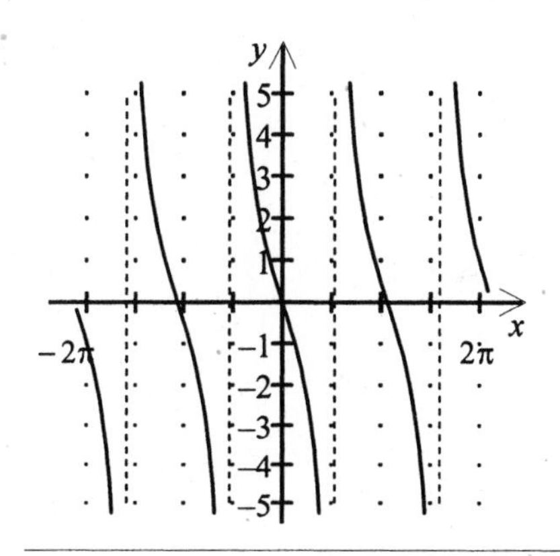

[80]

Objective 2: Sketch the graphs of cotangent functions

[81] (A)

[82] (C)

Answers may vary. Sample answer: $x = \dfrac{3\pi}{4}$, $x = \dfrac{5\pi}{4}$.

[83] 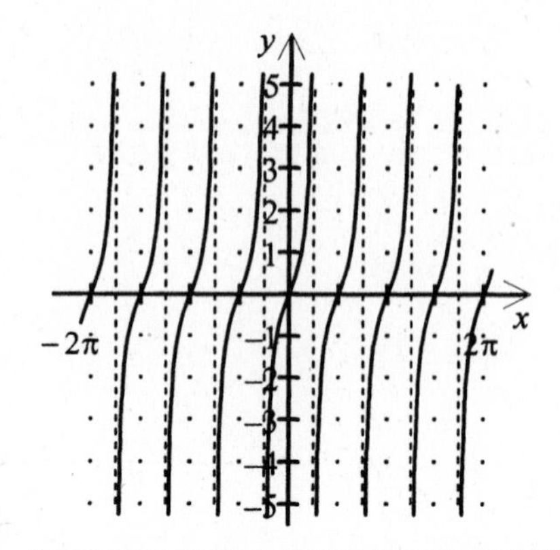

[84] 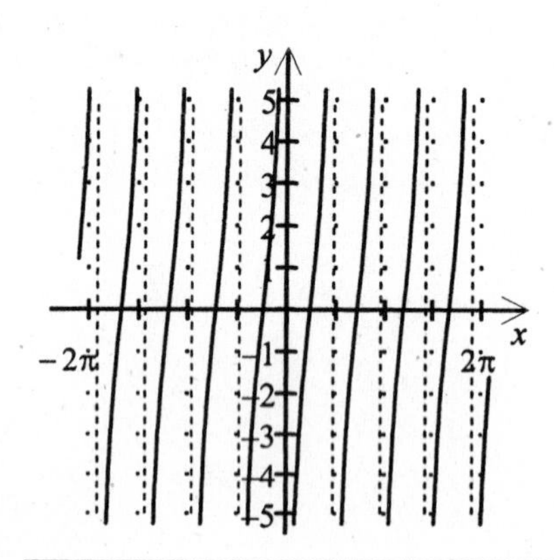

Objective 3: Sketch the graphs of secant and cosecant functions

[85] (A)

[86] (C)

Answers may vary. Sample answer: $x = -\frac{\pi}{36}$, $x = -\frac{17\pi}{36}$

[87] 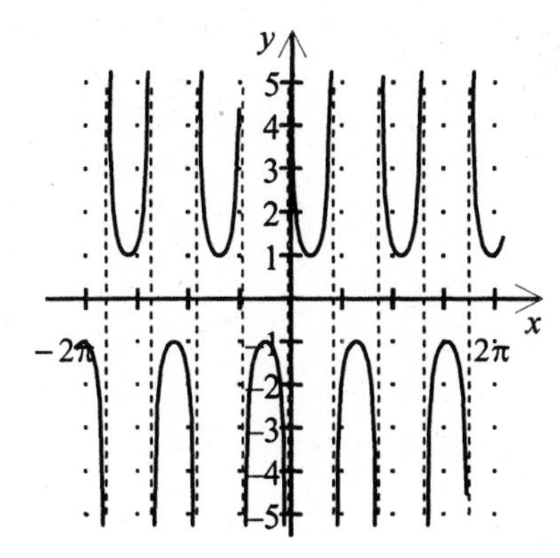

[88] period $= \frac{5\pi}{2}$, phase shift $= -\frac{7\pi}{4}$

Objective 4: Sketch the graphs of damped trigonometric functions

[89] (B)

[90] (B)

[91] 2^{-x}

[92] 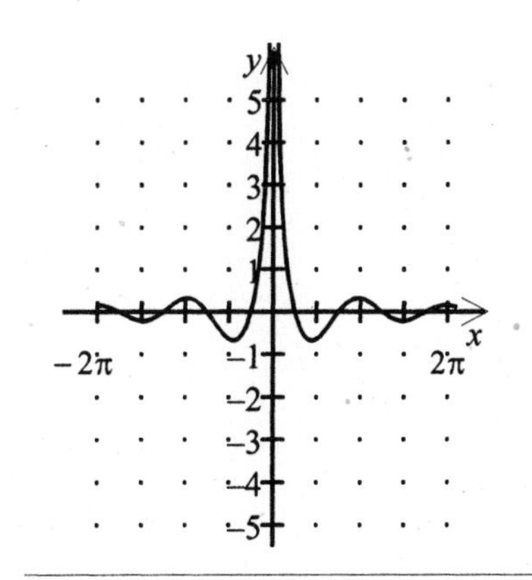

Section 4.7: Inverse Trigonometric Functions

Objective 1: Evaluate the inverse sine function

[93] (B)

[94] (C)

[95] $\left(-\frac{1}{2}\right)$

[96] $\theta = -\frac{\pi}{3}$

Objective 2: Evaluate the other inverse trigonometric functions

[97] (B)

[98] (A)

[99] $\theta = \frac{2\pi}{3}$

[100] $\frac{\pi}{6}$

Objective 3: Evaluate the compositions of trigonometric functions

[101] (D)

[102] (D)

[103] $\frac{11\pi}{12}$

[104] $\frac{\sqrt{2}}{2}$

Section 4.8: Applications and Models

Objective 1: Solve real-life problems involving right triangles

[105] (C)

[106] (C)

[107] roof line = 14.0 ft long, shed depth = 12.0 ft

[108] 2600 ft

Objective 2: Solve real-life problems involving directional bearings

[109] (D)

[110] (D)

[111] N 23.3° E, distance = 358.3 mi

[112] N 48° W

Objective 3: Solve real-life problems involving harmonic motion

[113] (B)

[114] (D)

[115] $d(t) = 7.2 \sin \frac{\pi}{4} t$

[116] $x(t) = 2.5 \cos 5.24\, t,\ x = -0.02$ cm

Chapter 5 Analytic Trigonometry

Section 5.1: Using Fundamental Identities

Objective 1: Recognize and write the fundamental trigonometric identities

1. Which expression completes the fundamental trigonometric identity? $\csc(-x)=$

(A) $-\csc x$ (B) $\sin x$ (C) $-\sin x$ (D) $\csc x$

2. Which is a trigonometric identity?

(A) $\cot u = -\dfrac{\sin u}{\cos u}$ (B) $\sec u = -\sin\dfrac{1}{u}$ (C) $\cot u = \dfrac{1}{\tan u}$ (D) $\tan^2 u + \cot^2 u = 1$

3. Is the equation a valid form of one of the Pythagorean trigonometric identities? $\cot^2 \beta = \csc^2 \beta + 1$

4. Write the radical form of one of the Pythagorean trigonometric identities.

Objective 2: Use the fundamental trigonometric identities to evaluate trigonometric functions, simplify trigonometric expressions, and rewrite trigonometric expressions

5. Factor the expression and use the fundamental identities to simplify.
$\cos^2 x \sec^2 x - \cos^2 x$

(A) 1 (B) $\sin^2 x$ (C) $\cos^2 x$ (D) $\cos^2 x \cot^2 x$

6. Use trigonometric substitution to write the algebraic expression as a trigonometric function of θ, where $0<\theta<\dfrac{\pi}{2}$.

$\sqrt{7x^2-567},\ x = 9\sec\theta$

(A) $9\sqrt{7}+9\cos\theta$ (B) $9\sqrt{7}\tan\theta$ (C) $18\sqrt{7}\sin\theta$ (D) $567\tan^2\theta$

7. Use the fundamental identities to write the expression in terms of a single trigonometric function.

$\dfrac{\sec x + \csc x}{1+\tan x}$

8. Perform the addition and use the fundamental identities to simplify.

$\dfrac{5+3\cos x}{\sin x}+\dfrac{3\sin x}{1+\cos x}$

Section 5.2: Verifying Trigonometric Identities

Objective 1: Plan a strategy for verifying trigonometric identities

9. Convert all of the terms to sines and cosines and simplify to find the expression that completes the identity.

$$\frac{\sec x}{\tan x \csc x} =$$

(A) $\frac{\sin x}{\cos^2 x}$ (B) $\sin x$ (C) 1 (D) $\frac{\sin^2 x}{\cos x}$

10. Add or subtract the fractions. Then simplify using the Pythagorean identities and factoring to find the expression that completes the identity.

$$\frac{\tan x}{\sec x - 1} + \frac{\sec x - 1}{\tan x}$$

(A) $2 \sec x$ (B) $2 \tan x$ (C) $2 \cot x$ (D) $2 \csc x$

11. Which fundamental identities could be used to verify the identity?

$$\cot\theta \sin\theta = \cos\theta$$

12. Convert all of the terms to sines and cosines and simplify.

$$\cot x \csc x$$

Objective 2: Verify trigonometric identities

Identify the expression that completes the equation so that it is an identity.

13. $\frac{\sin u}{1 - \cos u} - \frac{1 + \cos u}{\sin u} =$ (A) $\tan u$ (B) $-\sin u$ (C) $2 + \cos u$ (D) 0

14. $\frac{\cos^2 x}{1 - \sin x} =$ (A) $\frac{1 - \sin x}{\cos x}$ (B) $\sin x$ (C) $\frac{1 + \csc x}{\csc x}$ (D) $\frac{1 + \sin x}{\cos x}$

Verify the identity.

15. $\frac{\csc^2 x - \cot^2 x}{\sec x} = \cos x$

Verify the identity.

16. $\dfrac{\sin x}{1-\sin^2 x} = \sec x \tan x$

Section 5.3: Solving Trigonometric Equations

Objective 1: Use standard algebraic techniques to solve trigonometric equations

Identify the x-values that are solutions of the equation.

17. $9 \tan x - 2 = 11 \tan x$

(A) $x = \frac{\pi}{4}, \frac{7\pi}{4}$ (B) $x = \frac{5\pi}{4}, \frac{7\pi}{4}$ (C) $x = \frac{3\pi}{4}, \frac{7\pi}{4}$ (D) $x = \frac{\pi}{4}, \frac{5\pi}{4}$

18. $9\cot^2 x - 3 = 0$

(A) $\frac{\pi}{3}, \frac{2\pi}{3}, \frac{4\pi}{3}, \frac{5\pi}{3}$ (B) $\frac{\pi}{4}, \frac{\pi}{2}, \frac{2\pi}{3}, \frac{5\pi}{6}$

(C) $\frac{\pi}{6}, \frac{5\pi}{6}, \frac{7\pi}{6}, \frac{11\pi}{6}$ (D) $\frac{\pi}{4}, \frac{3\pi}{4}, \frac{5\pi}{4}, \frac{7\pi}{4}$

19. Solve the equation.
$4 \csc x + 6 = 7 \csc x$

20. Find all solutions of the equation in the interval $[0, 2\pi)$.
$2 \cos x \sin x = \sin x$

Objective 2: Solve trigonometric equations of quadratic type

Find all solutions of the equation in the interval $[0, 2\pi)$.

21. $\tan^2 \theta = \frac{3}{2} \sec \theta$ (A) $\frac{2\pi}{3}, \frac{4\pi}{3}$ (B) $\frac{\pi}{6}, \frac{11\pi}{6}$ (C) $\frac{\pi}{3}, \frac{5\pi}{3}$ (D) None of these

22. $2 \cot^2 x - 3 \csc x = 0$ (A) 0 (B) $\frac{7\pi}{6}, \frac{11\pi}{6}$ (C) $\frac{\pi}{6}, \frac{5\pi}{6}$ (D) $\frac{2\pi}{3}, \pi, \frac{4\pi}{3}$

Find all solutions of the equation in the interval $[0, 2\pi)$.

23. $-\sin x + 1 = 2\cos^2 x$

24. $\cot^2 x = 3$

Objective 3: Solve trigonometric equations involving multiple angles

Find all solutions of the equation in the interval $[0, 2\pi)$.

25. $2\tan 2x - 2\sqrt{3} = 0$

(A) $\frac{\pi}{6}, \frac{2\pi}{3}$ (B) $\frac{\pi}{6}, \frac{2\pi}{3}, \frac{7\pi}{6}$ (C) $\frac{\pi}{6}, \frac{2\pi}{3}, \frac{7\pi}{6}, \frac{5\pi}{3}$ (D) $\frac{\pi}{6}, \frac{5\pi}{12}, \frac{2\pi}{3}, \frac{7\pi}{6}, \frac{5\pi}{3}$

26. $4\sec^2\frac{x}{2} - 6\sec\frac{x}{2} - 4 = 0$ (A) 0 (B) 2π (C) $\frac{2\pi}{3}$ (D) $\frac{4\pi}{3}$

27. Find all solutions of the equation.

$\sin 3x = \frac{\sqrt{3}}{2}$

28. Find all solutions of the equation in the interval $[0, 2\pi)$.

$2\sin^2 6x - 1 = 0$

Objective 4: Use inverse trigonometric functions to solve trigonometric equations

Use inverse functions where needed to find all solutions of the equation in the interval $[0, 2\pi)$.

29. $6\sin^2 x - \left(-4 - 3\sqrt{2}\right)\cos x - 2\sqrt{2} - 6 = 0$

(A) $\arccos\left(-\frac{3}{2}\right), \arccos\left(-\frac{3}{2}\right) + \pi, \frac{\pi}{4}, \frac{7\pi}{4}$ (B) $\arccos\left(\frac{2}{3}\right), 2\pi - \arccos\left(\frac{2}{3}\right), \frac{\pi}{4}, \frac{7\pi}{4}$

(C) $\arccos\left(\frac{2}{3}\right), 2\pi - \arccos\left(\frac{2}{3}\right), \frac{3\pi}{4}, \frac{5\pi}{4}$ (D) $\arccos\left(-\frac{3}{2}\right), \arccos\left(-\frac{3}{2}\right) + \pi, \frac{3\pi}{4}, \frac{5\pi}{4}$

Use inverse functions where needed to find all solutions of the equation in the interval $[0, 2\pi)$.

30. $3\csc^2 x + \cot x - 5 = 0$

(A) $\operatorname{arccot}\left(\frac{2}{3}\right)$, $\operatorname{arccot}\left(\frac{2}{3}\right)+\pi$, $\frac{\pi}{4}$, $\frac{5\pi}{4}$

(B) $\operatorname{arccot}\left(-\frac{3}{2}\right)$, $\operatorname{arccot}\left(-\frac{3}{2}\right)+\pi$, $\frac{3\pi}{4}$, $\frac{7\pi}{4}$

(C) $\operatorname{arccot}\left(\frac{2}{3}\right)$, $\operatorname{arccot}\left(\frac{2}{3}\right)+\pi$, $\frac{3\pi}{4}$, $\frac{7\pi}{4}$

(D) $\operatorname{arccot}\left(-\frac{3}{2}\right)$, $\operatorname{arccot}\left(-\frac{3}{2}\right)+\pi$, $\frac{\pi}{4}$, $\frac{5\pi}{4}$

Use inverse functions where needed to find all solutions of the equation in the interval $[0, 2\pi)$.

31. $8\sin^2 x + \left(-2+4\sqrt{3}\right)\cos x + \sqrt{3} - 8 = 0$

32. $2\tan^2 x + \sec x - 8 = 0$

Section 5.4: Sum and Difference Formulas

Objective 1: Use sum and difference formulas to evaluate trigonometric functions, verify identities, and solve trigonometric equations

33. Find the exact value of $\cos(105°)$.

(A) $\frac{\sqrt{2}-\sqrt{6}}{4}$ (B) $\frac{\sqrt{6}-\sqrt{2}}{4}$ (C) $\frac{\sqrt{6}+\sqrt{2}}{4}$ (D) $\frac{-\sqrt{2}-\sqrt{6}}{4}$

34. Identify the expression that completes the identity.

$\sin\left(\frac{3\pi}{4} - x\right) =$

(A) $-\frac{\sqrt{2}}{2}(\cos x + \sin x)$

(B) $\frac{\sqrt{2}}{2}(\sin x - \cos x)$

(C) $\frac{\sqrt{2}}{2}(\cos x - \sin x)$

(D) $\frac{\sqrt{2}}{2}(\cos x + \sin x)$

35. Find the exact value of the expression.

$\sin\frac{\pi}{12}\cos\frac{5\pi}{12} + \cos\frac{\pi}{12}\sin\frac{5\pi}{12}$

36. Verify the identity.

$\cos\left(x - \frac{\pi}{2}\right) = \sin x$

Section 5.5: Multiple-Angle and Product-to-Sum Formulas

Objective 1: Use multiple-angle formulas to rewrite and evaluate trigonometric functions

37. Find the exact value of $\tan 2x$ using the double angle formula.

$\tan x = \frac{7}{11}, \quad \pi < x < \frac{3\pi}{2}$

(A) $-\frac{84\sqrt{2}}{121}$ (B) $\frac{77}{36}$ (C) $\frac{84\sqrt{2}}{121}$ (D) $\frac{23}{121}$

38. Find the exact solutions to the equation in the interval $[0, 2\pi)$.

$\sin 4x = 2\cos 2x$

(A) $\frac{\pi}{6}, \frac{5\pi}{6}, \frac{7\pi}{6}, 2\pi$ (B) $\frac{\pi}{4}, \frac{3\pi}{4}, \frac{5\pi}{4}, \frac{7\pi}{4}$

(C) $\frac{\pi}{6}, \frac{\pi}{8}, \frac{5\pi}{6}, \frac{3\pi}{4}, \frac{11\pi}{12}, \frac{5\pi}{4}, \frac{7\pi}{4}$ (D) $\frac{\pi}{8}, \frac{\pi}{4}, \frac{5\pi}{12}, \frac{3\pi}{4}, \frac{13\pi}{12}, \frac{5\pi}{4}, \frac{7\pi}{4}$

39. Use the figure to find the exact value of the trigonometric function.

$\cos 2\theta$

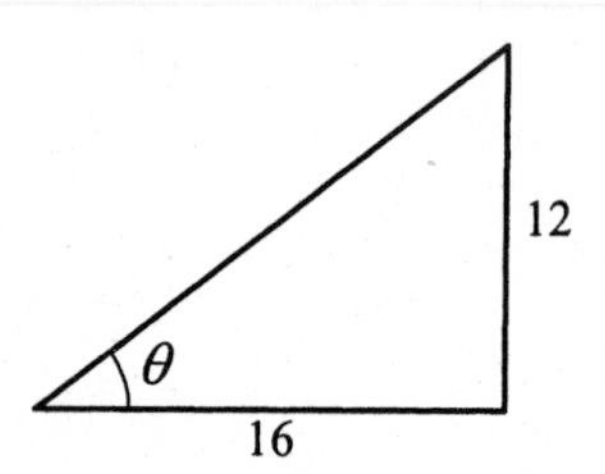

40. Use a double angle formula to rewrite the expression.

$\frac{10\tan x}{1 - \tan^2 x}$

Objective 2: Use power-reducing formulas to rewrite and evaluate trigonometric functions

Use the power-reducing formulas to find the exact value of the trigonometric function.

41. $\sin^2 157.5°$ (A) $\frac{2+\sqrt{2}}{4}$ (B) $\frac{3+\sqrt{2}}{4}$ (C) $\frac{3-\sqrt{2}}{4}$ (D) $\frac{2-\sqrt{2}}{4}$

42. $\cos^2 \frac{3\pi}{8}$ (A) $\frac{2+\sqrt{2}}{4}$ (B) $\frac{4-\sqrt{2}}{4}$ (C) $\frac{4+\sqrt{2}}{4}$ (D) $\frac{2-\sqrt{2}}{4}$

43. $\tan^2 22.5°$

44. $\sin^2 \frac{19\pi}{8}$

Objective 3: Use half-angle formulas to rewrite and evaluate trigonometric functions

45. Use the figure to find the exact value of the trigonometric function.

$\sin \frac{\theta}{2}$

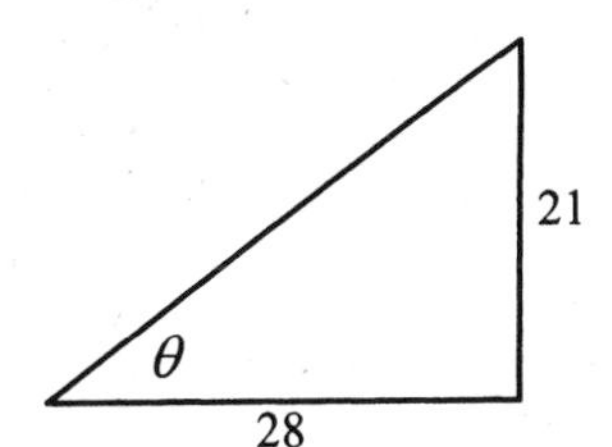

(A) 3 (B) $\frac{1}{3}$ (C) $\frac{\sqrt{10}}{3}$ (D) $\frac{\sqrt{10}}{10}$

46. Use the half-angle formulas to simplify the expression.

$-\sqrt{\frac{1+\cos 6x}{1-\cos 6x}}$

(A) $\tan 3x$ (B) $-|\cot 3x|$ (C) $\cot 3x$ (D) $-|\tan 3x|$

47. Find all solutions in the interval $[0,\ 2\pi)$.

$$\cos\frac{x}{2} - \sin x = 0$$

48. Find the exact value of $\cos 15°$.

Objective 4: Use product-to-sum and sum-to-product formulas to rewrite and evaluate trigonometric functions

49. Express $\sin 4x - \sin 2x$ as a product containing only sines and/or cosines.

(A) $2 \sin x \cos 3x$ (B) $2 \sin 3x \cos x$ (C) $-2 \sin 3x \cos x$ (D) $-2 \sin x \cos 3x$

Find all solutions in the interval $[0,\ 2\pi)$.

50. $\cos 3x + \cos x = 0$

(A) $0,\ \frac{\pi}{4},\ \frac{3\pi}{4},\ \pi,\ \frac{5\pi}{4},\ \frac{7\pi}{4}$

(B) $0,\ \frac{\pi}{2},\ \pi$

(C) $0,\ \frac{\pi}{2},\ \pi,\ \frac{3\pi}{2}$

(D) $\frac{\pi}{4},\ \frac{\pi}{2},\ \frac{3\pi}{4},\ \frac{5\pi}{4},\ \frac{3\pi}{2},\ \frac{7\pi}{4}$

51. $\sin^2 6x - \sin^2 4x = 0$

52. Express $\cos 6\theta \cos 2\theta$ as a sum containing only sines or cosines.

Answer Key for Chapter 5 Analytic Trigonometry

Section 5.1: Using Fundamental Identities

Objective 1: Recognize and write the fundamental trigonometric identities

[1] (A)

[2] (C)

[3] No

[4] Answers may vary. Sample answer: $\cos u = \pm\sqrt{1-\sin^2 u}$

Objective 2: Use the fundamental trigonometric identities to evaluate trigonometric functions, simplify trigonometric expressions, and rewrite trigonometric expressions

[5] (B)

[6] (B)

[7] $\csc x$

[8] Answers may vary. Sample answer: $\dfrac{8}{\sin x}$

Section 5.2: Verifying Trigonometric Identities

Objective 1: Plan a strategy for verifying trigonometric identities

[9] (C)

[10] (D)

[11] Quotient identity

[12] $\dfrac{\cos x}{\sin^2 x}$

Objective 2: Verify trigonometric identities

[13] (D)

[14] (C)

[15] Answers will vary.

[16] Answers will vary.

Section 5.3: Solving Trigonometric Equations

Objective 1: Use standard algebraic techniques to solve trigonometric equations

[17] (C)

[18] (A)

[19] $x = \frac{\pi}{6} + 2n\pi, \frac{5\pi}{6} + 2n\pi$

[20] $0, \frac{\pi}{3}, \pi, \frac{5\pi}{3}$

Objective 2: Solve trigonometric equations of quadratic type

[21] (C)

[22] (C)

[23] $x = \frac{7\pi}{6}, x = \frac{11\pi}{6}, x = \frac{\pi}{2}$

[24] $\frac{\pi}{6}, \frac{5\pi}{6}, \frac{7\pi}{6}, \frac{11\pi}{6}$

Objective 3: Solve trigonometric equations involving multiple angles

[25] (C)

[26] (C)

[27] $\frac{\pi}{9}+\frac{2}{3}n\pi,\ \frac{2\pi}{9}+\frac{2}{3}n\pi$

[28] $\frac{\pi}{24}, \frac{\pi}{8}, \frac{5\pi}{24}, \frac{7\pi}{24}, \frac{3\pi}{8}, \frac{11\pi}{24}, \frac{13\pi}{24}, \frac{5\pi}{8}, \frac{17\pi}{24}, \frac{19\pi}{24}, \frac{7\pi}{8}, \frac{23\pi}{24}, \frac{25\pi}{24}, \frac{9\pi}{8}, \frac{29\pi}{24}, \frac{31\pi}{24}, \frac{11\pi}{8}, \frac{35\pi}{24}, \frac{37\pi}{24}, \frac{13\pi}{8}, \frac{41\pi}{24}, \frac{43\pi}{24}, \frac{15\pi}{8}, \frac{47\pi}{24}$

Objective 4: Use inverse trigonometric functions to solve trigonometric equations

[29] (B)

[30] (C)

[31] $\arccos\left(-\frac{1}{4}\right),\ 2\pi-\arccos\left(-\frac{1}{4}\right),\ \frac{\pi}{6},\ \frac{11\pi}{6}$

[32] $\operatorname{arcsec}\left(-\frac{5}{2}\right),\ 2\pi-\operatorname{arcsec}\left(-\frac{5}{2}\right),\ \frac{\pi}{3},\ \frac{5\pi}{3}$

Section 5.4: Sum and Difference Formulas

Objective 1: Use sum and difference formulas to evaluate trigonometric functions, verify identities, and solve trigonometric equations

[33] (A)

[34] (D)

[35] 1

[36] Answers will vary.

Section 5.5: Multiple-Angle and Product-to-Sum Formulas

Objective 1: Use multiple-angle formulas to rewrite and evaluate trigonometric functions

[37] (B)

[38] (B)

[39] $\cos 2\theta = \frac{7}{25}$

[40] $5 \tan 2x$

Objective 2: Use power-reducing formulas to rewrite and evaluate trigonometric functions

[41] (D)

[42] (D)

[43] $3 - 2\sqrt{2}$

[44] $\frac{2+\sqrt{2}}{4}$

Objective 3: Use half-angle formulas to rewrite and evaluate trigonometric functions

[45] (D)

[46] (B)

[47] $\frac{1}{3}\pi, \pi, \frac{5}{3}\pi$

[48] $\frac{\sqrt{2+\sqrt{3}}}{2}$

Objective 4: Use product-to-sum and sum-to-product formulas to rewrite and evaluate trigonometric functions

[49] (A)

[50] (D)

[51] $0, \frac{\pi}{10}, \frac{\pi}{5}, \frac{3\pi}{10}, \frac{2\pi}{5}, \frac{\pi}{2}, \frac{3\pi}{5}, \frac{7\pi}{10}, \frac{4\pi}{5}, \frac{9\pi}{10}, \pi, \frac{11\pi}{10}, \frac{6\pi}{5}, \frac{13\pi}{10}, \frac{7\pi}{5}, \frac{3\pi}{2}, \frac{8\pi}{5}, \frac{17\pi}{10}, \frac{9\pi}{5}, \frac{19\pi}{10}$

[52] $\frac{1}{2}(\cos 8\theta + \cos 4\theta)$

Chapter 6 Additional Topics in Trigonometry

Section 6.1: Law of Sines

Objective 1: Use the Law of Sines to solve oblique triangles (AAS or ASA)

1. Use the information to solve the triangle.
$B = 38°$, $C = 35°$, and $a = 20$

(A) $A = 117°$, $b \approx 59.0$, $c \approx 31.1$

(B) $A = 117°$, $b \approx 13.8$, $c \approx 12.0$

(C) $A = 107°$, $b \approx 12.9$, $c \approx 12.0$

(D) $A = 107°$, $b \approx 59.0$, $c \approx 33.3$

2. Use the information to solve the triangle for the requested side (if possible). If two solutions exist, find both.
Find b.

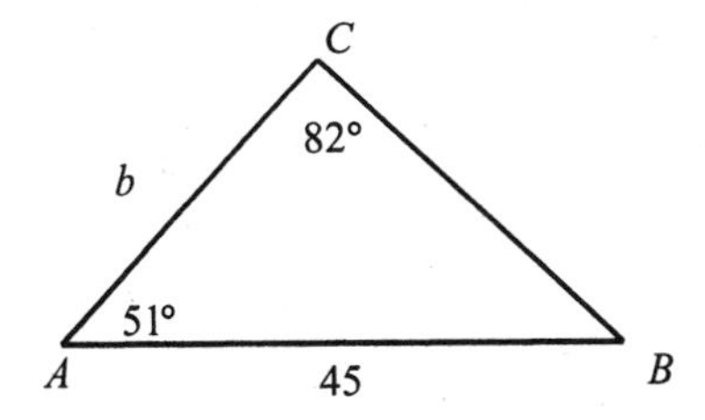

(A) $b \approx 33.2$

(B) $b \approx 33.2$ or $b \approx 34.2$

(C) $b \approx 32.2$

(D) No solution exists.

3. Use the Law of Sines to solve for a to the nearest tenth.

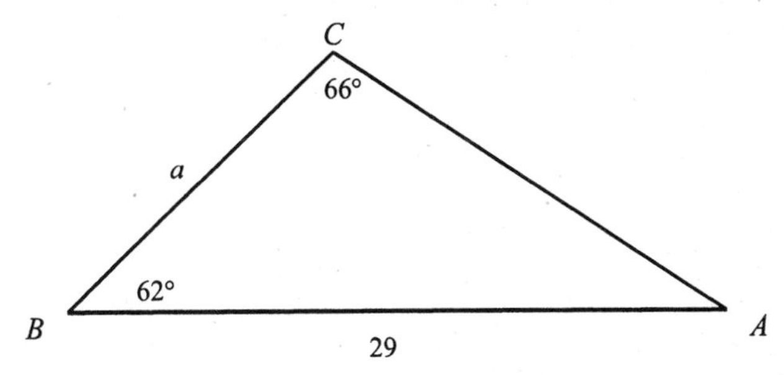

4. Given a triangle with $a = 13$, $A = 37°$, and $B = 19°$, what is the length of c? Round your answer to two decimal places.

Objective 2: Use the Law of Sines to solve oblique triangles (SSA)

5. Use the information to solve the triangle for the requested side (if possible). If two solutions exist, find both.
Find c if $A = 23°$, $a = 9$, and $b = 10$.

(A) $c \approx 17.31$ (B) $c \approx 13.45$ or $c \approx 1.54$ (C) $c \approx 17.31$ or $c \approx 1.10$ (D) No solution exists.

6. Use the Law of Sines to solve for $m\angle C$ to the nearest tenth.

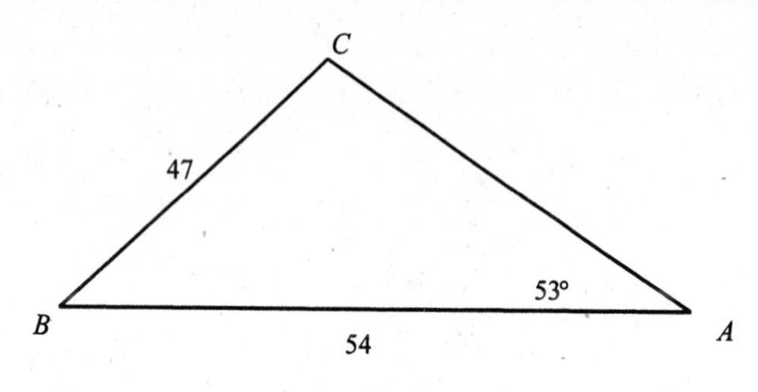

(A) 66.6 (B) 65.6 (C) 68.6 (D) 67.6

7. How many solutions exist for the triangle?
$a = 12.1,\ b = 3.4$, and $A = 30°$

8. Find a value for a such that the triangle has the number of solutions required, if possible.
One solution when $b = 3$ and $A = 22°$

Objective 3: Find the areas of oblique triangles

9. Find the area of the triangle having the indicated sides and angle.
$B = 20°\,10',\ a = 10,\ c = 4$

(A) 13.79 (B) 18.77 (C) 6.9 (D) None of these

10. Use the given measures to find the area of triangle ABC.
$A = 35°,\ a = 14.7,\ b = 14.7$

(A) 101.53 (B) 36.95 (C) 203.06 (D) 61.97

11. Triangle ABC has the given area, angle, and side. Find the length of the requested side. Round to the nearest integer.
Area = 3.64, $A = 29°$, $b = 3$. Find c.

12. Triangle ABC has the given measures. If the triangle has exactly one solution, give the area of the triangle to two decimal places. If the triangle does not have exactly one solution, determine how many solutions there are.
$A = 116°,\ a = 15.3,\ b = 11.6$

Objective 4: Use the Law of Sines to model and solve real-life problems

13. A loading dock ramp that is 19 feet long rises at an angle of 14.3° from the horizontal. Due to new design specifications, a longer ramp is to be used, so that the angle is reduced to 9.3°. How much farther out from the dock will that put the foot of the ramp?

(A) 10.2 ft (B) 9.2 ft (C) 35.2 ft (D) 29.0 ft

14. A pole 55 feet tall is situated at the bottom of a hill that slopes up at an angle of 12.2°. A guy wire from the top of the pole to the hillside forms an angle of 26° with the top of the pole. Find the distance from the base of the pole to the guy wire's point of attachment.

(A) 23.6 ft (B) 24.2 ft (C) 24.8 ft (D) 18.8 ft

15. A ship at sea, the Intrepid, spots two other ships, the Ranger and the Lancer, and measures the angle between them to be 49°. The distance between the Intrepid and the Ranger is 3650 meters. The Ranger measures an angle of 46° between the Intrepid and the Lancer. To the nearest meter, what is the distance between the Ranger and the Lancer?

16. An airplane left an airport and flew east for 169 miles. Then it turned northward to N 32° E. When it was 264 miles from the airport, there was an engine problem and it turned to take the shortest route back to the airport. Find θ, the angle through which the airplane turned.

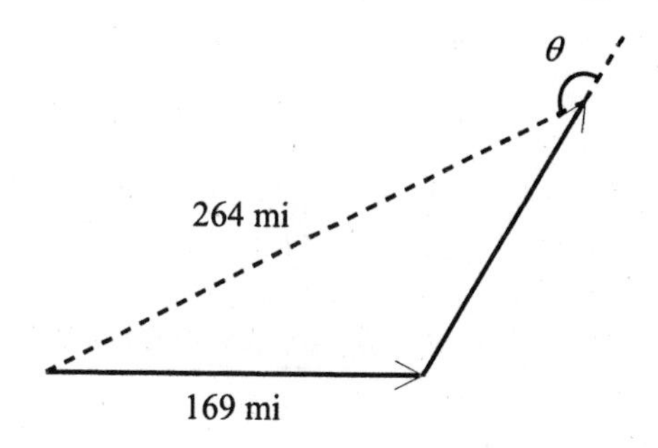

Section 6.2: Law of Cosines

Objective 1: Use the Law of Cosines to solve oblique triangles (SSS or SAS)

17. Use the given measures and the Law of Cosines to solve triangle *ABC*.
$a = 18,\ b = 16,\ c = 14$

(A) $A = 131.8°;\ B = 31.6°;\ C = 16.6°$
(B) $A = 58.4°;\ B = 73.4°;\ C = 48.2°$
(C) $A = 131.8°;\ B = 16.6°;\ C = 31.6°$
(D) $A = 73.4°;\ B = 58.4°;\ C = 48.2°$

18. Use the Law of Cosines to find the third side of the triangle.

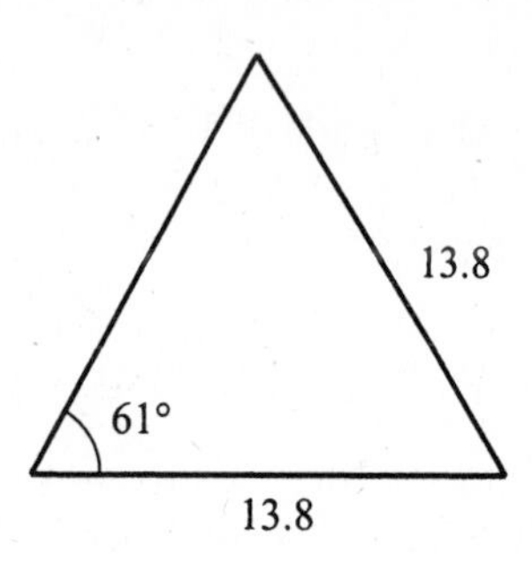

(A) 13.4
(B) 14
(C) 16.1
(D) 24.1

19. Use the Law of Cosines to find the length of the diagonal of the parallelogram.

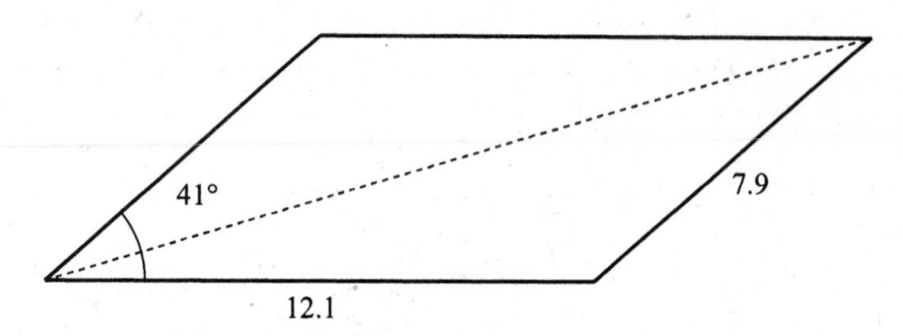

20. Use the Law of Cosines to find angle θ in the parallelogram. Round to the nearest tenth of a degree.

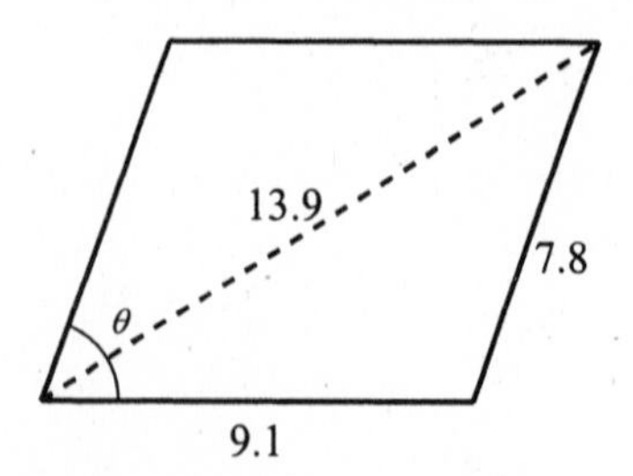

Objective 2: Use the Law of Cosines to model and solve real-life problems

21. Two ships leave a port at the same time. When ship A is 135 miles due east of the port, ship B is 145 miles from the port and 250 miles from ship A, in the direction shown below. What is ship B's bearing?

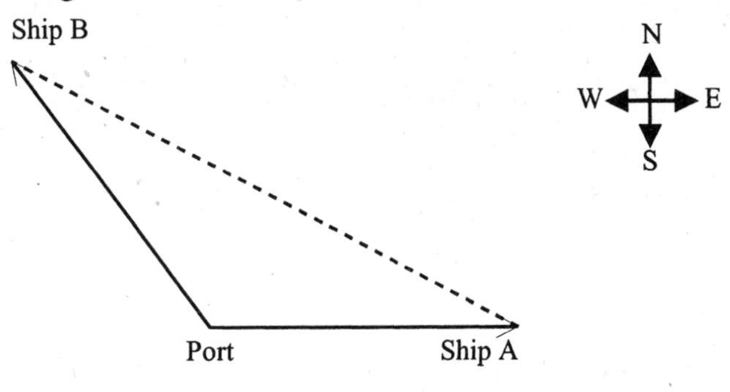

(A) N 36.4° W (B) N 53.6° W (C) N 25.8° W (D) N 64.2° W

22. A plane travels 120 miles at a heading of N 41° W. It then changes direction and travels 135 miles at a heading of N 55° W. How far is the plane from its original position?

(A) 255 mi (B) 240.5 mi (C) 227.8 mi (D) 253.1 mi

23. A ship travels due west for 91 miles. It then travels in a northern direction for 48 miles and ends up 117 miles from its original position. How many degrees did it turn when it changed direction? Round your answer to the nearest tenth.

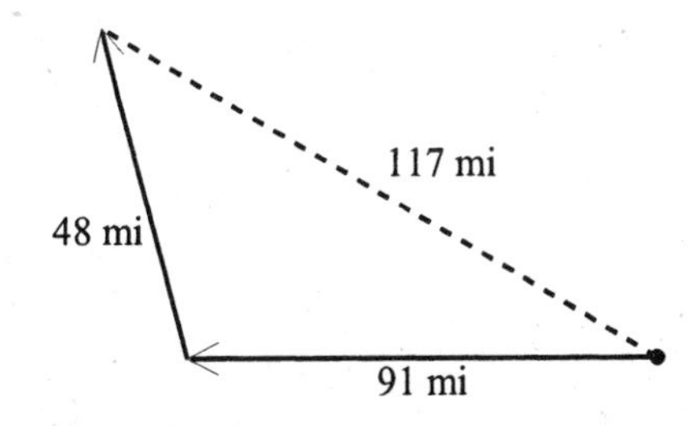

24. Cindy and Jane are golfing on a beautiful summer day. The fifth hole is 335 yards from the tee. On that hole, Cindy hit her drive to the left of the correct direction, as sketched below.

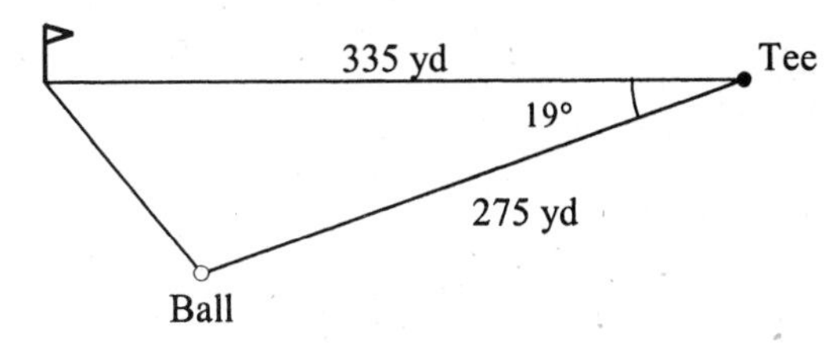

Write the Law of Cosines equation for the distance from Cindy's ball to the hole. Find the distance to the nearest tenth of a yard.

Objective 3: Use Heron's Area Formula to find the area of a triangle

25. Use Heron's Area Formula to find the area of the triangle with the given measures.
Equilateral triangle, with a perimeter of 21.6

(A) 22.4 (B) 6.2 (C) 254.0 (D) 720.1

26. Use Heron's Area Formula to find the value of x in the triangle with the given measures.
Area = 81.7

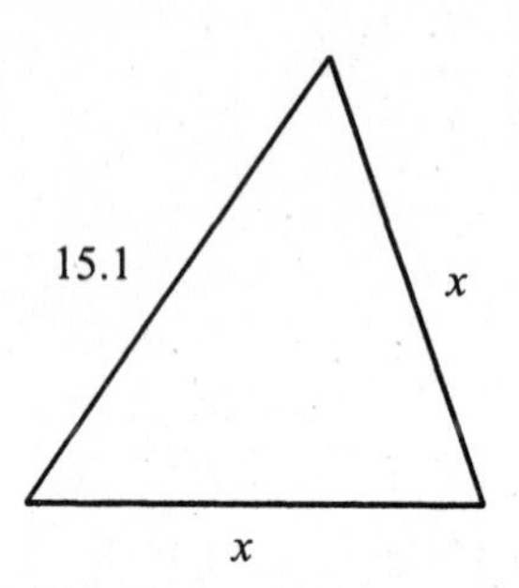

(A) $x = 29.8$ (B) $x = 13.2$ (C) $x = 17.9$ (D) $x = 13.8$

27. Use Heron's Area Formula to find the area of the triangle. Round the result to the nearest tenth.
$a = 18,\ b = 30,\ c = 33$

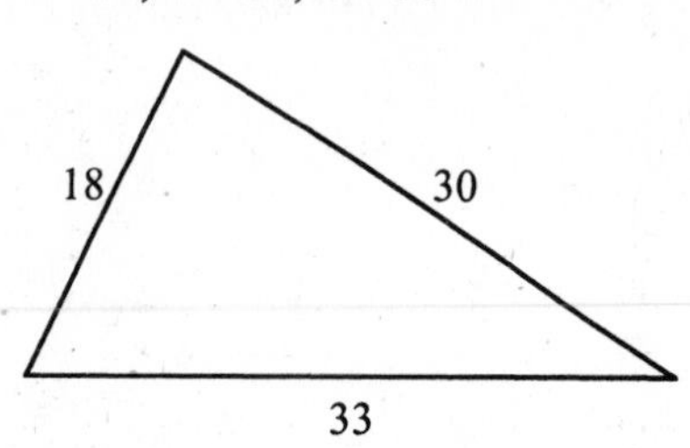

28. Use Heron's Area Formula to find the area of the triangle with the given measures.
Perimeter = 28.3

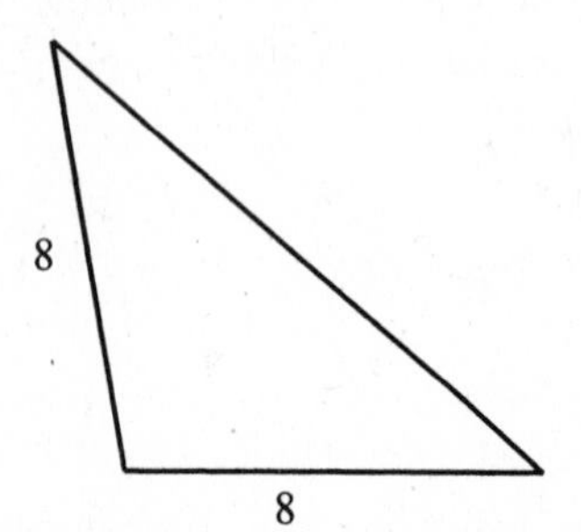

Section 6.3: Vectors in the Plane

Objective 1: Represent vectors as directed line segments

29. Identify the initial and terminal points of a vector that has the same direction as **v**.

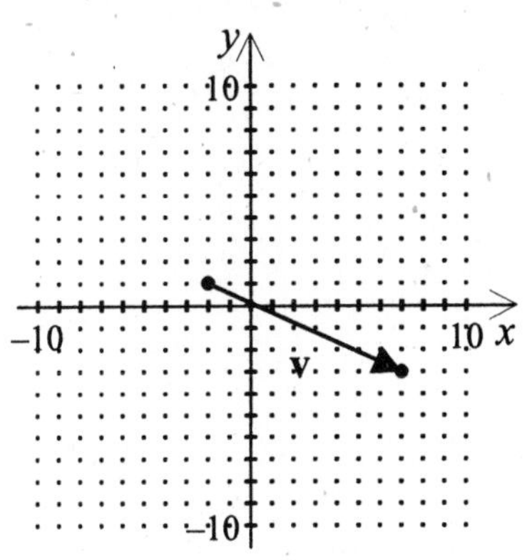

(A) $(-3, -3)$ to $(-12, -7)$

(B) $(1, 1)$ to $(-3, 10)$

(C) $(-3, -3)$ to $(1, -12)$

(D) $(1, 1)$ to $(10, -3)$

30. Identify the vector with the same magnitude as **v**.

$\mathbf{v} = \overrightarrow{AB}$ with $A = (5, 1)$ and $B = (9, 7)$.

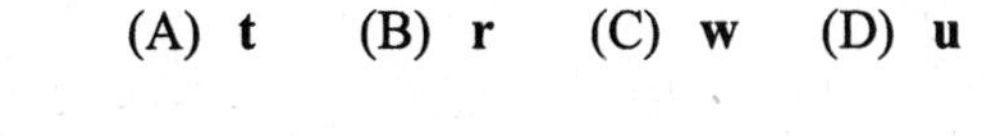

(A) **t** (B) **r** (C) **w** (D) **u**

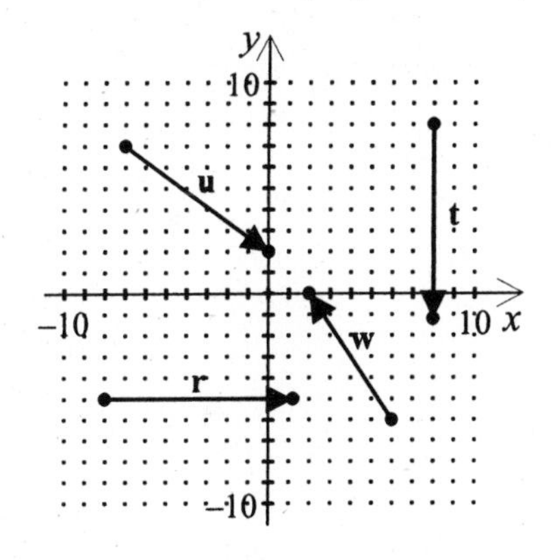

31. Show that **u** = **v**.

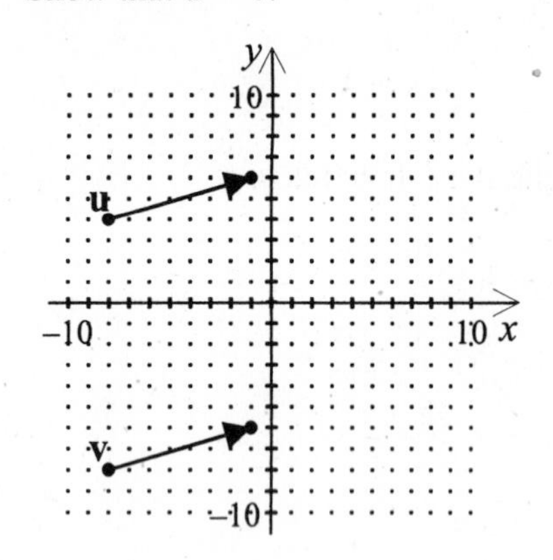

32. Find the value of k that makes **u** = **v** .

$\mathbf{u} = \overrightarrow{AB}$ with $A = (-7,\ 4)$, $B = (0,\ 6)$

$\mathbf{v} = \overrightarrow{MN}$ with $M = (-3,\ -8)$, $N = (k,\ -6)$

Objective 2: Write the component forms of vectors

33. Identify the pair of points that could be the initial and terminal points of the vector.

$\mathbf{u} = \langle -7, 6 \rangle$

(A) $(-1,\ -9)$ and $(5,\ -2)$

(B) $(5,\ -2)$ and $(12,\ 4)$

(C) $(4,\ -4)$ and $(-3,\ 2)$

(D) $(10,\ -11)$ and $(4,\ -4)$

34. Identify the initial point of vector **v**.

$\mathbf{v} = \langle -8,\ -6 \rangle$; terminal point is $(-11,\ -12)$.

(A) $(-19,\ -18)$ (B) $(-6,\ -3)$ (C) $(-3,\ -6)$ (D) $(5,\ 0)$

35. Give the component form of the vector $\mathbf{v} = \overrightarrow{EF}$ and sketch the vector in standard position.

$E = (0,\ 1)$ and $F = (8,\ -5)$

36. Find all possible values of the missing component of the vector **u**.

$\mathbf{u} = \langle 7,\ y \rangle$, $\|\mathbf{u}\| = \sqrt{149}$

Objective 3: Perform basic vector operations and represent them graphically

37. Use the figure to identify the graph of the result of the specified vector operation.
v – **w**

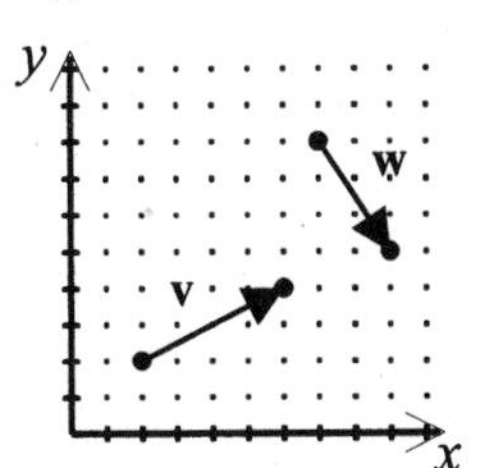

(A)

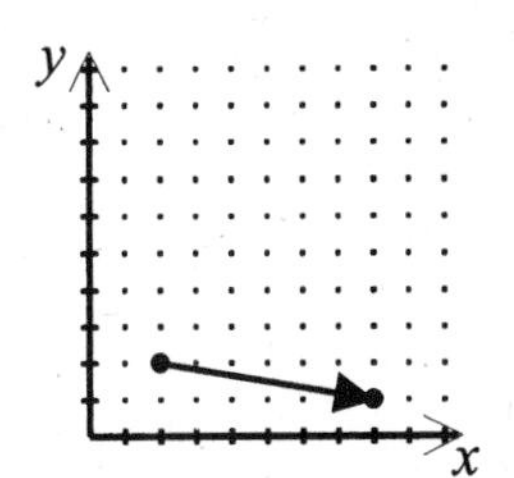

(B)

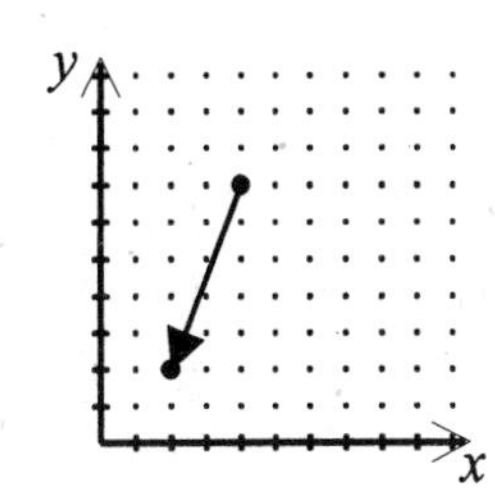

(C)

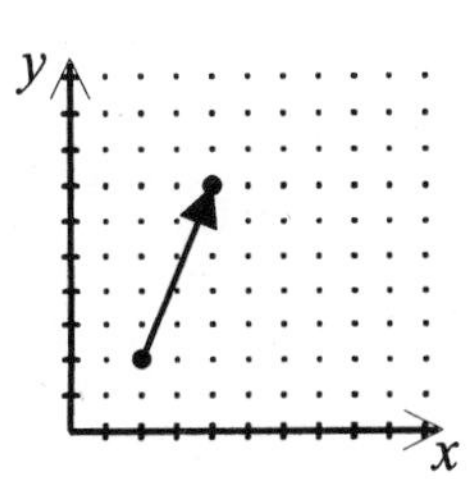

(D)

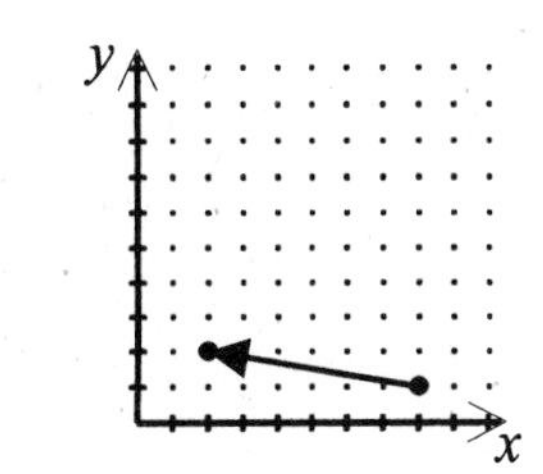

38. Identify the expression that is represented by the dashed segment in the graph.
$\mathbf{u} = \langle -3, -3 \rangle$, $\mathbf{v} = \langle -3, -5 \rangle$

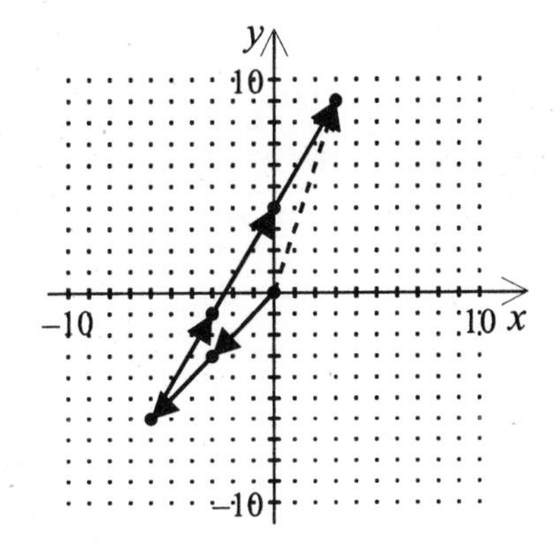

(A) 2**v** – 3**u** (B) 2**u** – 3**v** (C) 2**v** + 3**u** (D) 2**u** + 3**v**

39. Find the sum of the pair of vectors. Express the answer in component form.

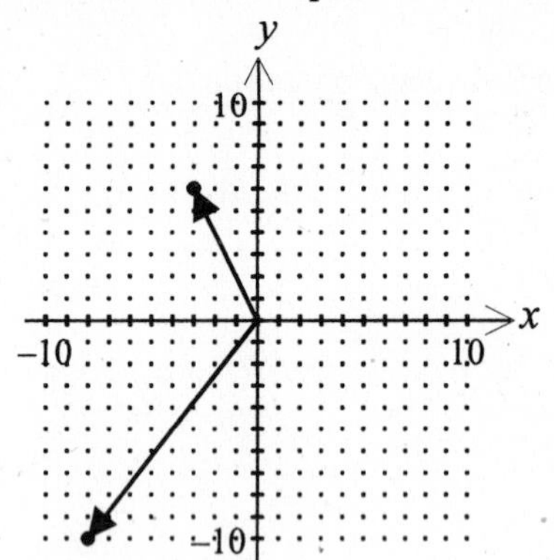

40. Give the component form of the vector **u** that has the magnitude described.

$\mathbf{v} = \langle -3,\ 1\rangle$, magnitude of $\mathbf{u} = \dfrac{1}{6}\|\mathbf{v}\|$

Objective 4: Write vectors as linear combinations of unit vectors

41. Let $\mathbf{u} = -2\mathbf{i} - 5\mathbf{j}$ and $\mathbf{v} = 2\mathbf{i} - 3\mathbf{j}$. Find $3\mathbf{u} + 2\mathbf{v}$.

(A) $-6\mathbf{i} - 6\mathbf{j}$ (B) $-10\mathbf{i} - 9\mathbf{j}$ (C) $-2\mathbf{i} - 21\mathbf{j}$ (D) $3\mathbf{i} - 6\mathbf{j}$

42. Find the number by which the components of the vector can be divided to find the unit vector in the same direction.
$\mathbf{u} = -2\mathbf{i} - 10\mathbf{j}$

(A) $2\sqrt{6}$ (B) 104 (C) 52 (D) $2\sqrt{26}$

43. Find the unit vector in the same direction as **u**.
$\mathbf{u} = 10\mathbf{i} + 11\mathbf{j}$

44. Express the vector as a combination of the standard unit vectors **i** and **j**.
$\mathbf{v} = \overrightarrow{AB}$ where $A = (-11,\ 2)$ and $B = (3,\ -9)$

Objective 5: Find the direction angles of vectors

Find the direction angle of the vector.

45.

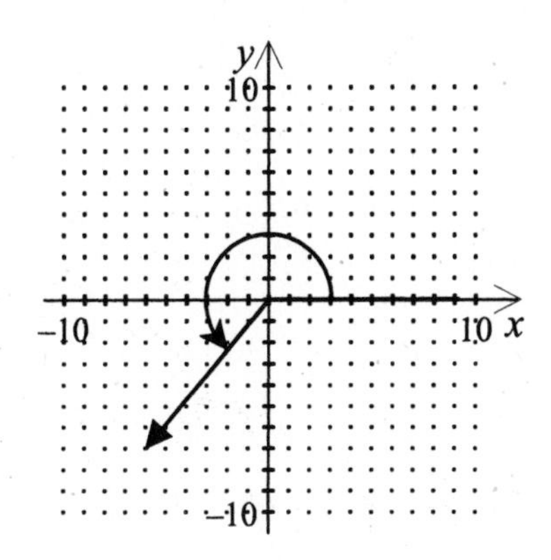

(A) $\theta = 229.4°$ (B) $\theta = 139.4°$ (C) $\theta = 140.4°$ (D) $\theta = 228.4°$

46. $\mathbf{v} = 3\mathbf{i} + 4\mathbf{j}$ (A) $\theta = 306.9°$ (B) $\theta = 36.9°$ (C) $\theta = 126.9°$ (D) $\theta = 53.1°$

47. $\mathbf{v} = \langle -7,\ 6 \rangle$

48. Find the missing component of the vector with the given direction angle.
$\mathbf{v} = \langle x,\ -3 \rangle$, with $\theta = 330°$

Objective 6: Use vectors to model and solve real-life problems

49. A force of 868 pounds is needed to push a stalled car up a hill inclined at an angle of 16° to the horizontal. Find the weight of the car. Ignore friction.

(A) 3149 lb (B) 2830 lb (C) 3027 lb (D) 3132 lb

50. A box is being pulled by means of two ropes. One rope exerts 75 pounds of force at 18° above the horizontal. The other rope exerts 50 pounds of force at 15° below the horizontal. Find the direction and magnitude of the resultant of the two forces.

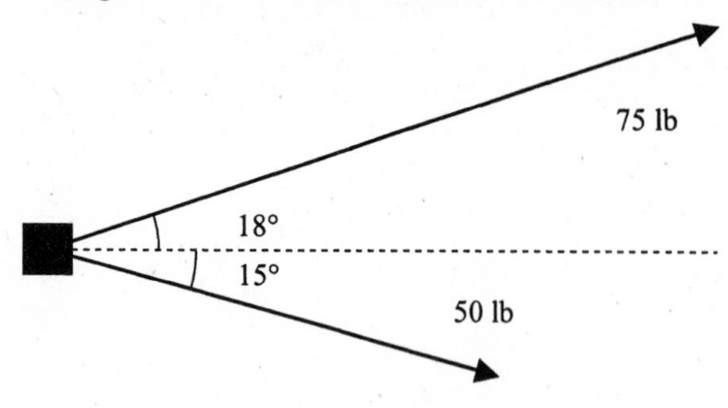

(A) 120.1 lbs at 4.9° above the horizontal

(B) 120.1 lbs at 4.9° below the horizontal

(C) 42.8 lbs at 3° above the horizontal

(D) 42.8 lbs at 3° below the horizontal

51. An airplane is flying due north at 350 miles per hour. A wind begins to blow in the direction N 16° W at 62 miles per hour. Find the bearing the pilot must fly the aircraft to continue traveling due north.

52. A pendulum bob is held stationary, as shown, by a force of 32.5 newtons. Find the tension in the pendulum string.

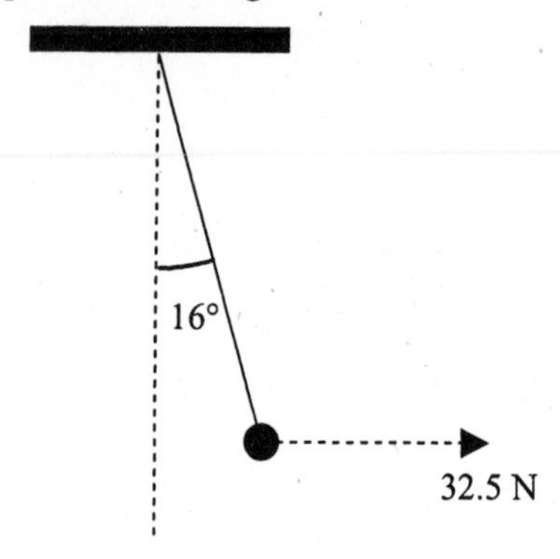

Section 6.4: Vectors and Dot Products

Objective 1: Find the dot product of two vectors and use the Properties of the Dot Product

53. Use the given vectors to find the indicated quantities.
$\mathbf{u} = \langle -1, -4 \rangle, \; \mathbf{v} = \langle -3, 5 \rangle, \; \mathbf{w} = \langle 2, 1 \rangle$
(a) $(\mathbf{u} \cdot \mathbf{v})\mathbf{w}$ (b) $5\mathbf{u} \cdot \mathbf{v}$

(A) (a) $\langle 46, 23 \rangle$ (b) 115
(B) (a) $\langle -34, -17 \rangle$ (b) -85
(C) (a) $\langle 14, 7 \rangle$ (b) 35
(D) (a) $\langle -22, -11 \rangle$ (b) -55

54. Find the indicated dot product.
$\mathbf{v} = -6\mathbf{i} + 4\mathbf{j}$ and $\mathbf{w} = -12\mathbf{i} - 18\mathbf{j}$, find $\mathbf{v} \cdot \mathbf{w}$

(A) 192 (B) 0 (C) –240 (D) 144

55. Use the properties of the dot product to find $\|\mathbf{u}\|$.
$\mathbf{u} \cdot \mathbf{u} = 881$

56. Use the properties of the dot product to evaluate the expression.
$\mathbf{u} = \langle 3, -4 \rangle, \; \mathbf{v} = \langle 1, 1 \rangle, \; \mathbf{w} = \langle -4, -4 \rangle$, find $\mathbf{v} \cdot (\mathbf{w} + \mathbf{u})$

Objective 2: Find the angle between two vectors and determine whether two vectors are orthogonal

57. Find the angle between the two vectors.
$$\mathbf{u} = \cos\left(\frac{2\pi}{3}\right)\mathbf{i} + \sin\left(\frac{2\pi}{3}\right)\mathbf{j}$$
$$\mathbf{v} = \cos\left(\frac{7\pi}{6}\right)\mathbf{i} + \sin\left(\frac{7\pi}{6}\right)\mathbf{j}$$

(A) $\theta = \frac{3\pi}{4}$ (B) $\theta = \frac{\pi}{2}$ (C) $\theta = \frac{\pi}{12}$ (D) $\theta = \frac{4\pi}{3}$

58. Find two vectors in opposite directions that are orthogonal to the vector $\mathbf{u}$.
$$\mathbf{u} = \left\langle \frac{20}{29}, \frac{21}{29} \right\rangle$$

(A) $\left\langle \frac{21}{29}, -\frac{20}{29} \right\rangle, \left\langle \frac{40}{29}, \frac{42}{29} \right\rangle$
(B) $\left\langle \frac{21}{29}, -\frac{20}{29} \right\rangle, \left\langle -\frac{42}{29}, \frac{40}{29} \right\rangle$
(C) $\left\langle \frac{21}{29}, -\frac{20}{29} \right\rangle, \left\langle \frac{42}{29}, -\frac{40}{29} \right\rangle$
(D) $\left\langle \frac{40}{29}, \frac{42}{29} \right\rangle, \left\langle -\frac{21}{29}, \frac{20}{29} \right\rangle$

59. Find the angle between the two vectors. Express your answer in radians rounded to the nearest tenth.

$\mathbf{u} = 9\mathbf{i} - 8\mathbf{j}$

$\mathbf{v} = 4\mathbf{i} + 2\mathbf{j}$

60. Determine if the vectors are orthogonal, parallel, or neither.

$-8\mathbf{i} + 20\mathbf{j}$ and $-15\mathbf{i} - 6\mathbf{j}$

Objective 3: Write a vector as the sum of two vector components

61. Identify the decomposition of **u** into the sum of two orthogonal vectors, one of which is $\text{proj}_{\mathbf{v}}\mathbf{u}$.

$\mathbf{u} = \langle -4, -5 \rangle,\ \mathbf{v} = \langle -5, -2 \rangle$

(A) $\mathbf{u} = \left\langle -\frac{146}{29}, -\frac{86}{29} \right\rangle + \left\langle \frac{30}{29}, -\frac{59}{29} \right\rangle$ (B) $\mathbf{u} = \left\langle -\frac{150}{29}, -\frac{60}{29} \right\rangle + \left\langle \frac{34}{29}, -\frac{85}{29} \right\rangle$

(C) $\mathbf{u} = \left\langle -\frac{146}{29}, -\frac{59}{29} \right\rangle + \left\langle \frac{30}{29}, -\frac{86}{29} \right\rangle$ (D) $\mathbf{u} = \left\langle \frac{34}{29}, -\frac{60}{29} \right\rangle + \left\langle -\frac{150}{29}, -\frac{85}{29} \right\rangle$

62. A car weighing 3000 pounds is parked on the side of a hill with a 20° slope. Assume the only force to overcome is gravity.
(a) Find the force needed to keep the car from rolling down the hill.
(b) Find the force perpendicular to the slope of the hill.

(A) (a) 1026 pounds (b) 2907 pounds (B) (a) 739 pounds (b) 2907 pounds (C) (a) 739 pounds (b) 2819 pounds (D) (a) 1026 pounds (b) 2819 pounds

63. Find the projection of $\mathbf{v} = 25\mathbf{i} + 20\mathbf{j}$ onto $\mathbf{u} = 2\mathbf{i} - \mathbf{j}$ and the vector component of **v** orthogonal to **u**.

64. Kim pushes a lawn spreader across a lawn by applying a force of 93 newtons along the handle, which makes an angle of 55° with the horizontal.
(a) What are the horizontal and vertical components of the force?
(b) The handle is lowered so it makes an angle of 45° with the horizontal. What are the horizontal and vertical components of this force?

Objective 4: Use vectors to find the work done by a force

65. A force of 30 pounds is applied at an angle of 40° above the horizontal to push a cart across the floor. Find the work done if the cart is moved 30 feet.

(A) 900 foot-pounds (B) 579 foot-pounds (C) 400 foot-pounds (D) 689 foot-pounds

66. A horizontal force of 120 pounds is applied to an object as it is pushed up a ramp that is 18 feet long. Find the work done if the ramp is inclined at an angle of 50° above the horizontal.

(A) 1388 foot-pounds (B) 1655 foot-pounds

(C) 1200 foot-pounds (D) 2160 foot-pounds

67. Find the work done in moving a particle from P to Q if the magnitude and direction of the force are given by $\mathbf{v}$.

$P=(5,\ 3),\ Q=(5,\ 8),\ \mathbf{v}=\langle 5,\ 2\rangle$

68. A force of 5 pounds is applied to a push lawn mower, at an angle of 55° with the horizontal. Find the work done if the lawn mower is moved 7 feet. Round your answer to the nearest tenth.

Section 6.5: Trigonometric Form of a Complex Number

Objective 1: Plot complex numbers in the complex plane

69. Which number is farthest from the origin in the complex plane?

(A) $-3-5i$ (B) $2-3i$ (C) $-8-5i$ (D) $-7-7i$

70. Match the complex number with its graph.

$-2-5i$

(A)

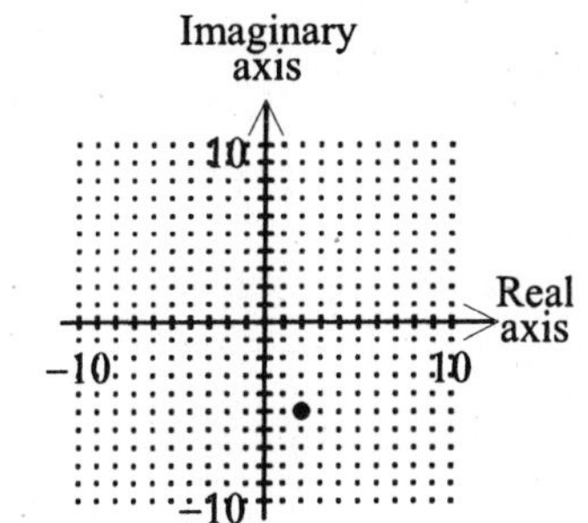

(B)

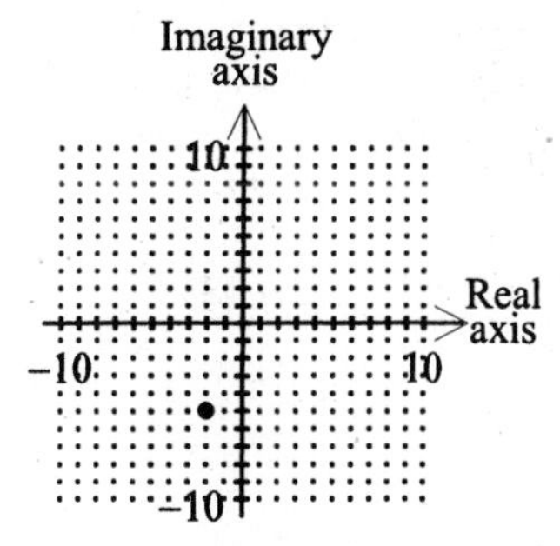

(C)

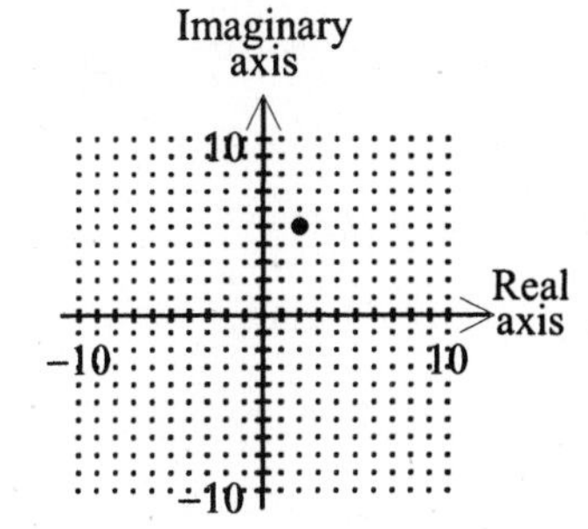

(D)

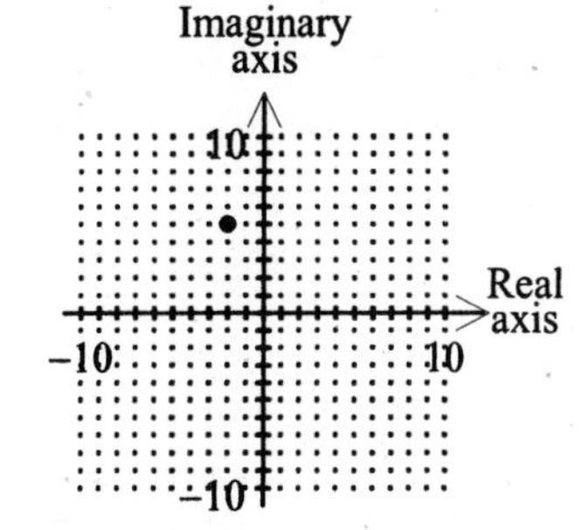

71. Plot the complex number.
$6-4i$

72. Write the complex number in standard form.

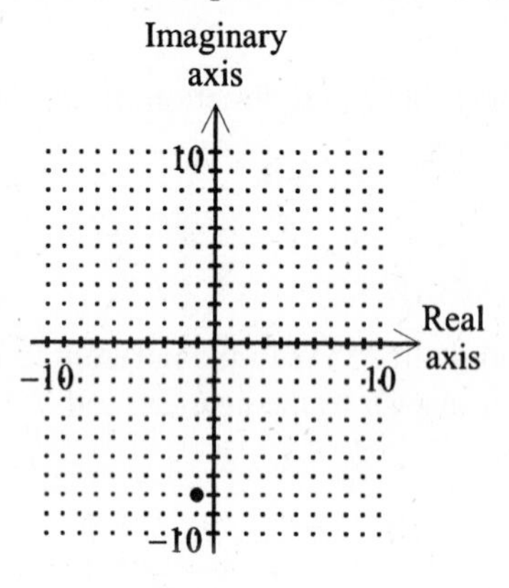

Objective 2: Write the trigonometric forms of complex numbers

73. Identify the standard form of the complex number.
$20(\cos 244° + i\sin 244°)$

(A) $-8.77-17.98i$ (B) $-8.77-0.44i$ (C) $-0.90-17.98i$ (D) $-0.90-0.44i$

74. Identify the trigonometric form of the complex number.
$7-5i$

(A) $8.6(\cos 324.5° + i\sin 324.5°)$ (B) $2(\cos 144.5° + i\sin 144.5°)$

(C) $8.6(\cos 324.5° - i\sin 324.5°)$ (D) $74(\cos 144.5° - i\sin 144.5°)$

75. Write the complex number in trigonometric form.

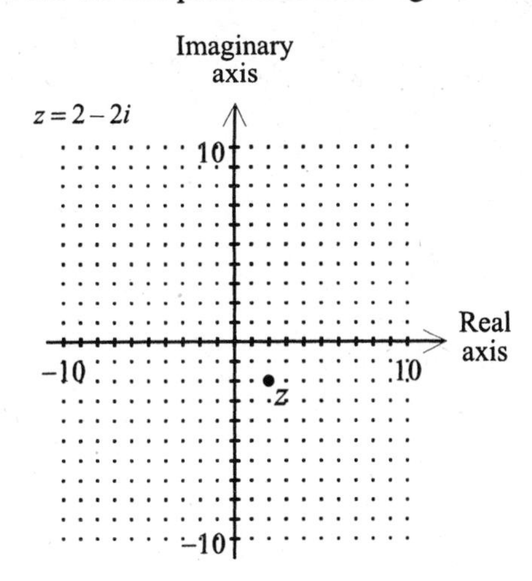

76. Find the exact modulus of the trigonometric form of the complex number z. Then find its argument to the nearest tenth of a degree.
$z = -4 + 12i$

Objective 3: Multiply and divide complex numbers written in trigonometric form

77. Find zw.
$z = 8\left(\cos\frac{7\pi}{18} + i\sin\frac{7\pi}{18}\right),\ w = 4\left(\cos\frac{7\pi}{36} + i\sin\frac{7\pi}{36}\right)$

(A) $12\left(\cos\frac{7\pi}{36} + i\sin\frac{7\pi}{36}\right)$

(B) $32\left(\cos\frac{7\pi}{12} + i\sin\frac{7\pi}{12}\right)$

(C) $32\left(\cos\frac{245\pi}{18} + i\sin\frac{245\pi}{18}\right)$

(D) None of these

78. Find the value of the expression.
$$\frac{(\cos 25° + i\sin 25°)}{(\cos 75° + i\sin 75°)}$$

(A) $-0.766 + 0.643i$ (B) $-0.643 - 0.766i$ (C) $0.643 - 0.766i$ (D) $0.766 - 0.643i$

79. Perform the multiplication. Give your answer in both trigonometric form and standard form. Do not estimate.
$3\left(\cos\frac{7\pi}{12} + i\sin\frac{7\pi}{12}\right)\cdot 4\left(\cos\frac{5\pi}{4} + i\sin\frac{5\pi}{4}\right)$

80. If $z = 6(\cos 90° + i\sin 90°)$ and $w = 3(\cos 15° + i\sin 15°)$, find the following. Give your answer in trigonometric form.
(a) zw
(b) $\frac{z}{w}$

Objective 4: Use DeMoivre's Theorem to find powers of complex numbers

Use DeMoivre's theorem to find the indicated power of the complex number.

81. $\left(-\sqrt{2}+\sqrt{2}i\right)^6$ (A) $64i$ (B) -64 (C) 64 (D) $-64i$

82. $\left(3+\sqrt{5}i\right)^6$

(A) $-2096-1770.97i$ (B) $-17.148-14.489i$
(C) $-14.489-17.148i$ (D) $-1770.97-2096i$

83. $(1-i)^4$

84. Use DeMoivre's theorem to find the indicated power of the complex number. Express the result in standard form. Do not approximate.
$\left[2(\cos 6° + i\sin 6°)\right]^5$

Objective 5: Find nth roots of complex numbers

Find the indicated roots of the complex number.

85. Fifth roots of $-32i$

(A) $2(\cos 36° + i\sin 36°)$, $2(\cos 108° + i\sin 108°)$, $2(\cos 108° + i\sin 108°)$, $2(\cos 252° + i\sin 252°)$, $2(\cos 324° + i\sin 324°)$

(B) $\sqrt[5]{2}(\cos 54° + i\sin 54°)$, $\sqrt[5]{2}(\cos 126° + i\sin 126°)$, $\sqrt[5]{2}(\cos 198° + i\sin 198°)$, $\sqrt[5]{2}(\cos 270° + i\sin 270°)$, $\sqrt[5]{2}(\cos 342° + i\sin 342°)$

(C) $32(\cos 54° + i\sin 54°)$, $32(\cos 126° + i\sin 126°)$, $32(\cos 198° + i\sin 198°)$, $32(\cos 270° + i\sin 270°)$, $32(\cos 54° + i\sin 54°)$

(D) None of these

86. Fourth roots of $81(\cos 55° + i\sin 55°)$

(A) $2.9140 + 0.7131i$, $-0.7131 + 2.9140i$, $-2.9140 - 0.7131i$, $0.7131 - 2.9140i$

(B) $0.7131 + 2.9140i$, $2.9140 - 0.7131i$, $-0.7131 - 2.9140i$, $-2.9140 + 0.7131i$

(C) $19.6697 + 4.8131i$, $-4.8131 + 19.6697i$, $-19.6697 - 4.8131i$, $4.8131 - 19.6697i$

(D) $4.8131 + 19.6697i$, $19.6697 - 4.8131i$, $-4.8131 - 19.6697i$, $-19.6697 + 4.8131i$

87. Find all the solutions of the equation.
$x^6 + 729 = 0$

88. Find the indicated roots of the complex number. Express the answers in trigonometric form, with θ in degrees.
Cube roots of $1 + i$

Answer Key for Chapter 6 Additional Topics in Trigonometry

Section 6.1: Law of Sines

Objective 1: Use the Law of Sines to solve oblique triangles (AAS or ASA)

[1] (C)

[2] (A)

[3] 25.0

[4] 17.91

Objective 2: Use the Law of Sines to solve oblique triangles (SSA)

[5] (C)

[6] (A)

[7] There is exactly 1 solution.

[8] Answers may vary. Sample answer: $a = 4$ or $a \geq 3$

Objective 3: Find the areas of oblique triangles

[9] (C)

[10] (A)

[11] $c = 5$

[12] 31.86

Objective 4: Use the Law of Sines to model and solve real-life problems

[13] (A)

[14] (C)

[15] 2765 m

[16] 147.1°

Section 6.2: Law of Cosines

Objective 1: Use the Law of Cosines to solve oblique triangles (SSS or SAS)

[17] (D)

[18] (A)

[19] 18.8

[20] 69.6°

Objective 2: Use the Law of Cosines to model and solve real-life problems

[21] (A)

[22] (D)

[23] 69.2°

[24] $d^2 = 275^2 + 335^2 - 2(275)(335)\cos 19°$; $d \approx 116.8$ yards

Objective 3: Use Heron's Area Formula to find the area of a triangle

[25] (A)

[26] (B)

[27] 267.9

[28] 31.5 square units

Section 6.3: Vectors in the Plane

Objective 1: Represent vectors as directed line segments

[29] (D)

[30] (C)

[31] $\|\mathbf{u}\| = \sqrt{(-8-(-1))^2 + (4-6)^2} = \sqrt{53}$, slope $= \frac{2}{7}$

$\|\mathbf{v}\| = \sqrt{(-8-(-1))^2 + (-8-(-6))^2} = \sqrt{53}$, slope $= \frac{2}{7}$

[32] $k = 4$

Objective 2: Write the component forms of vectors

[33] (C)

[34] (C)

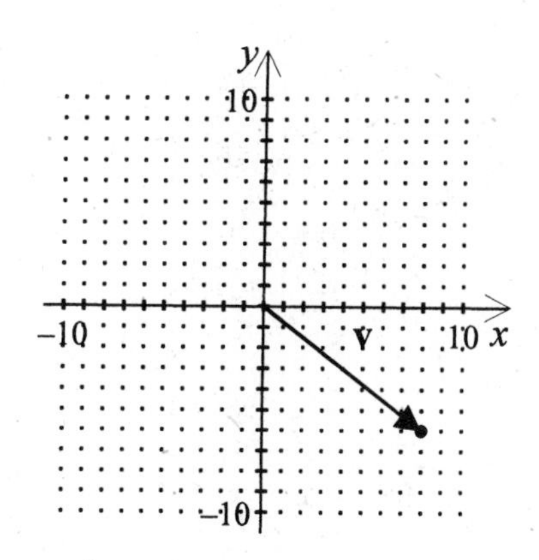

[35] $\mathbf{v} = \langle 8, -6 \rangle$

[36] $y = \pm 10$

Objective 3: Perform basic vector operations and represent them graphically

[37] (C)

[38] (B)

[39] $\langle -11, \ -4 \rangle$

[40] $\left\langle -\frac{1}{2}, \frac{1}{6} \right\rangle$

Objective 4: Write vectors as linear combinations of unit vectors

[41] (C)

[42] (D)

[43] $\frac{10\sqrt{221}}{221}\mathbf{i} + \frac{11\sqrt{221}}{221}\mathbf{j}$

[44] $14\mathbf{i} - 11\mathbf{j}$

Objective 5: Find the direction angles of vectors

[45] (A)

[46] (D)

[47] $\theta = 139.4°$

[48] $x = 3\sqrt{3}$

Objective 6: Use vectors to model and solve real-life problems

[49] (A)

[50] (A)

[51] N 2.8° E

[52] 118 N

Section 6.4: Vectors and Dot Products

Objective 1: Find the dot product of two vectors and use the Properties of the Dot Product

[53] (B)

[54] (B)

[55] $\|\mathbf{u}\| = \sqrt{881}$

[56] -9

Objective 2: Find the angle between two vectors and determine whether two vectors are orthogonal

[57] (B)

[58] (B)

[59] 1.2 radians

[60] Orthogonal

Objective 3: Write a vector as the sum of two vector components

[61] (B)

[62] (D)

[63] $12\mathbf{i} - 6\mathbf{j}$; $13\mathbf{i} + 26\mathbf{j}$

[64] (a) 53 N, 76 N
(b) 66 N, 66 N

Objective 4: Use vectors to find the work done by a force

[65] (D)

[66] (A)

[67] 10

[68] 20.1 foot-pounds

Section 6.5: Trigonometric Form of a Complex Number

Objective 1: Plot complex numbers in the complex plane

[69] (D)

[70] (B)

[71] 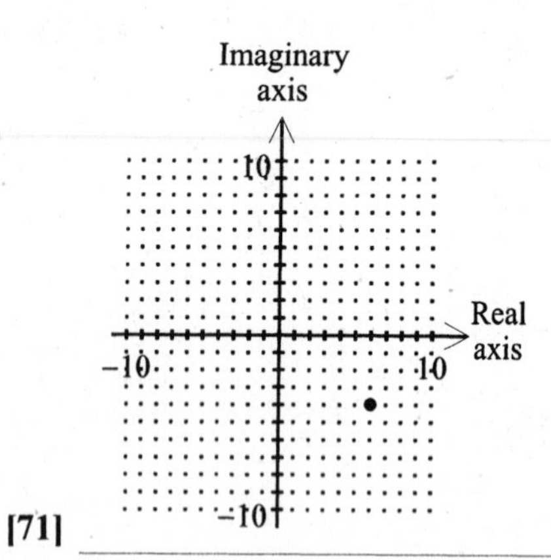

[72] $-1-8i$

Objective 2: Write the trigonometric forms of complex numbers

[73] (A)

[74] (A)

[75] $2\sqrt{2}\left(\cos\frac{7\pi}{4}+i\sin\frac{7\pi}{4}\right)$

[76] Modulus $=4\sqrt{10}$; argument $=108.4°$

Objective 3: Multiply and divide complex numbers written in trigonometric form

[77] (B)

[78] (C)

[79] $12\left(\cos\frac{11\pi}{6}+i\sin\frac{11\pi}{6}\right)=6\sqrt{3}-6i$

[80] (a) $zw=18\left(\cos 105°+i\sin 105°\right)$

(b) $\frac{z}{w}=2\left(\cos 75°+i\sin 75°\right)$

Objective 4: Use DeMoivre's Theorem to find powers of complex numbers

[81] (A)

[82] (A)

[83] -4

[84] $16\sqrt{3}+16i$

Objective 5: Find nth roots of complex numbers

[85] (D)

[86] (A)

[87] $2.5981 + 1.5i,\ \ 3i,\ \ -2.5981 + 1.5i,\ \ -2.5981 - 1.5i,\ \ -3i,\ \ 2.5981 - 1.5i$

[88] $\sqrt[6]{2}(\cos 15° + i \sin 15°),\ \sqrt[6]{2}(\cos 135° + i \sin 135°),\ \sqrt[6]{2}(\cos 255° + i \sin 255°)$

Chapter 7 Systems of Equations and Inequalities

Section 7.1: Solving Systems of Equations

Objective 1: Use the method of substitution to solve systems of equations

Solve the system by the method of substitution.

1. $\begin{cases} 3x+4y=-11 \\ -x+y=6 \end{cases}$ (A) $(-4,\ 10)$ (B) $(-5,\ -1)$ (C) $(-5,\ 1)$ (D) $\left(1,\ -\frac{7}{2}\right)$

2. $\begin{cases} 5x-2y=-24 \\ 2x-7y=9 \end{cases}$ (A) $(-3,\ -6)$ (B) $(-6,\ -3)$ (C) $(-24,\ 9)$ (D) $(9,\ -24)$

3. $\begin{cases} y=x^2-x-3 \\ y=x^2-9 \end{cases}$

4. $\begin{cases} y+2x=-4 \\ x^2+y^2=16 \end{cases}$

Objective 2: Use a graphical approach to solve systems of equations in two variables

5. Solve the system graphically.

$\begin{cases} x+y=-2 \\ 2x-y=5 \end{cases}$

(A)

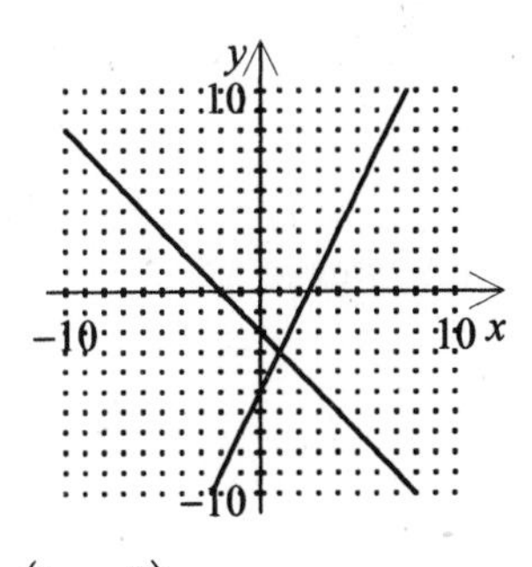

$(1,\ -3)$

(B)

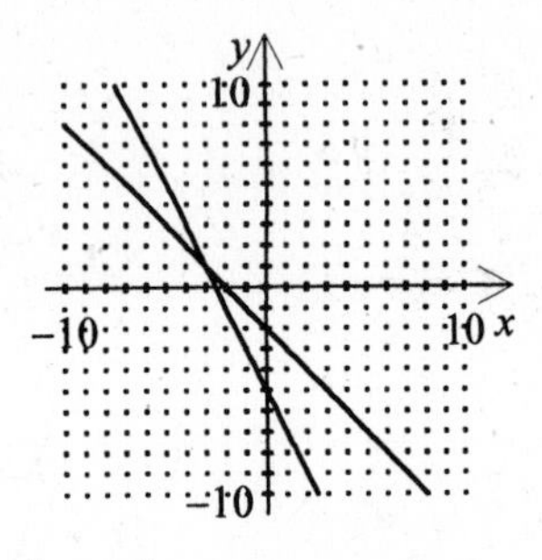

$(-3, 1)$

(C)

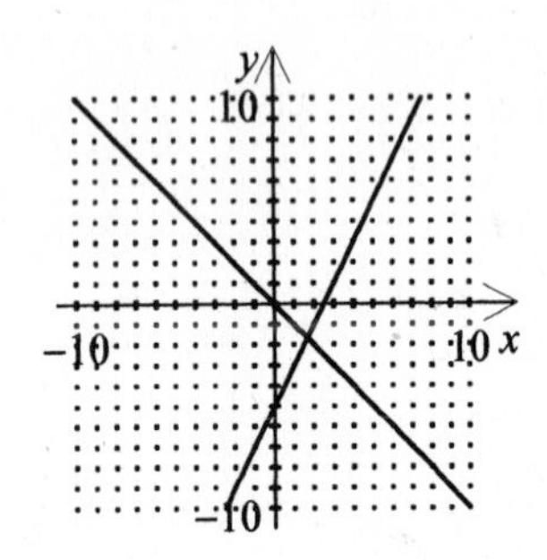

$\left(\frac{5}{3}, -\frac{5}{3}\right)$

(D)

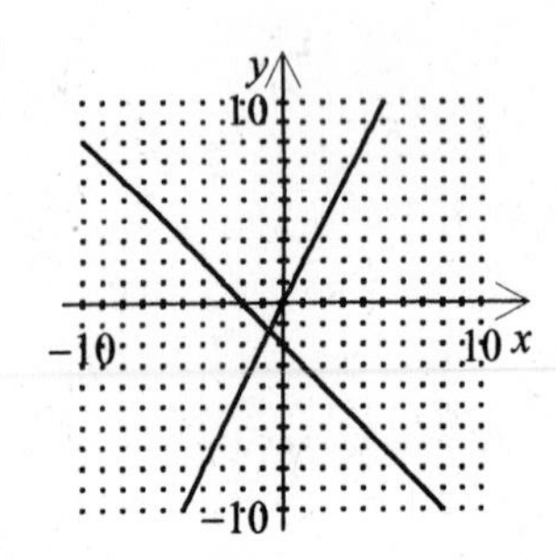

$\left(-\frac{2}{3}, -\frac{4}{3}\right)$

(5.)

6. Use a graphing utility to approximate all points of intersection of the graphs.

$$\begin{cases} -2x - y = 3 \\ -\frac{1}{2}x^2 - y - 5 = 0 \end{cases}$$

(A)

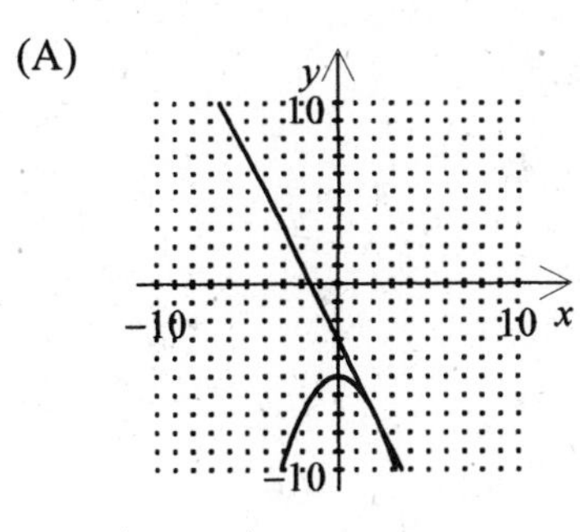

$(2,\ -7)$

(B)

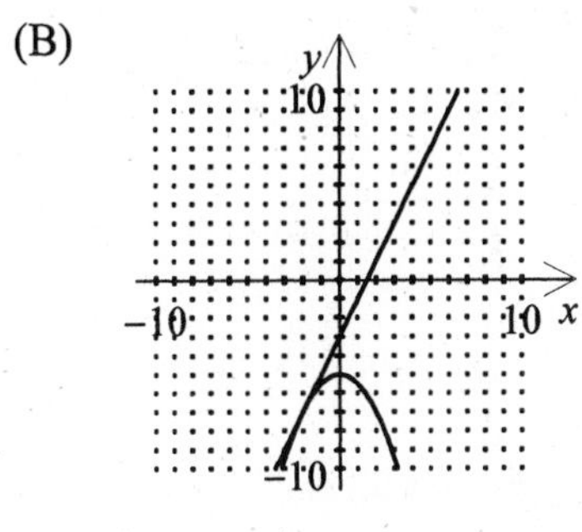

$(-2,\ -7)$

(C)

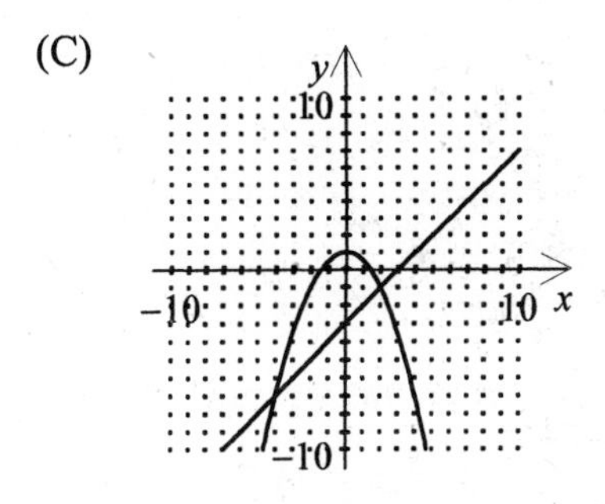

$(2,\ -1),\ (-4,\ -7)$

(D)

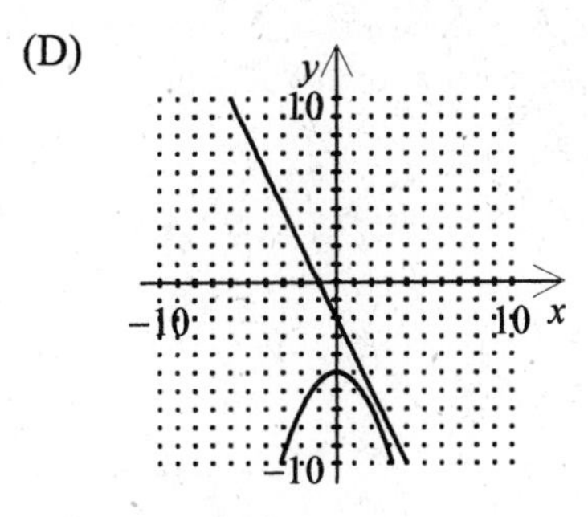

No solution

(6.)

Solve the system graphically.

7. $\begin{cases} 5x+6y=2 \\ x-9y=-20 \end{cases}$

8. $\begin{cases} x^2+y=-3 \\ x^2-y=5 \end{cases}$

Objective 3: Use systems of equations to model and solve real-life problems

9. The perimeter of a rectangular concrete slab is 90 feet and its area is 450 square feet. What is the length of the longer side of the slab?

(A) 32 ft (B) 31 ft (C) 33 ft (D) 30 ft

10. Sue bought some saltwater fish for \$3 each and some freshwater fish for \$2 each for her two new aquariums. If she bought a total of 20 fish and spent a total of \$51, how many freshwater fish did she buy?

(A) 8 (B) 9 (C) 11 (D) 10

11. The cost C for a company to produce, and the revenue R from the sale of, x units of a very expensive computer microprocessor are given by

$C=12.4x+135,\ R=36.9x-0.4x^2.$

How many microprocessors should the company produce for the venture to be profitable $(R>C)$?

12. At the local ballpark, the team charges \$7.75 for each ticket and averages \$2205 of income per game from concessions. The team must pay its players a total of \$3465 and all other workers a total of \$2025. Each fan gets a free bat, which costs the team \$3.25 per bat. How many tickets must be sold for the team to break even on holding a game?

Section 7.2: Two-Variable Linear Systems

Objective 1: Use the method of elimination to solve systems of linear equations in two variables

Solve the system by elimination.

13. $\begin{cases} 8x-4y=-28 \\ 3x+4y=-38 \end{cases}$ (A) $(-28, -5)$ (B) $(a, 3a+4)$ (C) $(-6, -5)$ (D) $(a, 2a+2)$

14. $\begin{cases} 12x+y=59 \\ 13x+2y=63 \end{cases}$ (A) $(17, 17)$ (B) $(5, -1)$ (C) $(9, -5)$ (D) $(2, -5)$

15. $\begin{cases} 4x-5y=-3 \\ x-2y=-4 \end{cases}$

16. $\begin{cases} 3x+\frac{1}{2}y=-12 \\ \frac{7}{2}x+3y=\frac{1}{2} \end{cases}$

Objective 2: Interpret graphically the numbers of solutions of systems of equations in two variables

Identify the graph of the system. Use the graph to determine if the system is consistent or inconsistent. If the system is consistent, determine the number of solutions.

17. $\begin{cases} y-x=1 \\ 3y-3x=6 \end{cases}$

(A)

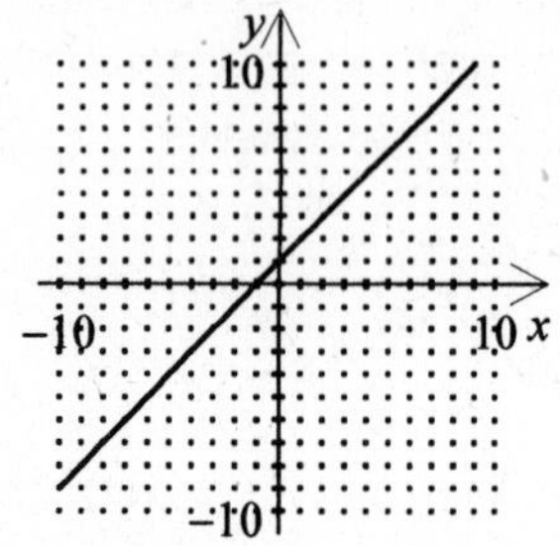

Inconsistent

(B)

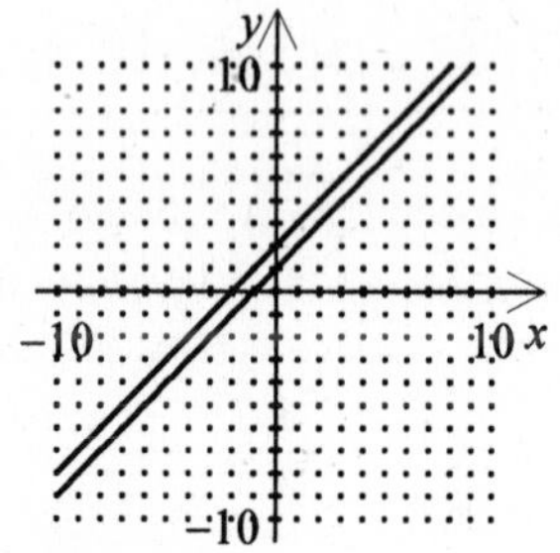

Inconsistent

(C)

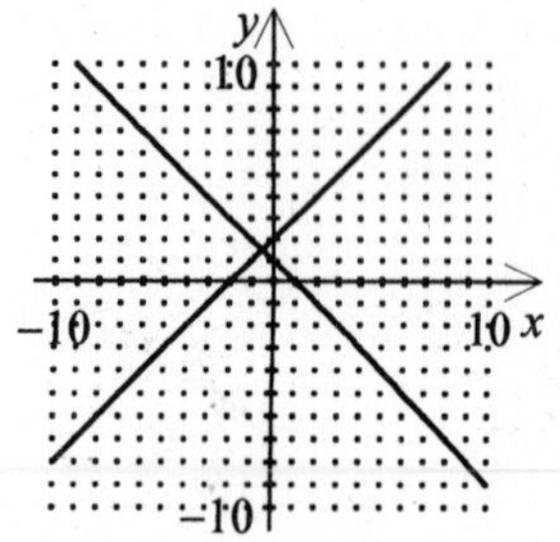

Consistent, infinitely many solutions

(D)

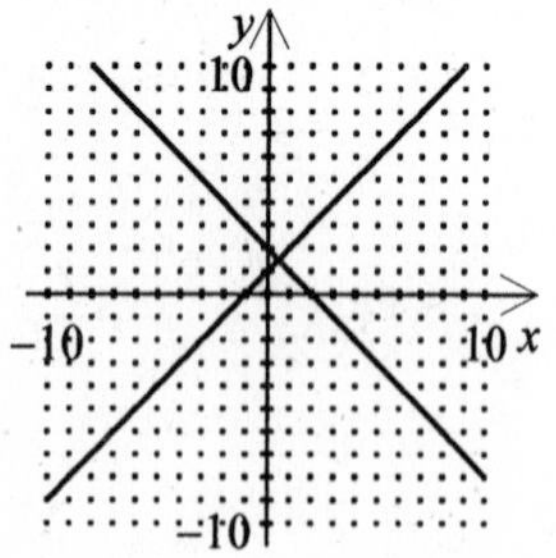

Consistent, one solution

(17.)

Identify the graph of the system. Use the graph to determine if the system is consistent or inconsistent. If the system is consistent, determine the number of solutions.

18. $\begin{cases} -2x + y = -2 \\ x = 1 + \dfrac{1}{2}y \end{cases}$

(A)

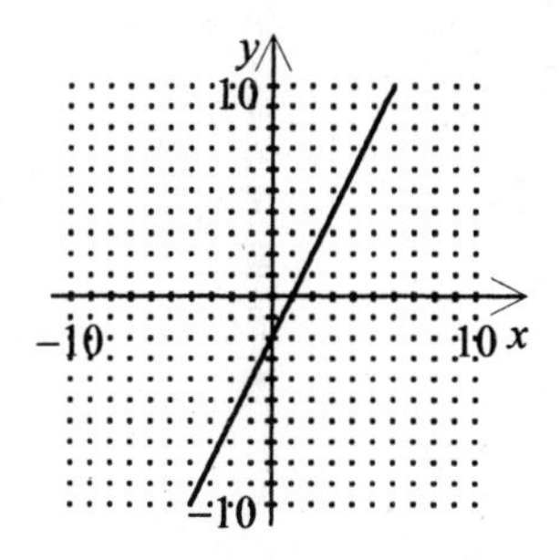

Consistent, infinitely many solutions

(B)

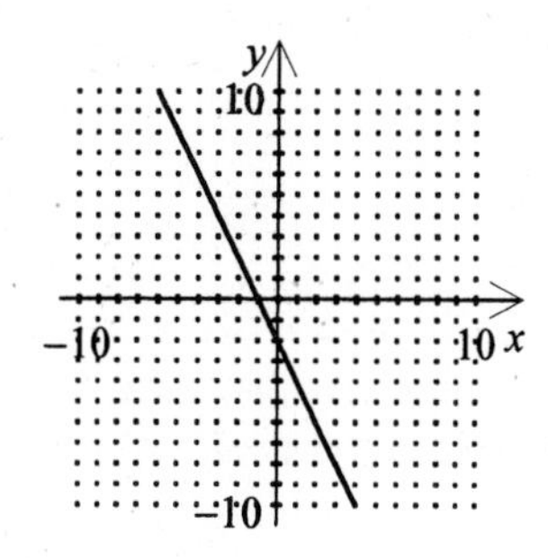

Inconsistent

(C)

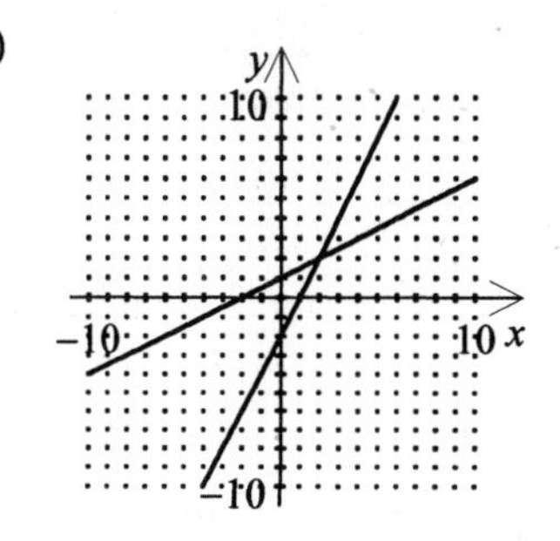

Consistent, one solution

(D)

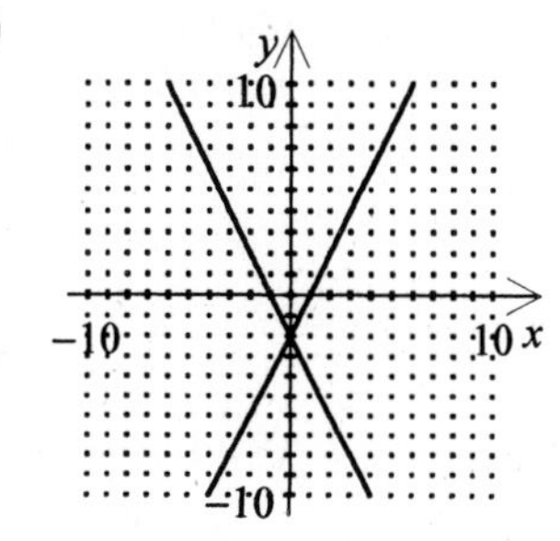

Inconsistent

Graph the lines in the system. Use the graph to determine if the system is consistent or inconsistent. If the system is consistent, determine the number of solutions.

19. $\begin{cases} 6x - 4y = -24 \\ -9x + 6y = 48 \end{cases}$

20. $\begin{cases} 4x + 6y = -16 \\ 2x + 3y = -8 \end{cases}$

Objective 3: Use systems of equations in two variables to model and solve real-life problems

21. Leslie and Kathy drove a total of 654 miles in 11.4 hours. Leslie drove the first part of the trip and averaged 55 miles per hour. Kathy drove the remainder of the trip and averaged 65 miles per hour. For what length of the time did Leslie drive?

(A) 2.7 hours (B) 8.7 hours (C) 1.6 hours (D) 9.8 hours

22. How much pure water must be mixed with 9 pints of 60% citric acid to produce a mixture that is 25% citric acid?

(A) $21\frac{3}{5}$ pints (B) $12\frac{3}{5}$ pints (C) $21\frac{6}{25}$ pints (D) $12\frac{6}{25}$ pints

23. A motorboat can go 10 miles downstream on a river in 20 minutes. It takes 30 minutes for this boat to go back upstream the same 10 miles. Find the speed of the current.

24. Mr. Simon invests a total of $9959 in two savings accounts. One account yields 7% simple interest and the other 8% simple interest. He would like to find the amount placed in each account if a total of $773.26 in interest is received after one year. Write a system of linear equations to express this problem.

Section 7.3: Multivariable Linear Systems

Objective 1: Use back-substitution to solve linear systems in row-echelon form

Use back-substitution to solve the system of linear equations.

25. $$\begin{cases} -2x + 3y - 10z = -6 \\ 2y + 9z = 2 \\ z = -2 \end{cases}$$

(A) $\left(0, -\frac{26}{3}, -2\right)$ (B) $(-28, 10, -2)$ (C) $(13, 0, -2)$ (D) $(28, 10, -2)$

26. $$\begin{cases} 5x + 1.5y - z = 10 \\ 4.5y + 6.5z = -4.5 \\ z = -0.5 \end{cases}$$

(A) $(-4, 19.67, -0.5)$ (B) $(-1.98, -0.28, -0.5)$

(C) $(1.98, -0.28, -0.5)$ (D) $(3.1, -4, -0.5)$

Use back-substitution to solve the system of linear equations.

27. $\begin{cases} -8x+8y-3z=2 \\ 3y-7z=-8 \\ z=2 \end{cases}$

28. $\begin{cases} x+\frac{3}{4}y+z=1 \\ -\frac{1}{2}y+2z=-1 \\ z=-2 \end{cases}$

Objective 2: Use Gaussian elimination to solve systems of linear equations

Solve the system of linear equations using Gaussian elimination.

29. $\begin{cases} -x+y+z=3 \\ -3x+2y+5z=9 \\ -5x+3y+9z=15 \end{cases}$

(A) $(-3, -1, 5)$ (B) $(3a-3, 2a, a)$ (C) $(5, -5, -3)$ (D) No solution

30. $\begin{cases} 3x+10y=5 \\ -6x-5y=0 \end{cases}$

(A) $\left(\frac{2}{3}, -\frac{5}{9}\right)$ (B) $\left(\frac{1}{450}, -\frac{1}{1800}\right)$ (C) $\left(-\frac{5}{9}, \frac{2}{3}\right)$ (D) $\left(-\frac{1}{1800}, \frac{1}{450}\right)$

31. $\begin{cases} x-2y+z=5 \\ 3x-5y-6z=-3 \\ 2x-6y+21z=7 \end{cases}$

32. $\begin{cases} w+13x-y-4z=60 \\ -2w-5x-10y+7z=31 \\ -5w-3x-y+10z=14 \\ -w-5x+y-6z=-40 \end{cases}$

Objective 3: Solve nonsquare systems of linear equations

Solve the system of linear equations.

33. $\begin{cases} 5x-5y+2z=-2 \\ -25x-10y+6z=4 \end{cases}$

(A) $\left(\frac{2}{35}a+\frac{8}{35},\ \frac{16}{35}a-\frac{6}{35},\ a\right)$

(B) $\left(-\frac{2}{35}a-\frac{8}{35},\ -\frac{16}{35}a+\frac{6}{35},\ a\right)$

(C) $\left(\frac{2}{35}a-\frac{8}{35},\ \frac{16}{35}a+\frac{6}{35},\ a\right)$

(D) $\left(-\frac{2}{35}a+\frac{8}{35},\ -\frac{16}{35}a-\frac{6}{35},\ a\right)$

34. $\begin{cases} -2x-\frac{8}{3}y-\frac{2}{3}z=\frac{5}{3} \\ \frac{15}{2}x-4y-\frac{3}{2}z=-\frac{5}{2} \end{cases}$

(A) $\left(-\frac{1}{21}a+\frac{10}{21},\ \frac{2}{7}a+\frac{15}{56},\ a\right)$

(B) $\left(\frac{1}{21}a-\frac{10}{21},\ -\frac{2}{7}a-\frac{15}{56},\ a\right)$

(C) $\left(-\frac{1}{21}a-\frac{10}{21},\ \frac{2}{7}a-\frac{15}{56},\ a\right)$

(D) $\left(\frac{1}{21}a+\frac{10}{21},\ -\frac{2}{7}a+\frac{15}{56},\ a\right)$

35. $\begin{cases} -8x+y+5z=-9 \\ -32x-3y-45z=45 \end{cases}$

36. $\begin{cases} \frac{3}{2}x+\frac{9}{2}y+2z=-5 \\ 3x-\frac{27}{4}y-9z=\frac{25}{2} \end{cases}$

Objective 4: Use systems of linear equations in three or more variables to model and solve application problems

37. In a certain country, the official unit of currency is the copper leaf. One silver leaf is worth 15 copper leaves, and one gold leaf is worth 48 copper leaves. Lily returns from her vacation to this country and finds that she has 12 more copper leaves than silver leaves, and a total of 33 leaves. If her money is worth 460 copper leaves, how many leaves of each kind does she have?

(A) 19 copper leaves
7 silver leaves
7 gold leaves

(B) 16 copper leaves
7 silver leaves
10 gold leaves

(C) 19 copper leaves
4 silver leaves
10 gold leaves

(D) 16 copper leaves
4 silver leaves
13 gold leaves

38. Find the partial fraction decomposition for the rational expression.

$$\frac{-7x^2-60x-63}{(x+6)(x+3)(x+9)}=\frac{A}{x+6}+\frac{B}{x+3}+\frac{C}{x+9}$$

(A) $-\frac{5}{x+6}+\frac{3}{x+3}-\frac{5}{x+9}$

(B) $\frac{5}{x+6}-\frac{3}{x+3}+\frac{5}{x+9}$

(C) $-\frac{5}{x+6}+\frac{3}{x+3}+\frac{5}{x+9}$

(D) $\frac{5}{x+6}+\frac{3}{x+3}-\frac{5}{x+9}$

39. The sum of the measures of the angles of a triangle is $180°$. The middle-sized angle measures $6°$ more than 2 times the smallest angle. The middle-sized angle measures $78°$ less than the largest angle. Find the measure of each angle.

40. Homer has three investments totaling $100,000. These investments earn interest at 5%, 7%, and 9% respectively. Homer's total income from these investments is $7800. The income from the 9% investment exceeds the total income from the other two investments by $1200. Find how much Homer has invested at 5%.

Section 7.4: Systems of Inequalities

Objective 1: Sketch the graphs of inequalities in two variables

Identify the graph of the inequality.

41. $y \le 4x + 4$

(A)

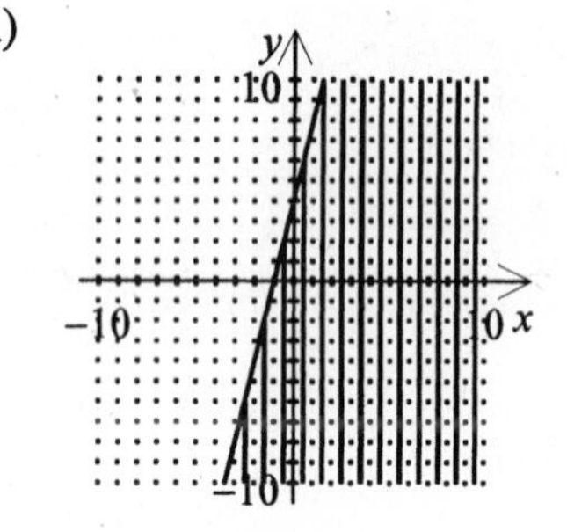

(B)

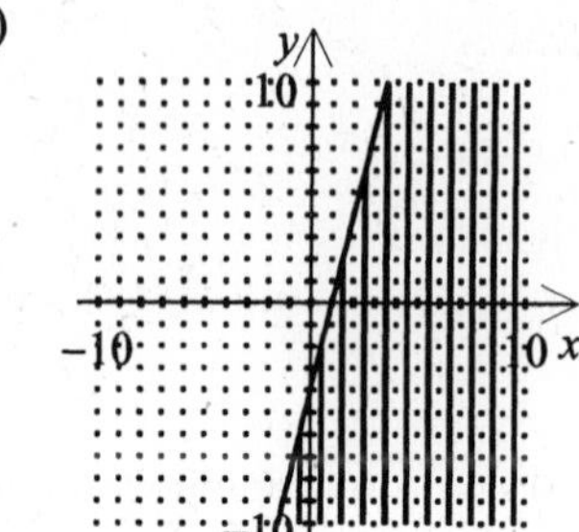

(C)

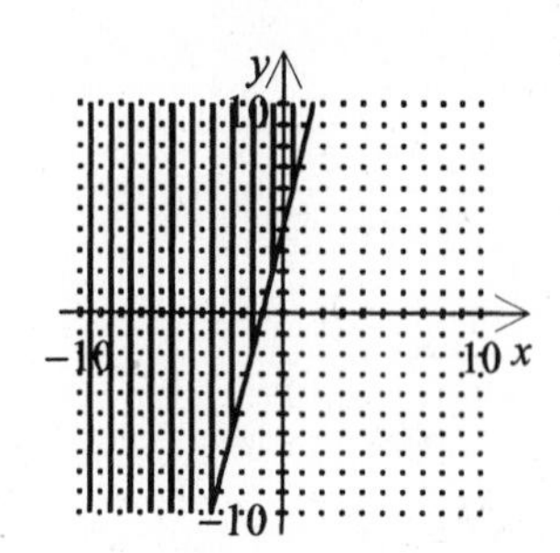

(D)

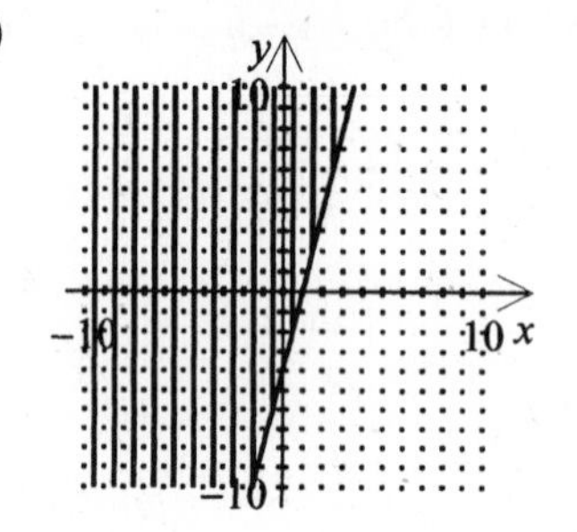

Identify the graph of the inequality.

42. $y \geq 5^x$

(A)

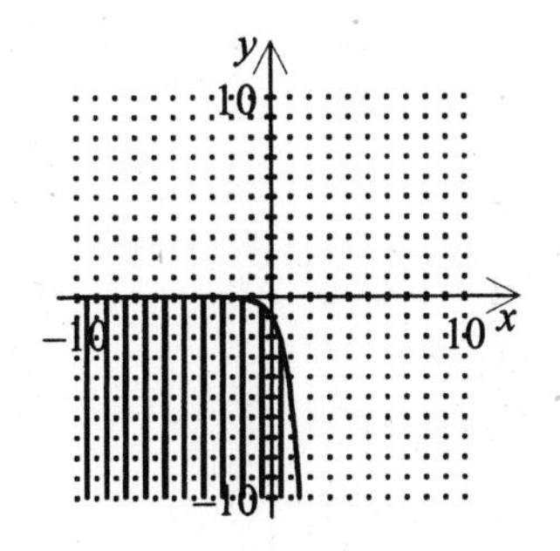

(B)

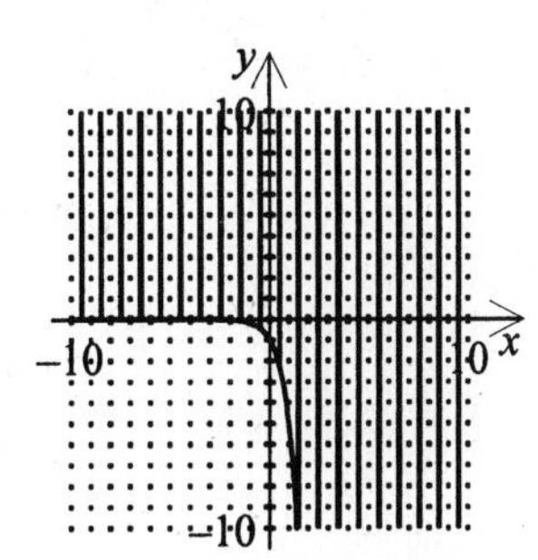

(C)

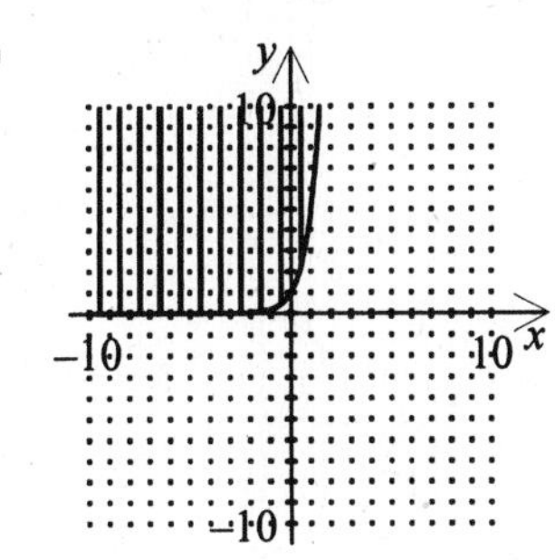

(D)

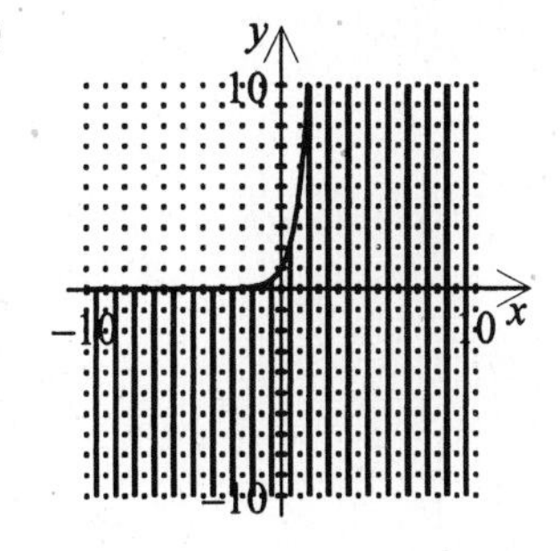

Sketch the graph of the inequality.

43. $x^2 + y^2 \geq 36$

44. $y > -x^2 + 5x - 6$

Objective 2: Solve systems of inequalities

45. Which is the shaded region representing the solution of the system?

$$\begin{cases} 10x + 2y \le 26 \\ -8x + 5y \le -34 \end{cases}$$

(A)

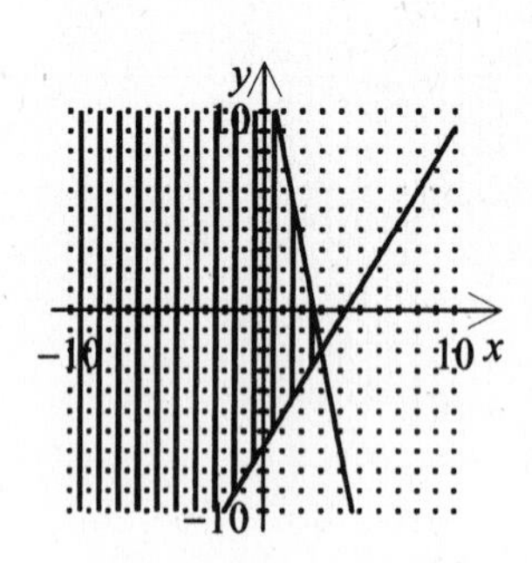

(B)

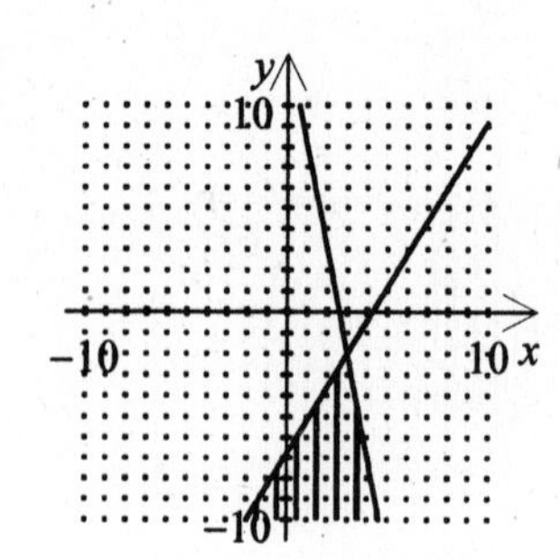

(C)

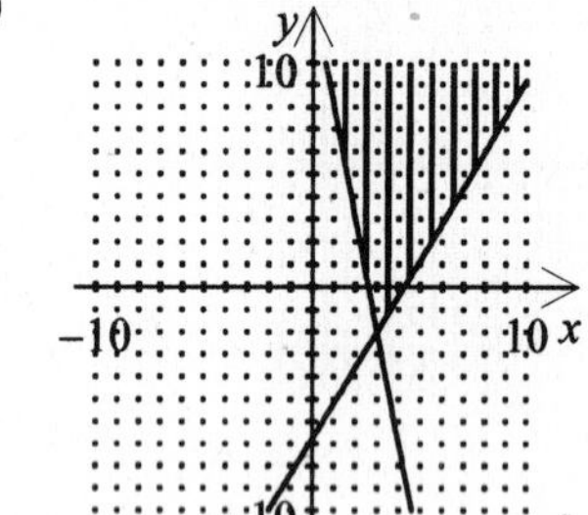

(D)

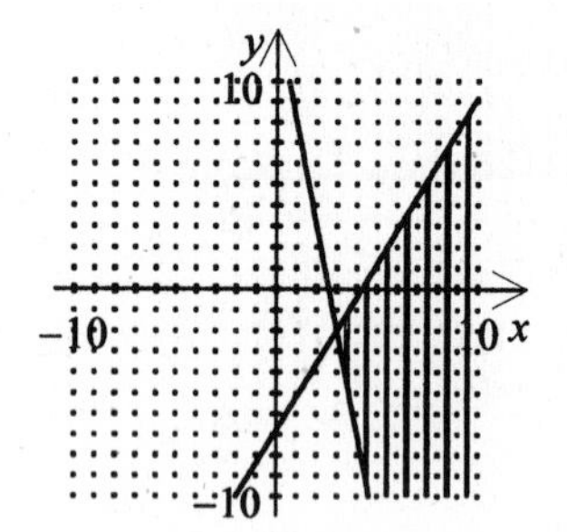

46. Which is the shaded region representing the solution of the system?

$$\begin{cases} x^2 + y^2 \le 64 \\ y > 8 - x^2 \end{cases}$$

(A)

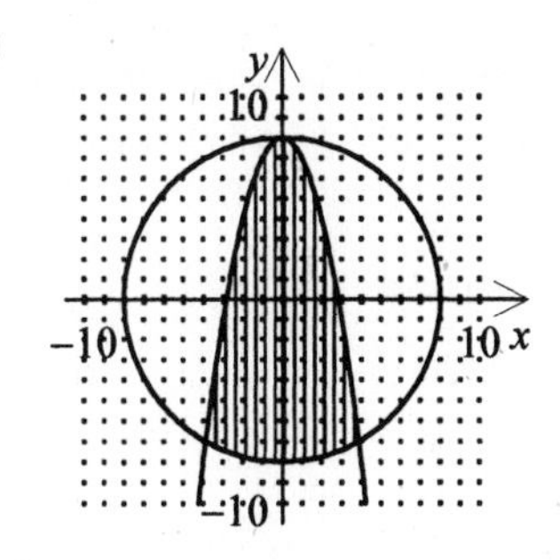

(B)

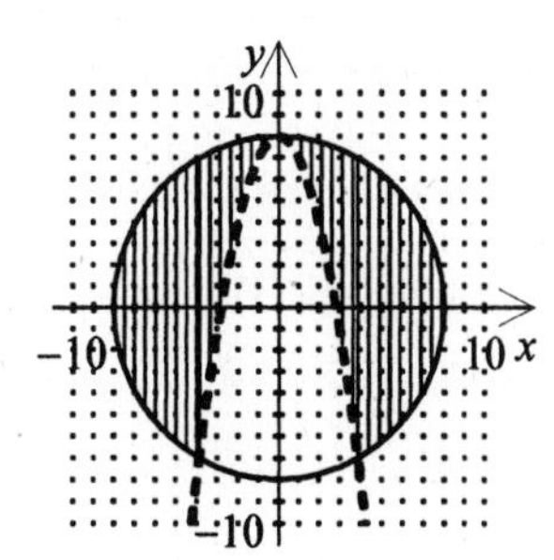

(C)

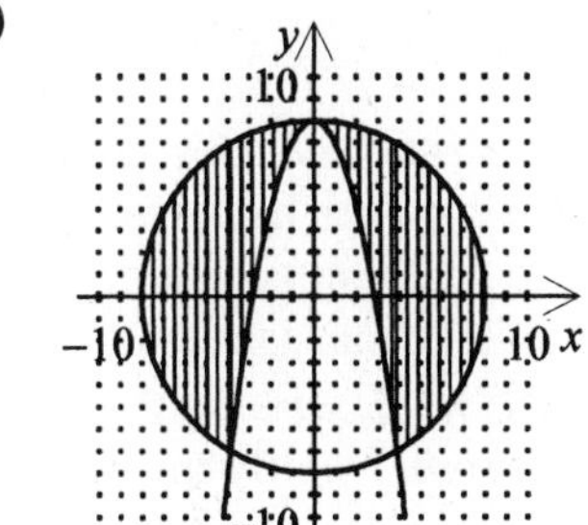

(D)

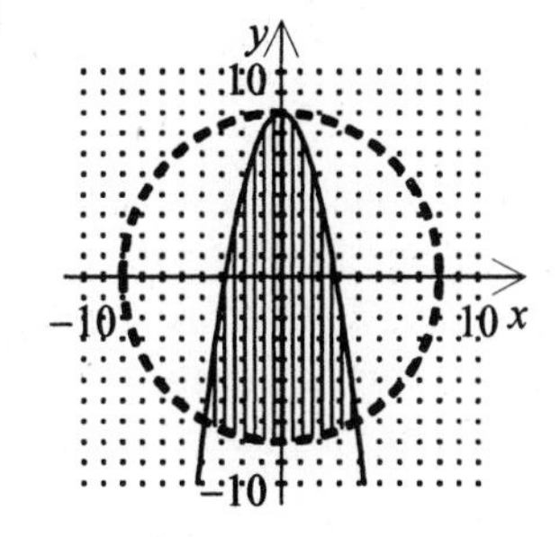

Graph the inequalities. Shade the region representing the solution of the system.

47. $\begin{cases} 3x - 4y \le -12 \\ x + y \ge 2 \\ x \le 8 \end{cases}$

48. $\begin{cases} y \ge (x+1)^2 - 8 \\ y \le x - 5 \end{cases}$

Objective 3: Use systems of inequalities in two variables to model and solve real-life problems

49. A large investment firm wants to invest in two companies. The firm has budgeted enough capital to buy up to a total of 46,000 shares. To protect their investment, they want to own at least 8000 shares in each company. Because one of the companies looks more promising than the other one, they also want to be sure they have at least three times as many shares in that company as in the other company. Find a system of inequalities that describes the number of shares that the investment firm may purchase from each company.

(A) $\begin{cases} x+y \geq 46{,}000 \\ x \leq 8000 \\ y \leq 8000 \\ x \leq 3y \end{cases}$ (B) $\begin{cases} x+y \geq 46{,}000 \\ x \geq 8000 \\ y \geq 8000 \\ x \leq 3y \end{cases}$ (C) $\begin{cases} x+y \leq 46{,}000 \\ x \geq 8000 \\ y \geq 8000 \\ x \geq 3y \end{cases}$ (D) $\begin{cases} x+y \leq 46{,}000 \\ x \leq 8000 \\ y \leq 8000 \\ x \geq 3y \end{cases}$

50. A pharmacist is preparing prescription medication. The smallest amounts of three substances that she needs in order to mix the medication properly are 150 units of substance A, 300 units of substance B, and 200 units of substance C. She must purchase these supplies from two different vendors. One package from vendor x contains 20 units of substance A, 25 units of substance B, and 30 units of substance C. One package from vendor y contains 30 units of substance A, 10 units of substance B, and 15 units of substance C. Write a system of inequalities describing the different numbers of packages the pharmacist can purchase from the two vendors. Identify the graph of the system.

(A) $\begin{cases} 20x+30y \geq 150 \\ 25x+10y \geq 300 \\ 30x+15y \geq 200 \\ x \geq 0 \\ y \geq 0 \end{cases}$

(B) 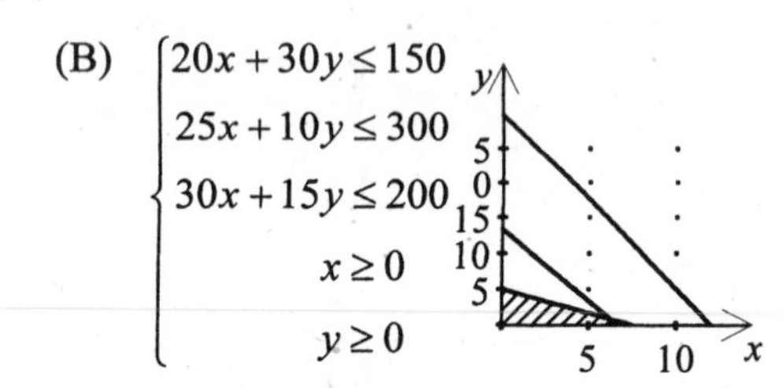

$\begin{cases} 20x+30y \leq 150 \\ 25x+10y \leq 300 \\ 30x+15y \leq 200 \\ x \geq 0 \\ y \geq 0 \end{cases}$

(C) $\begin{cases} 20x+30y \leq 150 \\ 25x+10y \leq 300 \\ 30x+15y \leq 200 \\ x \geq 0 \\ y \geq 0 \end{cases}$

(D) 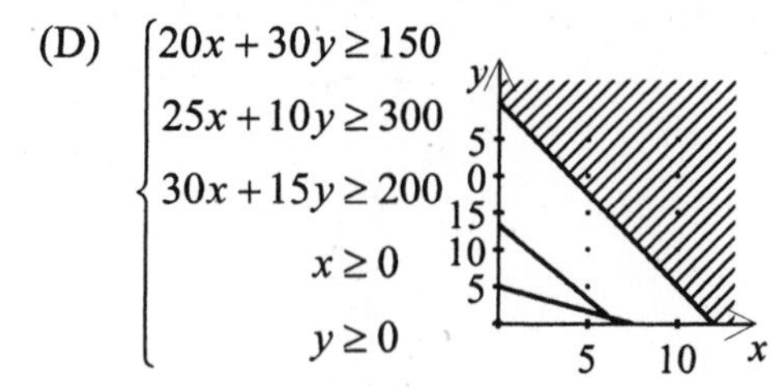

$\begin{cases} 20x+30y \geq 150 \\ 25x+10y \geq 300 \\ 30x+15y \geq 200 \\ x \geq 0 \\ y \geq 0 \end{cases}$

51. Pioneer Math Club must sell at least 10 school jackets and at least 19 caps during a fundraiser. The club will make \$14 profit on every jacket sold and \$3 profit on every cap sold.
(a) Write a system of inequalities that shows how many jackets and caps the club members need to sell to make a profit of at least \$280.
(b) Graph the system of inequalities from part (a).
(c) Will the club meet its goal of \$280 profit if it sells 13 jackets and 36 caps? How much above or below its goal will the club be?

52. A contractor has entered into an agreement with a development company to build a 100-unit apartment building. The agreement states that the contractor must install at least $\frac{1}{2}$ as many fire extinguishers and at least twice as many fire alarms as there are units in the building. Each extinguisher costs $50 and each alarm costs $30 to purchase and install. The most that the contractor can spend on the fire safety equipment is $15,000. Write a system of inequalities that represents the numbers of fire extinguishers and fire alarms that may be purchased and installed to fulfill the contract. Graph the system.

Section 7.5: Linear Programming

Objective 1: Solve linear programming problems

53. Find the minimum and maximum values of the objective function and where they occur, subject to the indicated constraints.

Objective function:

$z = 3x + 5y$

Constraints:

$$x + y \geq 2$$
$$6x - 6y \leq 12$$
$$-4x + 8y \leq 16$$

(A) Minimum at $(0, 2)$: 6
Maximum at $(8, 4)$: 44

(B) Minimum at $(0, 2)$: 6
Maximum at $(6, 8)$: 54

(C) Minimum at $(2, 0)$: 6
Maximum at $(8, 6)$: 54

(D) Minimum at $(2, 0)$: 6
Maximum at $(8, 4)$: 44

54. Find the minimum and maximum values of the objective function and where they occur, subject to the indicated constraints. The graph of the region determined by the constraints is provided.

Objective function:

$z = 3x + 5y$

Constraints:

$x + y \geq 2$

$3x - 2y \leq 6$

$-x + 4y \leq 8$

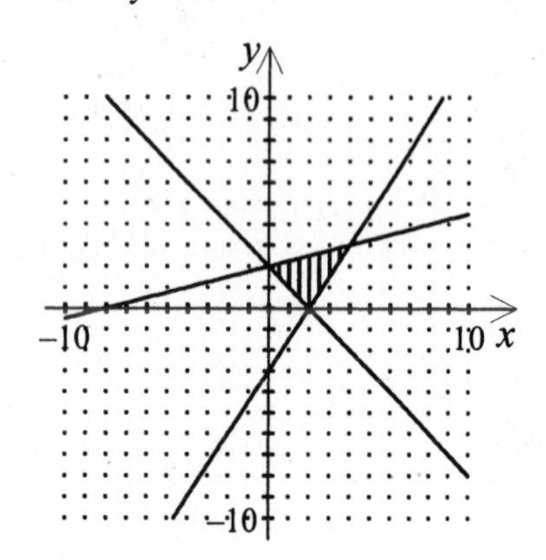

(A) Minimum at $(0, 2)$: 6
Maximum at $(3, 4)$: 27

(B) Minimum at $(2, 0)$: 6
Maximum at $(4, 3)$: 27

(C) Minimum at $(2, 0)$: 6
Maximum at $(4, 6)$: 42

(D) Minimum at $(0, 2)$: 6
Maximum at $(4, 6)$: 42

55. Find the minimum and maximum values of the objective function and where they occur, subject to the indicated constraints.

Objective function:

$z = 2x + 6y$

Constraints:

$x + y \geq 3$

$4x - 2y \leq 12$

$-x + 5y \leq 15$

56. Sketch the region determined by the constraints. Then find the minimum and maximum values of the objective function and where they occur, subject to the constraints.
Objective function:

$z = 2x + 6y$

Constraints:

$x + y \geq 2$

$3x - 2y \leq 6$

$-x + 4y \leq 8$

Objective 2: Use linear programming to model and solve real-life problems

57. Your computer supply store sells two types of laser printers. The first, type A, costs \$137 and you make a \$40 profit on each one. The second, type B, costs \$120 and you make a \$35 profit on each one. You expect to sell at least 100 laser printers this month and you need to make at least \$3675 profit on them. If you must order at least one of each type of printer, how many of each type of printer should you order if you want to minimize your cost?

(A) 30 of type A
70 of type B

(B) 35 of type A
65 of type B

(C) 70 of type A
30 of type B

(D) 65 of type A
35 of type B

58. A factory can produce two products, x and y, with a profit approximated by $P = 11x + 21y - 600$. The production of y can exceed x by no more than 200 units. Moreover, production levels are limited by the formula $x + 2y \leq 1000$. What production levels yield maximum profit?

(A) $x = 0$
$y = 0$

(B) $x = 1000$
$y = 0$

(C) $x = 200$
$y = 400$

(D) $x = 0$
$y = 200$

59. A company makes two explosives: Type I and Type II. Due to storage problems, a maximum of 100 pounds of Type I and 150 pounds of Type II can be mixed and packaged each year. One pound of Type I takes 62 hours to mix and 78 hours to package; one pound of Type II takes 42 hours to mix and 48 hours to package. The mixing department has at most 6920 man-hours available each year, and packaging has at most 8280 man-hours available. If the profit for one pound of Type I is \$76 and for one pound of Type II is \$48, what is the maximum profit possible each year?

60. Roland's Boat Tours sells deluxe and economy seats for each tour it conducts. In order to complete a tour, at least 16 economy seats must be sold and at least 6 deluxe seats must be sold. The maximum number of passengers allowed on each boat is 35. Roland's Boat Tours makes \$50 profit for each economy seat sold and \$45 profit for each deluxe seat sold. What is the maximum profit from one tour?

Answer Key for Chapter 7 Systems of Equations and Inequalities

Section 7.1: Solving Systems of Equations

Objective 1: Use the method of substitution to solve systems of equations

[1] (C)

[2] (B)

[3] $(6, 27)$

[4] $(0, -4), \left(-\frac{16}{5}, \frac{12}{5}\right)$

Objective 2: Use a graphical approach to solve systems of equations in two variables

[5] (A)

[6] (A)

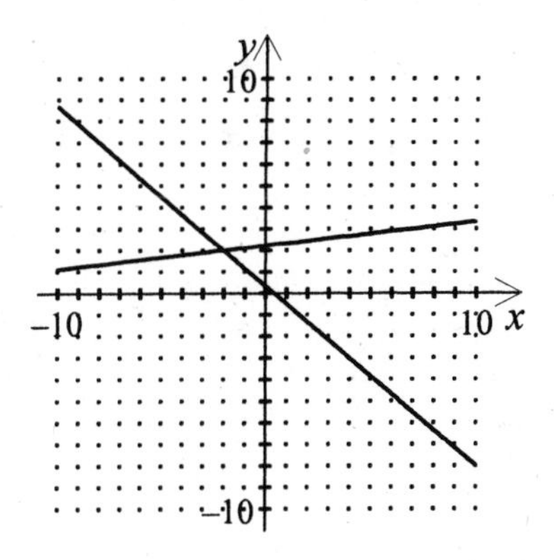

[7] $(-2, 2)$

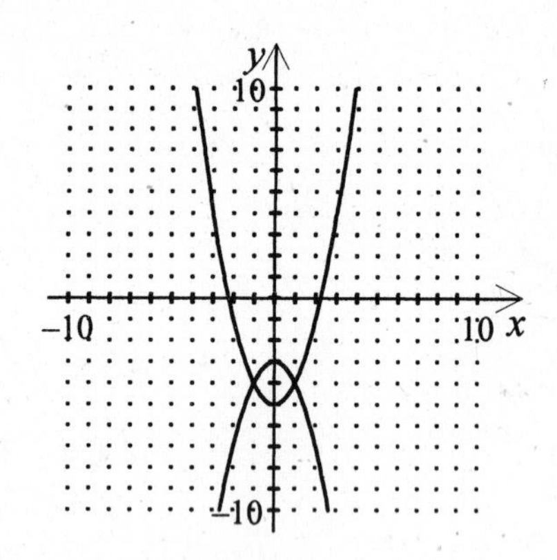

[8] $(-1, -4), (1, -4)$

Objective 3: Use systems of equations to model and solve real-life problems

[9] (D)

[10] (B)

[11] Between 7 and 55 microprocessors, inclusive

[12] 730 tickets

Section 7.2: Two-Variable Linear Systems

Objective 1: Use the method of elimination to solve systems of linear equations in two variables

[13] (C)

[14] (B)

[15] $\left(\frac{14}{3}, \frac{13}{3}\right)$

[16] $(-5, 6)$

Objective 2: Interpret graphically the numbers of solutions of systems of equations in two variables

[17] (B)

[18] (A)

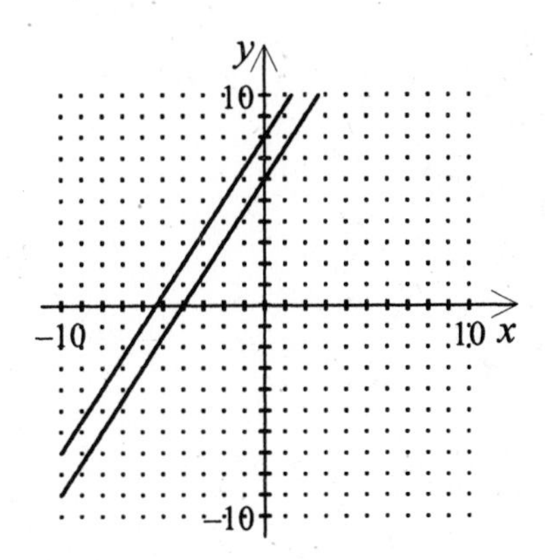

[19] Inconsistent

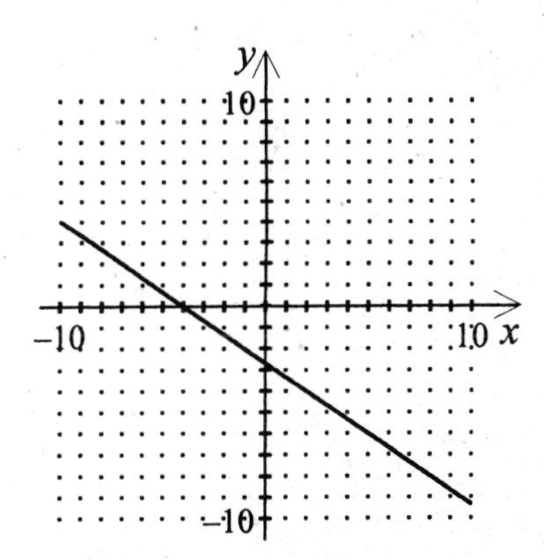

[20] Consistent, infinitely many solutions

Objective 3: Use systems of equations in two variables to model and solve real-life problems

[21] (B)

[22] (B)

[23] 5 mph

[24] $\begin{cases} x + y = 9959 \\ 0.07x + 0.08y = 773.26 \end{cases}$

Section 7.3: Multivariable Linear Systems

Objective 1: Use back-substitution to solve linear systems in row-echelon form

[25] (D)

[26] (C)

[27] $(1,\ 2,\ 2)$

[28] $\left(\frac{15}{2},\ -6,\ -2\right)$

Objective 2: Use Gaussian elimination to solve systems of linear equations

[29] (B)

[30] (C)

[31] $(-694,\ -369,\ -39)$

[32] $(-1,\ 5,\ -4,\ 2)$

Objective 3: Solve nonsquare systems of linear equations

[33] (C)

[34] (B)

[35] $\left(-\frac{15}{28}a-\frac{9}{28},\ -\frac{65}{7}a-\frac{81}{7},\ a\right)$

[36] $\left(\frac{8}{7}a+\frac{20}{21},\ -\frac{52}{63}a-\frac{10}{7},\ a\right)$

Objective 4: Use systems of linear equations in three or more variables to model and solve application problems

[37] (A)

[38] (A)

[39] 18°, 42°, 120°

[40] $10,000

Section 7.4: Systems of Inequalities

Objective 1: Sketch the graphs of inequalities in two variables

[41] (A)

[42] (C)

[43]

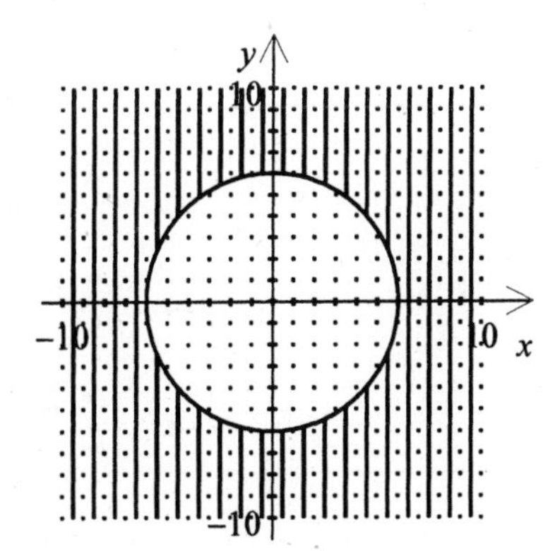

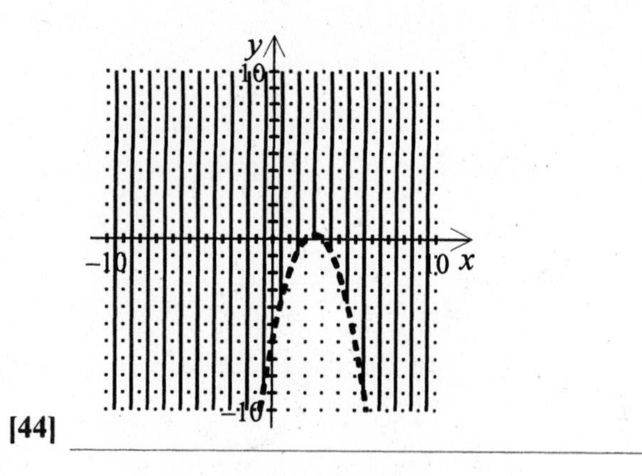

[44] ______

Objective 2: Solve systems of inequalities

[45] (B)

[46] (B)

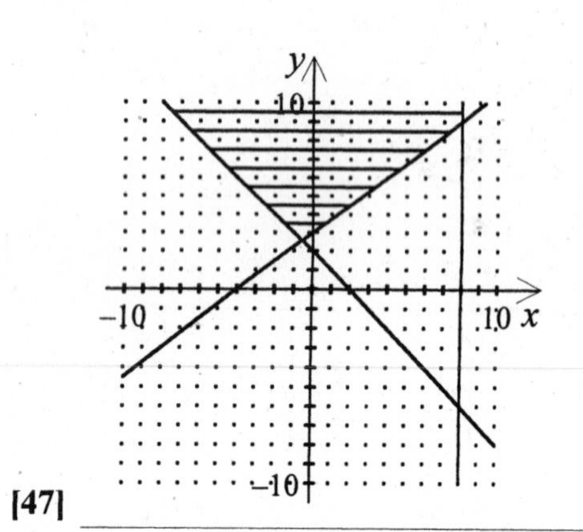

[47] ______

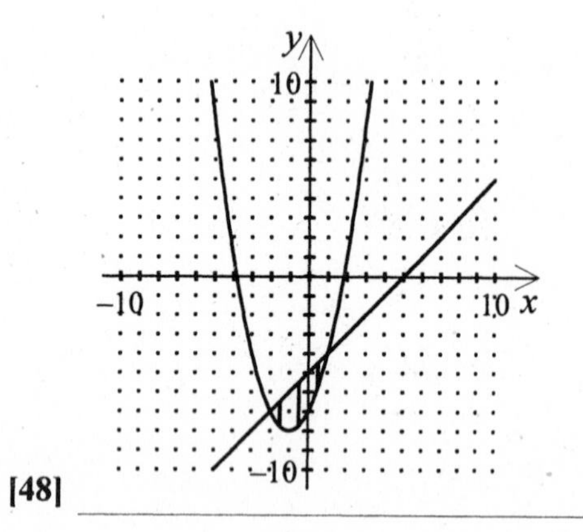

[48] ______

Objective 3: Use systems of inequalities in two variables to model and solve real-life problems

[49] (C)

[50] (D)

(a) $\begin{cases} 14x + 3y \geq 280 \\ x \geq 10 \\ y \geq 19 \end{cases}$

(b)

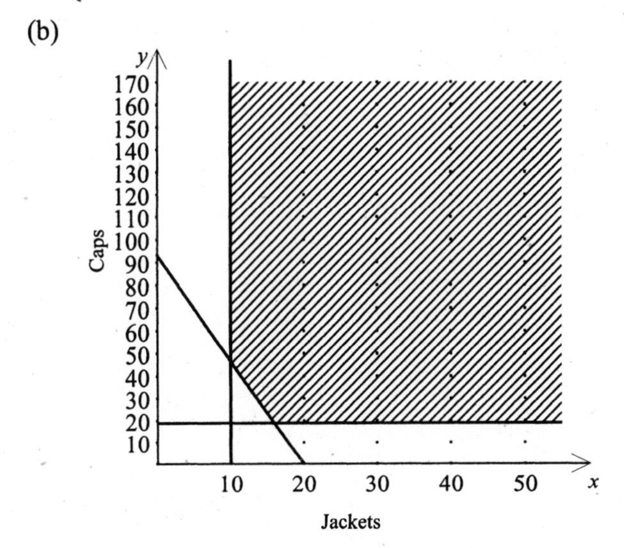

[51] (c) Yes; $10 above

$$\begin{cases} 50x + 30y \le 15{,}000 \\ x \ge 50 \\ y \ge 200 \end{cases}$$

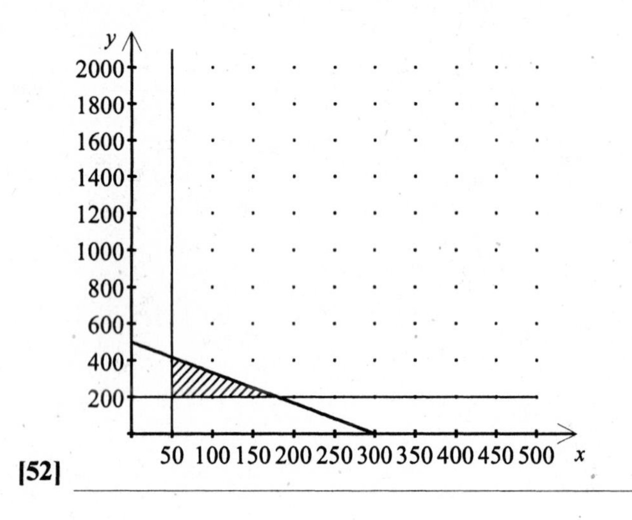

[52]

Section 7.5: Linear Programming

Objective 1: Solve linear programming problems

[53] (C)

[54] (B)

[55] Minimum at $(3, 0)$: 6
Maximum at $(5, 4)$: 34

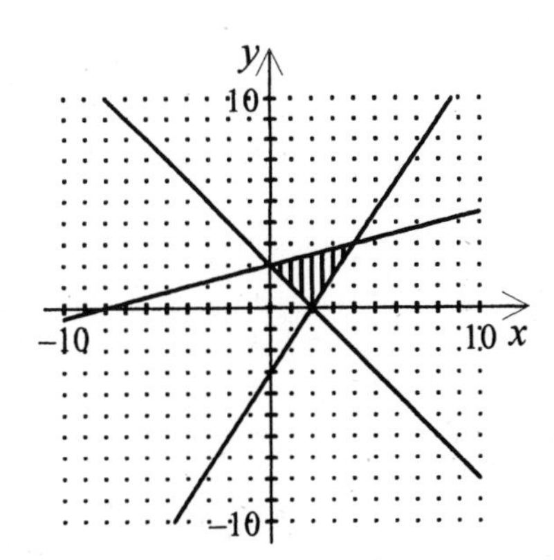

Minimum at $(2, 0)$: 4

[56] Maximum at $(4, 3)$: 26

Objective 2: Use linear programming to model and solve real-life problems

[57] (B)

[58] (B)

[59] \$8176

[60] \$1720

Chapter 8 Matrices and Determinants

Section 8.1: Matrices and Systems of Equations

Objective 1: Write a matrix and identify its order

Determine the order of the matrix.

1. $\begin{bmatrix} 7 & -0.13 & 0.5 \\ -0.8 & 9 & 0 \\ 0.86 & -7 & 4 \end{bmatrix}$

(A) 3×3 (B) 2×3 (C) 5×2 (D) 4×3

2. $\begin{bmatrix} -3 & 0 & -\sqrt{5} & -2 & 0.88 \\ 5 & -\sqrt{3} & -5 & 2 & 0.5 \\ -2 & 0 & -0.88 & 2 & -2.5 \\ -2 & 2.33 & 3 & 0 & -0.75 \end{bmatrix}$

(A) 5×5 (B) 6×4 (C) 5×4 (D) 4×5

3. $\begin{bmatrix} 4 & 7 & 6 \\ 5 & 1 & 5 \\ 8 & 3 & 2 \\ 2 & 9 & 1 \\ 0 & 7 & 4 \\ 1 & 2 & 0 \\ 6 & 0 & 6 \\ 9 & 4 & 8 \\ 3 & 8 & 5 \end{bmatrix}$

4. $\begin{bmatrix} 18 & -4 & 1 \\ 12 & 3 & 6 \end{bmatrix}$

Objective 2: Perform elementary row operations on matrices

5. Perform the row operation on the matrix.
Multiply R_1 by $-\frac{3}{5}$.

$$\begin{bmatrix} -9 & -5 & 8 & -8 \\ -1 & 9 & -6 & 6 \\ 4 & 0 & 3 & 2 \end{bmatrix}$$

(A) $\begin{bmatrix} -9 & -5 & 8 & -8 \\ -1 & 9 & -6 & 6 \\ -\frac{12}{5} & 0 & -\frac{9}{5} & -\frac{6}{5} \end{bmatrix}$

(B) $\begin{bmatrix} -9 & -5 & 8 & -8 \\ \frac{3}{5} & -\frac{27}{5} & \frac{18}{5} & -\frac{18}{5} \\ 4 & 0 & 3 & 2 \end{bmatrix}$

(C) $\begin{bmatrix} \frac{27}{5} & -5 & 8 & -8 \\ \frac{3}{5} & 9 & -6 & 6 \\ -\frac{12}{5} & 0 & 3 & 2 \end{bmatrix}$

(D) $\begin{bmatrix} \frac{27}{5} & 3 & -\frac{24}{5} & \frac{24}{5} \\ -1 & 9 & -6 & 6 \\ 4 & 0 & 3 & 2 \end{bmatrix}$

6. Perform the sequence of row operations on the matrix.
(a) Swap R_1 and R_2.
(b) Replace R_2 by R_2 plus 5 times R_1.
(c) Replace R_3 by R_3 plus -4 times R_1.

$$\begin{bmatrix} -5 & 5 & -3 \\ 1 & 4 & -8 \\ 4 & -2 & -9 \end{bmatrix}$$

(A) $\begin{bmatrix} 9 & -10 & 7 \\ 3 & -18 & 23 \\ -1 & 25 & -43 \end{bmatrix}$

(B) $\begin{bmatrix} 1 & 4 & -8 \\ 0 & 25 & -43 \\ 0 & -18 & 23 \end{bmatrix}$

(C) $\begin{bmatrix} 1 & -9 & -1 \\ 0 & -11 & 29 \\ 0 & 18 & -49 \end{bmatrix}$

(D) $\begin{bmatrix} 9 & -10 & 7 \\ 3 & 18 & -49 \\ -1 & -11 & 29 \end{bmatrix}$

7. Perform the row operation on the matrix.
Replace row R_1 by row R_1 plus -4 times row R_3.

$$\begin{bmatrix} 7 & 2 & -1 \\ 8 & -7 & -4 \\ 4 & -9 & -6 \end{bmatrix}$$

8. Perform the sequence of row operations on the matrix.

(a) Swap R_1 and R_3.

(b) Replace R_2 by R_2 plus -5 times R_1.

(c) Replace R_3 by R_3 plus 2 times R_1.

$$\begin{bmatrix} -2 & -8 & 5 \\ 5 & 10 & -3 \\ 1 & 7 & -7 \end{bmatrix}$$

Objective 3: Use matrices and Gaussian elimination to solve systems of linear equations

If possible, solve the system of equations using Gaussian elimination with back-substitution.

9. $\begin{cases} 2x+4y+10=0 \\ x-4y=7 \end{cases}$

(A) $(-10,\ -2)$ (B) $(a,\ a+5)$ (C) $(-1,\ -2)$ (D) $(a,\ 4a+2)$

10. $\begin{cases} 6x-7y=14 \\ -12x+14y=-28 \end{cases}$

(A) $(16,\ -23)$ (B) $(-12,\ 29)$ (C) $\left(a,\ \frac{6}{7}a-2\right)$ (D) Inconsistent

11. $\begin{cases} x+3y+z=20 \\ 2x+7y-4z=39 \\ x-y-4z=-5 \end{cases}$

12. $\begin{cases} x-2y+z=6 \\ 3x-5y-18z=-3 \\ 2x-6y+45z=4 \end{cases}$

Objective 4: Use matrices and Gauss-Jordan elimination to solve systems of linear equations

If possible, solve the system of equations using Gauss-Jordan elimination.

13. $\begin{cases} -3x-4y+2z=7 \\ x-3y+4z=5 \\ 13x+13y-4z=-23 \end{cases}$

(A) $(7, -4, -2)$ (B) $(-2, 8, 7)$ (C) $\left(-\frac{10}{13}a-\frac{1}{13}, \frac{14}{13}a-\frac{22}{13}, a\right)$ (D) Inconsistent

14. $\begin{cases} 2x-6y=-9 \\ 10x+9y=0 \end{cases}$

(A) $\left(-\frac{27}{26}, \frac{15}{13}\right)$ (B) $\left(-\frac{1}{1620}, -\frac{1}{5400}\right)$ (C) $\left(-\frac{1}{5400}, -\frac{1}{1620}\right)$ (D) $\left(\frac{15}{13}, -\frac{27}{26}\right)$

15. $\begin{cases} x-2y+z=-3 \\ 3x-5y+7z=-6 \\ 2x-6y-5z=-4 \end{cases}$

16. $\begin{cases} -6x-4y-4z+9w=26 \\ -8x-5y-3z+4w=33 \\ -9x-7y+5z+3w=26 \\ -6x-5y-6z-8w=26 \end{cases}$

Section 8.2: Operations with Matrices

Objective 1: Decide whether two matrices are equal

17. Find x, y, and z.

$$\begin{bmatrix} 2x+4 & -9 & 7 \\ -5 & \frac{y}{3}-1 & -2 \\ -1 & -4 & 11z-5 \end{bmatrix} = \begin{bmatrix} 4x+2 & -9 & 7 \\ -5 & 3y+5 & -2 \\ -1 & -4 & 44z+10 \end{bmatrix}$$

(A) $x=1$, $y=-\frac{9}{4}$, $z=-\frac{5}{11}$ (B) $x=3$, $y=9$, $z=\frac{5}{11}$

(C) $x=-1$, $y=-\frac{9}{4}$, $z=-\frac{3}{11}$ (D) $x=3$, $y=-9$, $z=-\frac{3}{11}$

18. Find t, x, y, and z.

$$\begin{bmatrix} -7 & 8 & -2 & 3t+4 \\ 4x-3 & 3 & -5 & -4 \\ -6 & 2y+2 & -1 & z-8 \end{bmatrix} = \begin{bmatrix} -7 & 8 & -2 & 2t-5 \\ x+9 & 3 & -5 & -4 \\ -6 & 5y-1 & -1 & 3z-6 \end{bmatrix}$$

(A) $t=-9$, $x=4$, $y=1$, $z=-1$

(B) $t=-\frac{10}{3}$, $x=-\frac{27}{4}$, $y=-\frac{1}{5}$, $z=16$

(C) $t=-\frac{1}{5}$, $x=\frac{6}{5}$, $y=\frac{1}{7}$, $z=-\frac{7}{2}$

(D) $t=9$, $x=-4$, $y=-1$, $z=1$

19. Find x and y.

$$\begin{bmatrix} 3 & 2 \\ -5 & y \end{bmatrix} = \begin{bmatrix} x & 2 \\ -5 & -3 \end{bmatrix}$$

20. Find t, x, y, and z.

$$\begin{bmatrix} \frac{5}{2} & -9 & -\frac{1}{2} & 1 \\ 5x+4 & -\frac{8}{3} & -\frac{7}{4} & -1 \\ -3 & \frac{3}{4} & -4 & 3y+1 \end{bmatrix} = \begin{bmatrix} \frac{5}{2} & -9 & -\frac{1}{2} & t-6 \\ -\frac{7}{2} & -\frac{8}{3} & -\frac{7}{4} & -1 \\ -3 & 4z+2 & -4 & \frac{2}{3} \end{bmatrix}$$

Objective 2: Add and subtract matrices and multiply matrices by real numbers

Evaluate the expression.

21. $A-B$

$$A=\begin{bmatrix} -\frac{23}{3} & \frac{29}{3} \\ \frac{5}{2} & -\frac{8}{9} \end{bmatrix},\quad B=\begin{bmatrix} -\frac{33}{4} & 8 \\ -\frac{7}{2} & -\frac{21}{8} \end{bmatrix}$$

(A) $\begin{bmatrix} \frac{7}{12} & -\frac{11}{2} \\ \frac{79}{6} & \frac{125}{72} \end{bmatrix}$

(B) $\begin{bmatrix} \frac{7}{12} & \frac{79}{6} \\ -\frac{11}{2} & \frac{125}{72} \end{bmatrix}$

(C) $\begin{bmatrix} -\frac{191}{12} & \frac{53}{3} \\ -1 & -\frac{253}{72} \end{bmatrix}$

(D) $\begin{bmatrix} \frac{7}{12} & \frac{5}{3} \\ 6 & \frac{125}{72} \end{bmatrix}$

Evaluate the expression.

22. $-4A + 6B$

$$A = \begin{bmatrix} -4 & -2 & -3 \\ 0 & -5 & -7 \\ -10 & 10 & 2 \end{bmatrix}, \quad B = \begin{bmatrix} -6 & 9 & -4 \\ 6 & 2 & -1 \\ 1 & 7 & 0 \end{bmatrix}$$

(A) $\begin{bmatrix} 0 & -48 & -2 \\ -24 & -38 & -38 \\ -64 & 32 & 12 \end{bmatrix}$

(B) $\begin{bmatrix} -48 & 24 & -34 \\ 24 & -22 & -46 \\ -56 & 88 & 12 \end{bmatrix}$

(C) $\begin{bmatrix} -20 & 62 & -12 \\ 36 & 32 & 22 \\ 46 & 2 & -8 \end{bmatrix}$

(D) $\begin{bmatrix} 52 & -46 & 36 \\ -36 & 8 & 34 \\ 34 & -82 & -8 \end{bmatrix}$

23. $A - B$

$$A = \begin{bmatrix} -2 & -3 & 0 \\ 6 & -9 & 1 \\ 3 & -1 & -7 \end{bmatrix}, \quad B = \begin{bmatrix} -7 & -9 & -3 \\ 5 & 6 & 4 \\ 7 & -4 & -2 \end{bmatrix}$$

24. $2A$

$$A = \begin{bmatrix} -7 & -5 & 0 \\ -4 & -8 & -3 \\ -9 & -1 & 3 \end{bmatrix}$$

Objective 3: Multiply two matrices

Find the product, if possible.

25. AB, if $A = \begin{bmatrix} 0 & 2 & 1 \\ -3 & -1 & 0 \end{bmatrix}$, $B = \begin{bmatrix} 1 & 6 \\ 0 & 1 \\ -2 & -1 \end{bmatrix}$

(A) $\begin{bmatrix} -2 & -3 \\ 1 & -19 \end{bmatrix}$ (B) $\begin{bmatrix} 0 & -18 \\ 0 & -1 \\ 0 & 0 \end{bmatrix}$ (C) $\begin{bmatrix} -2 & 1 \\ -3 & -19 \end{bmatrix}$ (D) $\begin{bmatrix} -18 & -4 & 1 \\ -3 & -1 & 0 \\ 3 & -3 & 0 \end{bmatrix}$

26. AB, if $A = \begin{bmatrix} -6 & 7 & 5 \\ 9 & -1 & 2 \end{bmatrix}$, $B = \begin{bmatrix} -9 \\ 6 \\ -1 \end{bmatrix}$

(A) $\begin{bmatrix} 91 \\ -89 \end{bmatrix}$ (B) $\begin{bmatrix} -89 \\ 91 \end{bmatrix}$ (C) $\begin{bmatrix} 91 & -89 \end{bmatrix}$ (D) Not possible

27. BA, if $A = \begin{bmatrix} -10 & 3 \\ 5 & 4 \end{bmatrix}$, $B = \begin{bmatrix} 10 & -4 \\ 3 & 10 \end{bmatrix}$

28. AB, if $A = \begin{bmatrix} 2 & -3 & 1 \\ 4 & 5 & -1 \\ -1 & -2 & 3 \end{bmatrix}$, $B = \begin{bmatrix} 3 & -\frac{4}{3} & 3 \\ 1 & -2 & -4 \\ 5 & 1 & -1 \end{bmatrix}$

Objective 4: Use matrix operations to model and solve real-life problems

29. Three departments at a high school have requested administrative leave for some of their teachers to attend mini-training courses on classroom organization, using computers in the classroom, and standards interpretation. The number of teachers in each department requesting leave for each course is listed below.

Department	Organization	Computers	Standards
Math	11	5	6
Science	8	8	3
Language Arts	9	2	2

If the Organization course takes 2.25 hours, the Computers course takes 4.5 hours, and the Standards course takes 1.5 hours, which of the following correctly shows the use of matrices to find the number of hours of administrative leave each department requested for course attendance?

(A) $\begin{bmatrix} 11 & 5 & 6 \\ 8 & 8 & 3 \\ 9 & 2 & 2 \end{bmatrix}\begin{bmatrix} 2.25 & 4.5 & 1.5 \end{bmatrix} = \begin{bmatrix} 150.00 \end{bmatrix}$

(B) $\begin{bmatrix} 11 & 5 & 6 \\ 8 & 8 & 3 \\ 9 & 2 & 2 \end{bmatrix}\begin{bmatrix} 2.25 \\ 4.5 \\ 1.5 \end{bmatrix} = \begin{bmatrix} 56.25 \\ 58.50 \\ 32.25 \end{bmatrix}$

(C) $\begin{bmatrix} 11 & 5 & 6 \\ 8 & 8 & 3 \\ 9 & 2 & 2 \end{bmatrix}\begin{bmatrix} 2.25 \\ 4.5 \\ 1.5 \end{bmatrix} = \begin{bmatrix} 147.00 \end{bmatrix}$

(D) $\begin{bmatrix} 2.25 & 4.5 & 1.5 \end{bmatrix}\begin{bmatrix} 11 & 5 & 6 \\ 8 & 8 & 3 \\ 9 & 2 & 2 \end{bmatrix} = \begin{bmatrix} 63.00 & 72.00 & 15.00 \end{bmatrix}$

30. A supermarket chain sells oranges, apples, peaches, and bananas in three stores located throughout a large metropolitan area. The average number of pounds sold per day in each store is summarized in matrix M. "In season" and "out of season" prices, per pound, of each fruit are given in matrix N. What is the total, for the three stores, of "in season" daily revenue for the four fruits? To the nearest whole percent, what percentage of the daily total "out of season" revenues for store 3 does the "out of season" peach sales represent?

Fruit

$$M = \begin{array}{c} \\ \end{array}\begin{matrix} \text{Oranges} & \text{Apples} & \text{Peaches} & \text{Bananas} \end{matrix}$$

$$M = \begin{bmatrix} 25 & 50 & 20 & 45 \\ 50 & 30 & 65 & 40 \\ 50 & 40 & 40 & 45 \end{bmatrix} \begin{matrix} \text{Store 1} \\ \text{Store 2} \\ \text{Store 3} \end{matrix}$$

Price

"In Season" "Out of Season"

$$N = \begin{bmatrix} \$1.10 & \$1.26 \\ \$1.20 & \$1.50 \\ \$1.00 & \$1.30 \\ \$1.20 & \$1.62 \end{bmatrix} \begin{matrix} \text{Oranges} \\ \text{Apples} \\ \text{Peaches} \\ \text{Bananas} \end{matrix}$$

(A) \$711.23; 21% (B) \$711.23; 34% (C) \$562.50; 21% (D) \$654.10; 65%

31. The art department and the homecoming committee at a local school are ordering supplies. The supplies they need are listed in matrix A.

Paint Brushes Paper Glue Sticks Tape

$$A = \begin{bmatrix} 11 & 14 & 4 & 15 & 5 \\ 8 & 14 & 9 & 16 & 8 \end{bmatrix} \begin{matrix} \text{Art Department} \\ \text{Homecoming Committee} \end{matrix}$$

The costs of these items are listed in matrix B.

$$B = \begin{bmatrix} \$4 \\ \$2 \\ \$8 \\ \$3 \\ \$2 \end{bmatrix} \begin{matrix} \text{Bottle of Paint} \\ \text{Paint Brush} \\ \text{Ream of Colored Paper} \\ \text{Box of Glue Sticks} \\ \text{Roll of Tape} \end{matrix}$$

Compute AB and interpret the result.

32. In a certain town, the probability that a certain day is rainy, sunny, or cloudy during the summer months is determined by the weather on the preceding day. For example, if today is rainy, tomorrow will be rainy $\frac{1}{3}$ of the time, sunny none of the time, and cloudy $\frac{2}{3}$ of the time. Below is a table of the probabilities:

Tomorrow

		Rainy	Sunny	Cloudy
Today	Rainy	$\frac{1}{3}$	0	$\frac{2}{3}$
	Sunny	$\frac{1}{3}$	$\frac{1}{3}$	$\frac{1}{3}$
	Cloudy	0	$\frac{1}{2}$	$\frac{1}{2}$

Similarly, the weather forecasters use this table to predict more than one day in advance. By multiplying the matrix form of the table by itself n times, you will get a new table that allows you to predict the weather n days in advance from today's weather conditions. Use this method to determine the probability that it is cloudy 2 days from now given that today is rainy.

Section 8.3: The Inverse of a Square Matrix

Objective 1: Verify that two matrices are inverses of each other

Determine whether or not matrix B is the inverse of matrix A.

33. $A = \begin{bmatrix} 1 & 0 & -4 \\ 2 & -5 & -9 \\ -8 & 4 & -1 \end{bmatrix}$, $B = \begin{bmatrix} -\frac{3}{2} & \frac{3}{2} & \frac{7}{4} \\ \frac{9}{4} & \frac{3}{4} & -\frac{1}{2} \\ 2 & -\frac{7}{4} & -\frac{3}{4} \end{bmatrix}$

(A) $AB = \begin{bmatrix} -\frac{19}{2} & \frac{17}{2} & \frac{19}{4} \\ -\frac{129}{4} & 15 & \frac{51}{4} \\ 19 & -\frac{29}{4} & -\frac{61}{4} \end{bmatrix}$
B is not the inverse of A.

(B) $AB = \begin{bmatrix} -\frac{19}{2} & \frac{17}{2} & \frac{19}{4} \\ -\frac{129}{4} & 15 & \frac{51}{4} \\ 19 & -\frac{29}{4} & -\frac{61}{4} \end{bmatrix}$
B is the inverse of A.

(C) $AB = \begin{bmatrix} 1 & 0 & 0 \\ 0 & 1 & 0 \\ 0 & 0 & 1 \end{bmatrix}$ and $BA = \begin{bmatrix} 1 & 0 & 0 \\ 0 & 1 & 0 \\ 0 & 0 & 1 \end{bmatrix}$
B is the inverse of A.

(D) $AB = \begin{bmatrix} 1 & 0 & 0 \\ 0 & 1 & 0 \\ 0 & 0 & 1 \end{bmatrix}$ and $BA = \begin{bmatrix} 1 & 0 & 0 \\ 0 & 1 & 0 \\ 0 & 0 & 1 \end{bmatrix}$
B is not the inverse of A.

Determine whether or not matrix B is the inverse of matrix A.

34. $$A = \begin{bmatrix} 4 & 1 & -6 & 3 \\ 5 & -8 & -2 & -7 \\ -9 & -1 & 2 & -3 \end{bmatrix}, \quad B = \begin{bmatrix} -4 & 7 \\ -6 & 9 \\ 8 & -5 \\ 5 & 7 \end{bmatrix}$$

(A) B is the inverse of A.

$$AB = \begin{bmatrix} 1 & 0 \\ 0 & 1 \\ 0 & 0 \end{bmatrix} \text{ and } BA = \begin{bmatrix} 1 & 0 & 0 & 0 \\ 0 & 1 & 0 & 0 \\ 0 & 0 & 1 & 0 \\ 0 & 0 & 0 & 1 \end{bmatrix}$$

(B) B is not the inverse of A.

$$AB = \begin{bmatrix} -55 & 88 \\ -23 & -76 \\ 43 & -103 \end{bmatrix} \text{ and } BA = \begin{bmatrix} 9 & 2 & 7 & 5 \\ 5 & 2 & 1 & 1 \\ 2 & 2 & 6 & 1 \\ 5 & 4 & 9 & 8 \end{bmatrix}$$

(C) AB is not defined because the dimensions of A and B do not allow multiplication. B is not the inverse of A.

(D) BA is not defined because the dimensions of A and B do not allow multiplication. B is not the inverse of A.

Show that matrix B is the inverse of matrix A.

35. $$A = \begin{bmatrix} 1 & -2 & 0 \\ 19 & -35 & 6 \\ 4 & -8 & 1 \end{bmatrix}, \quad B = \begin{bmatrix} \frac{13}{3} & \frac{2}{3} & -4 \\ \frac{5}{3} & \frac{1}{3} & -2 \\ -4 & 0 & 1 \end{bmatrix}$$

36. $$A = \begin{bmatrix} 9 & -2 \\ 8 & 3 \end{bmatrix}, \quad B = \begin{bmatrix} \frac{3}{43} & \frac{2}{43} \\ -\frac{8}{43} & \frac{9}{43} \end{bmatrix}$$

Objective 2: Use Gauss-Jordan elimination to find the inverses of matrices

Use Gauss-Jordan elimination to find the inverse of the matrix (if it exists).

37. $\begin{bmatrix} -5 & 5 & 1 \\ 0 & -2 & 3 \\ 0 & 0 & 2 \end{bmatrix}$

(A) $\begin{bmatrix} -\frac{1}{5} & 0 & 0 \\ -\frac{1}{2} & -\frac{1}{2} & 0 \\ \frac{17}{20} & \frac{3}{4} & \frac{1}{2} \end{bmatrix}$ (B) $\begin{bmatrix} -\frac{1}{5} & -\frac{1}{2} & \frac{17}{20} \\ 0 & -\frac{1}{2} & \frac{3}{4} \\ 0 & 0 & \frac{1}{2} \end{bmatrix}$ (C) $\begin{bmatrix} 0 & -\frac{1}{5} & -\frac{1}{2} \\ 0 & 0 & \frac{17}{20} \\ -\frac{1}{2} & 0 & 0 \end{bmatrix}$ (D) Does not exist

38. $\begin{bmatrix} -3 & 0 & -5 \\ 3 & 1 & 4 \\ 5 & -4 & 2 \end{bmatrix}$

(A) $\begin{bmatrix} \frac{18}{31} & \frac{20}{31} & \frac{5}{31} \\ \frac{14}{31} & \frac{19}{31} & -\frac{3}{31} \\ -\frac{17}{31} & -\frac{12}{31} & -\frac{3}{31} \end{bmatrix}$ (B) $\begin{bmatrix} \frac{18}{31} & -\frac{20}{31} & \frac{5}{31} \\ -\frac{14}{31} & \frac{19}{31} & \frac{3}{31} \\ -\frac{17}{31} & \frac{12}{31} & -\frac{3}{31} \end{bmatrix}$

(C) $\begin{bmatrix} \frac{18}{31} & \frac{14}{31} & -\frac{17}{31} \\ \frac{20}{31} & \frac{19}{31} & -\frac{12}{31} \\ \frac{5}{31} & -\frac{3}{31} & -\frac{3}{31} \end{bmatrix}$ (D) Does not exist

39. $\begin{bmatrix} -5 & 2 & 0 \\ -3 & 3 & 5 \\ 4 & 1 & -2 \end{bmatrix}$

40. $\begin{bmatrix} -1 & 2 & 4 \\ 1 & -1 & -2 \\ 1 & 0 & 1 \end{bmatrix}$

Objective 3: Use a formula to find the inverses of 2 by 2 matrices

Use the formula for finding the inverse of a 2×2 matrix to find the inverse of the matrix (if it exists).

41. $\begin{bmatrix} -1 & 0 \\ 3 & -4 \end{bmatrix}$ (A) $\begin{bmatrix} -4 & 0 \\ -3 & -1 \end{bmatrix}$ (B) $\begin{bmatrix} -1 & 0 \\ -\frac{3}{4} & -\frac{1}{4} \end{bmatrix}$ (C) $\begin{bmatrix} \frac{1}{4} & \frac{3}{4} \\ 0 & 1 \end{bmatrix}$ (D) Does not exist

42. $\begin{bmatrix} -1 & -4 \\ -2 & -3 \end{bmatrix}$

(A) $\begin{bmatrix} -3 & -2 \\ -4 & -1 \end{bmatrix}$ (B) $\begin{bmatrix} \frac{1}{5} & -\frac{2}{5} \\ -\frac{4}{5} & \frac{3}{5} \end{bmatrix}$ (C) $\begin{bmatrix} \frac{3}{5} & -\frac{4}{5} \\ -\frac{2}{5} & \frac{1}{5} \end{bmatrix}$ (D) Does not exist

43. $\begin{bmatrix} -2 & -4 \\ 4 & 1 \end{bmatrix}$

44. $\begin{bmatrix} 0 & -1 \\ 1 & -5 \end{bmatrix}$

Objective 4: Use inverse matrices to solve systems of linear equations

If possible, solve the system of equations using an inverse matrix.

45. $\begin{cases} x - 4y + z = 9 \\ x + 3y + 2z = -7 \\ 4x - 2y + 6z = -5 \end{cases}$

(A) $(5, -4, 1)$ (B) $(0, -2, -3)$ (C) $(0, -3, 1)$ (D) No solution

46. $\begin{cases} 6x = 10 \\ -10x - 7y = -1 \end{cases}$ (A) $\left(-\frac{47}{21}, \frac{5}{3}\right)$ (B) $\left(-\frac{5}{3}, \frac{47}{21}\right)$ (C) $\left(\frac{5}{3}, -\frac{47}{21}\right)$ (D) No solution

47. $\begin{cases} x + 2y + z = 10 \\ 3x + 7y + 4z = 32 \\ x - y - 4z = 6 \end{cases}$

If possible, solve the system of equations using an inverse matrix.

48. $\begin{cases} 3x+4y=7 \\ 7x-2y=-29 \end{cases}$

Section 8.4: The Determinant of a Square Matrix

Objective 1: Find the determinants of 2 by 2 matrices

Find the determinant of the matrix.

49. $\begin{bmatrix} 8 & 4 \\ -10 & -8 \end{bmatrix}$ (A) 6 (B) –24 (C) –104 (D) 24

50. $\begin{bmatrix} \frac{4}{7} & \frac{3}{7} \\ -\frac{2}{7} & -\frac{8}{7} \end{bmatrix}$ (A) $-\frac{4}{49}$ (B) $-\frac{38}{49}$ (C) $\frac{4}{7}$ (D) $-\frac{26}{49}$

51. $\begin{bmatrix} 4 & -1 \\ 9 & -9 \end{bmatrix}$

52. $\begin{bmatrix} \frac{1}{12} & -\frac{1}{12} \\ -\frac{1}{6} & -\frac{1}{2} \end{bmatrix}$

Objective 2: Find minors and cofactors of square matrices

53. Find the cofactor C_{32} of the matrix A. (A) –8 (B) 8 (C) 29 (D) –29

$$A=\begin{bmatrix} 5 & 9 & 8 \\ -3 & -7 & 1 \\ -4 & -2 & 6 \end{bmatrix}$$

54. Find the three cofactors of the third row of the matrix.

$$\begin{bmatrix} -9 & -4 & -2 \\ -8 & -7 & 5 \\ 6 & -3 & 1 \end{bmatrix}$$

(A) $C_{31} = 34,\ C_{32} = -61,\ C_{33} = -31$ (B) $C_{31} = -34,\ C_{32} = 61,\ C_{33} = 31$

(C) $C_{31} = -66,\ C_{32} = 51,\ C_{33} = -31$ (D) $C_{31} = 66,\ C_{32} = -51,\ C_{33} = 31$

55. Find the minor M_{21} of the matrix A.

$$A = \begin{bmatrix} 3 & -8 & -2 \\ -6 & -9 & 4 \\ 1 & 7 & -5 \end{bmatrix}$$

56. Find the cofactor C_{22} of the matrix A.

$$A = \begin{bmatrix} 9 & 5 & -6 \\ -1 & 7 & 2 \\ 8 & -4 & -3 \end{bmatrix}$$

Objective 3: Find the determinants of square matrices

Find the determinant of the matrix.

57. $\begin{bmatrix} 3 & 5 & 5 \\ 4 & 1 & 1 \\ 2 & 2 & 3 \end{bmatrix}$ (A) -23 (B) 23 (C) -17 (D) 17

58. $\begin{bmatrix} 10 & -8 & -6 \\ -5 & -5 & 5 \\ -9 & -5 & -8 \end{bmatrix}$ (A) -1450 (B) 1450 (C) -1110 (D) 1110

59. $\begin{bmatrix} -1 & -2 & -3 \\ 5 & 3 & 4 \\ 4 & 5 & 2 \end{bmatrix}$

Find the determinant of the matrix.

60. $\begin{bmatrix} 1 & 0 & 9 \\ -10 & -2 & 10 \\ -1 & -7 & 5 \end{bmatrix}$

Section 8.5: Applications of Matrices and Determinants

Objective 1: Use Cramer's Rule to solve systems of linear equations

Use Cramer's Rule to solve (if possible) the system of equations.

61. $\begin{cases} 2x + y = -3 \\ 2x - 4y = 0 \end{cases}$ (A) $\left(\frac{3}{5}, -\frac{6}{5}\right)$ (B) $\left(-\frac{6}{5}, -\frac{3}{5}\right)$ (C) $\left(-\frac{3}{5}, -\frac{6}{5}\right)$ (D) $\left(-\frac{6}{5}, \frac{3}{5}\right)$

62. $\begin{cases} 5x + 8y = 43 \\ 6x + 5y = 47 \end{cases}$ (A) $(-7, 1)$ (B) $(9, 3)$ (C) $(7, 1)$ (D) $(6, 0)$

63. $\begin{cases} 9x + 4y = 31 \\ 3x + 7y = 16 \end{cases}$

64. $\begin{cases} 3x + 2y + z = -13 \\ 3x - 2y - z = -11 \\ 3x + 2y - z = -3 \end{cases}$

Objective 2: Use determinants to find the areas of triangles

65. Use a determinant and the vertices of the triangle to find the area of the triangle.

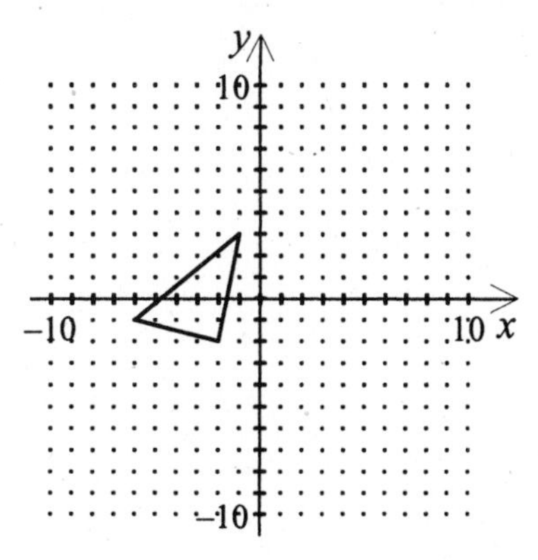

(A) 9 (B) $\frac{21}{2}$ (C) $\frac{19}{2}$ (D) 11

66. The vertices of a triangle are given below. Use a determinant and the vertices of the triangle to find the area of the triangle.

$(8, -2),\ (-4, -6),\ (1, 5)$

(A) 56 (B) $\frac{109}{2}$ (C) 55 (D) $\frac{113}{2}$

67. The vertices of a triangle are given below. Use a determinant and the vertices of the triangle to find the area of the triangle.

$(1, -1),\ (-9.5, -8),\ (-2.5, -0.5)$

68. A triangle has the vertices $(8, 3.5)$, $(-0.5, 8.5)$, and $(-1.5, -4)$. Use a determinant and the vertices of the triangle to find the area of the triangle.

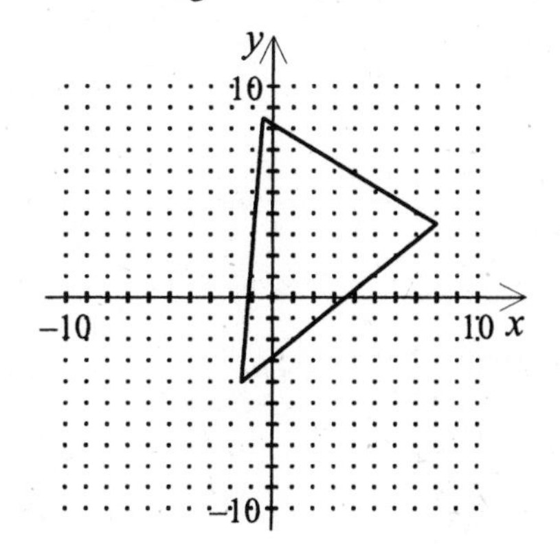

Objective 3: Use a determinant to test for collinear points and find an equation of a line passing through two points

69. Use a determinant to find an equation of the line through the points.

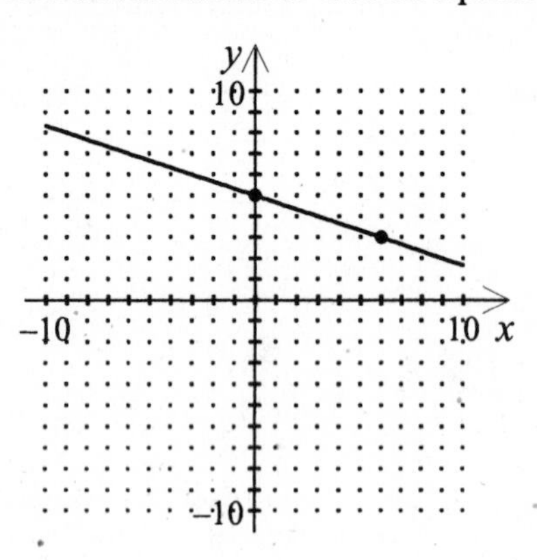

(A) $y=-\frac{1}{3}x+5$ (B) $y=\frac{1}{3}x+5$ (C) $y=-3x+5$ (D) $y=3x+5$

70. Use a determinant to find an equation of the line through the points listed below.

$(0,\ 1),\ (10,\ 9)$

(A) $-5x+4y-4=0$ (B) $5x+4y-4=0$ (C) $4x+5y-5=0$ (D) $4x-5y+5=0$

71. The line below passes through the points $\left(\frac{3}{2},\ \frac{9}{2}\right)$ and $(0,\ 3)$. Use a determinant to find an equation of the line through the points.

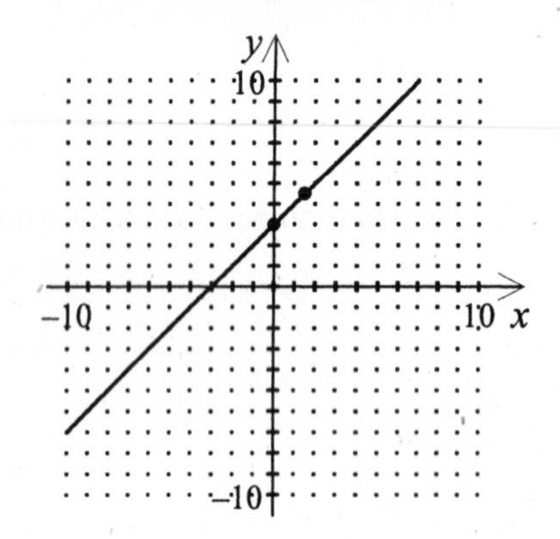

72. Use a determinant to find an equation of the line through the points listed below.

$\left(\frac{1}{2},\ -3\right),\ \left(\frac{3}{2},\ -\frac{3}{2}\right)$

Objective 4: Use matrices to code and decode messages

Assign a number to each letter of the alphabet and a blank space as shown by the table below:

0 = _	1 = A	2 = B	3 = C	4 = D	5 = E	6 = F	7 = G	8 = H
9 = I	10 = J	11 = K	12 = L	13 = M	14 = N	15 = O	16 = P	17 = Q
18 = R	19 = S	20 = T	21 = U	22 = V	23 = W	24 = X	25 = Y	26 = Z

73. Which is a cryptogram for "FEET ON THE GROUND" made using the matrix A?

$$A = \begin{bmatrix} -1 & 2 \\ 1 & 2 \end{bmatrix}$$

(A) –1 23 15 51 15 29 –14 29 –12 57 –5 9 11 49 6 73 –10 35

(B) –1 22 15 50 15 30 –14 28 –12 56 –5 10 11 50 6 72 –10 36

(C) 4 17 35 44 30 31 –14 13 –4 35 –5 4 29 44 27 56 –6 23

(D) 4 16 35 45 30 30 –14 14 –4 36 –5 5 29 43 27 57 –6 22

74. Which is a cryptogram for "CAN OF WORMS" made using the matrix A?

$$A = \begin{bmatrix} 2 & 1 & -1 \\ -1 & 1 & 0 \\ 0 & 2 & -1 \end{bmatrix}$$

(A) –7 –2 –12 9 15 24 8 23 31 30 –5 7

(B) 5 32 –17 –15 27 –6 –23 53 –15 23 69 –37

(C) 5 31 –17 –15 26 –6 –23 54 –15 23 68 –37

(D) –7 –1 –12 9 14 24 8 24 31 30 –6 7

75. Write a cryptogram for "FEW AND FAR BETWEEN" using the matrix A.

$$A = \begin{bmatrix} -1 & 0 & 1 \\ 1 & 0 & 0 \\ -1 & 2 & -1 \end{bmatrix}$$

76. Matrix A was used to encode the cryptogram below.

$$A = \begin{bmatrix} 0 & -1 & 1 \\ 1 & 1 & 0 \\ 1 & 0 & -1 \end{bmatrix}$$

8 –14 16 18 –1 11 35 20 –15 29 14 –15 5 –9 14

Use A^{-1} to decode the cryptogram.

Answer Key for Chapter 8 Matrices and Determinants

Section 8.1: Matrices and Systems of Equations

Objective 1: Write a matrix and identify its order

[1] (A)

[2] (D)

[3] 9×3

[4] 2×3

Objective 2: Perform elementary row operations on matrices

[5] (D)

[6] (B)

[7] $\begin{bmatrix} -9 & 38 & 23 \\ 8 & -7 & -4 \\ 4 & -9 & -6 \end{bmatrix}$

[8] $\begin{bmatrix} 1 & 7 & -7 \\ 0 & -25 & 32 \\ 0 & 6 & -9 \end{bmatrix}$

Objective 3: Use matrices and Gaussian elimination to solve systems of linear equations

[9] (C)

[10] (C)

[11] $(4, 5, 1)$

[12] $(-2086, -1071, -50)$

Objective 4: Use matrices and Gauss-Jordan elimination to solve systems of linear equations

[13] (C)

[14] (A)

[15] $(-69, -29, 8)$

[16] $(-5, 2, -1, 0)$

Section 8.2: Operations with Matrices

Objective 1: Decide whether two matrices are equal

[17] (A)

[18] (A)

[19] $x = 3, y = -3$

[20] $t = 7, x = -\frac{3}{2}, y = -\frac{1}{9}, z = -\frac{5}{16}$

Objective 2: Add and subtract matrices and multiply matrices by real numbers

[21] (D)

[22] (C)

[23] $\begin{bmatrix} 5 & 6 & 3 \\ 1 & -15 & -3 \\ -4 & 3 & -5 \end{bmatrix}$

[24] $\begin{bmatrix} -14 & -10 & 0 \\ -8 & -16 & -6 \\ -18 & -2 & 6 \end{bmatrix}$

Objective 3: Multiply two matrices

[25] (C)

[26] (A)

[27] $\begin{bmatrix} -120 & 14 \\ 20 & 49 \end{bmatrix}$

[28] $\begin{bmatrix} 8 & \frac{13}{3} & 17 \\ 12 & -\frac{49}{3} & -7 \\ 10 & \frac{25}{3} & 2 \end{bmatrix}$

Objective 4: Use matrix operations to model and solve real-life problems

[29] (B)

[30] (C)

[31] $\begin{bmatrix} 11 & 14 & 4 & 15 & 5 \\ 8 & 14 & 9 & 16 & 8 \end{bmatrix} \begin{bmatrix} \$4 \\ \$2 \\ \$8 \\ \$3 \\ \$2 \end{bmatrix} = \begin{bmatrix} \$159 \\ \$196 \end{bmatrix}$

The entries in AB are the costs to purchase all of the supplies needed by the two groups. The art department needs $159 and the homecoming committee needs $196.

[32] $\frac{5}{9}$

Section 8.3: The Inverse of a Square Matrix

Objective 1: Verify that two matrices are inverses of each other

[33] (A)

[34] (D)

[35] Student must show that $AB = I$ and $BA = I$.

$$\begin{bmatrix} 1 & -2 & 0 \\ 19 & -35 & 6 \\ 4 & -8 & 1 \end{bmatrix}\begin{bmatrix} \frac{13}{3} & \frac{2}{3} & -4 \\ \frac{5}{3} & \frac{1}{3} & -2 \\ -4 & 0 & 1 \end{bmatrix} = \begin{bmatrix} 1 & 0 & 0 \\ 0 & 1 & 0 \\ 0 & 0 & 1 \end{bmatrix} \text{ and } \begin{bmatrix} \frac{13}{3} & \frac{2}{3} & -4 \\ \frac{5}{3} & \frac{1}{3} & -2 \\ -4 & 0 & 1 \end{bmatrix}\begin{bmatrix} 1 & -2 & 0 \\ 19 & -35 & 6 \\ 4 & -8 & 1 \end{bmatrix} = \begin{bmatrix} 1 & 0 & 0 \\ 0 & 1 & 0 \\ 0 & 0 & 1 \end{bmatrix}$$

[36] Student must show that $AB = I$ and $BA = I$.

$$\begin{bmatrix} 9 & -2 \\ 8 & 3 \end{bmatrix}\begin{bmatrix} \frac{3}{43} & \frac{2}{43} \\ -\frac{8}{43} & \frac{9}{43} \end{bmatrix} = \begin{bmatrix} 1 & 0 \\ 0 & 1 \end{bmatrix} \text{ and } \begin{bmatrix} \frac{3}{43} & \frac{2}{43} \\ -\frac{8}{43} & \frac{9}{43} \end{bmatrix}\begin{bmatrix} 9 & -2 \\ 8 & 3 \end{bmatrix} = \begin{bmatrix} 1 & 0 \\ 0 & 1 \end{bmatrix}$$

Objective 2: Use Gauss-Jordan elimination to find the inverses of matrices

[37] (B)

[38] (A)

[39]
$$\begin{bmatrix} -\frac{11}{83} & \frac{4}{83} & \frac{10}{83} \\ \frac{14}{83} & \frac{10}{83} & \frac{25}{83} \\ -\frac{15}{83} & \frac{13}{83} & -\frac{9}{83} \end{bmatrix}$$

[40]
$$\begin{bmatrix} 1 & 2 & 0 \\ 3 & 5 & -2 \\ -1 & -2 & 1 \end{bmatrix}$$

Objective 3: Use a formula to find the inverses of 2 by 2 matrices

[41] (B)

[42] (C)

[43]
$$\begin{bmatrix} \frac{1}{14} & \frac{2}{7} \\ -\frac{2}{7} & -\frac{1}{7} \end{bmatrix}$$

[44]
$$\begin{bmatrix} -5 & 1 \\ -1 & 0 \end{bmatrix}$$

Objective 4: Use inverse matrices to solve systems of linear equations

[45] (D)

[46] (C)

[47] $(5,\ 3,\ -1)$

[48] $(-3,\ 4)$

Section 8.4: The Determinant of a Square Matrix

Objective 1: Find the determinants of 2 by 2 matrices

[49] (B)

[50] (D)

[51] –27

[52] $-\frac{1}{18}$

Objective 2: Find minors and cofactors of square matrices

[53] (D)

[54] (B)

[55] 54

[56] 21

Objective 3: Find the determinants of square matrices

[57] (C)

[58] (B)

[59] -37

[60] 672

Section 8.5: Applications of Matrices and Determinants

Objective 1: Use Cramer's Rule to solve systems of linear equations

[61] (B)

[62] (C)

[63] $(3, 1)$

[64] $(-4, 2, -5)$

Objective 2: Use determinants to find the areas of triangles

[65] (B)

[66] (A)

[67] 14.875

[68] 55.625

Objective 3: Use a determinant to test for collinear points and find an equation of a line passing through two points

[69] (A)

[70] (D)

[71] $y = x + 3$

[72] $6x - 4y = 15$

Objective 4: Use matrices to code and decode messages

[73] (B)

[74] (B)

[75] –24 46 –17 –13 28 –14 –10 12 –2 17 0 1 –17 40 –18 –23 10 18 –14 0 14

[76] SECOND TO NONE

Chapter 9 Sequences, Series, and Probability

Section 9.1: Sequences and Series

Objective 1: Use sequence notation to write the terms of a sequence

Write the first five terms of the sequence. Assume that n begins with 1.

1. $a_n = 5\left(\frac{1}{2}\right)^n$

(A) $5, \frac{7}{4}, \frac{9}{8}, \frac{11}{12}, \frac{13}{16}$

(B) $\frac{5}{2}, \frac{5}{4}, \frac{5}{8}, \frac{5}{16}, \frac{5}{32}$

(C) $5, \frac{5}{2}, \frac{5}{4}, \frac{5}{8}, \frac{5}{16}$

(D) $5, \frac{5}{2}, \frac{10}{3}, 5, 8$

2. $a_n = \frac{(2n)!}{(n+1)!}$

(A) $1, \frac{1}{2}, \frac{1}{6}, \frac{1}{24}, \frac{1}{120}$

(B) $\frac{1}{2}, \frac{2}{3}, \frac{3}{8}, \frac{2}{15}, \frac{5}{144}$

(C) $1, \frac{1}{2}, \frac{2}{3}, \frac{3}{2}, \frac{24}{5}$

(D) $1, 4, 30, 336, 5040$

3. Find the indicated term of the sequence.

$a_n = \frac{n(n+3)}{3}$

$a_{21} = \square$

4. Write the first five terms of the sequence. Assume that n begins with 1.

$a_n = -8n + 31$

Objective 2: Use factorial notation

Simplify the ratio of factorials.

5. $\frac{7!}{3!(7-3)!}$

(A) 210 (B) 5040 (C) 35 (D) 6

Simplify the ratio of factorials.

6. $\dfrac{(n+4)!}{(n+3)!}$ (A) $(n+4)$ (B) $(n+4)(n+3)$ (C) $(n+3)$ (D) $(n+5)(n+4)$

7. $\dfrac{8!}{3!(8-3)!}$

8. $\dfrac{(n+3)!}{(n+2)!}$

Objective 3: Use summation notation to write sums

Use sigma notation to write the sum.

9. $-5.4-6.8-8.2-9.6-11-12.4$

(A) $\sum_{i=0}^{5}(-5.4i-1.4)$ (B) $\sum_{i=1}^{6}(-5.4i-1.4)$ (C) $\sum_{i=0}^{5}(-1.4i-5.4)$ (D) $\sum_{i=1}^{6}(-1.4i-5.4)$

10. $7+13+19+25+31+37$

(A) $\sum_{i=1}^{6}(7+6(i-1))$ (B) $\sum_{i=0}^{6}(7+6i)$ (C) $\sum_{i=0}^{5}(7+6(i-1))$ (D) $\sum_{i=1}^{6}(7+6i)$

11. For the sequence defined by a_n, use sigma notation to write the sum of the first six terms of the sequence.

$a_n = \dfrac{n^2}{2}$

12. Use sigma notation to write the sum.

$-\dfrac{x^5}{9}+\dfrac{x^6}{10}-\dfrac{x^7}{11}+\dfrac{x^8}{12}-\dfrac{x^9}{13}$

Objective 4: Find the sum of an infinite series

Find the sum of the infinite series.

13. $\sum_{i=1}^{\infty} 2\left(\frac{1}{5}\right)^i$ (A) $\frac{20}{3}$ (B) $\frac{1}{3}$ (C) $\frac{1}{2}$ (D) $\frac{5}{3}$

14. $\sum_{k=1}^{\infty} 3\left(\frac{1}{6}\right)^k$ (A) $\frac{18}{7}$ (B) $\frac{3}{7}$ (C) 12 (D) $\frac{3}{5}$

15. $\sum_{k=1}^{\infty} 7\left(\frac{2}{7}\right)^k$

16. $\sum_{i=2}^{\infty} \left(\frac{5}{3^i} + \frac{2}{4^i}\right)$

Objective 5: Use sequences and series to model and solve real-life problems

17. The average price of a loaf of bread n years after 1950 is approximated by the sequence

$a_n = 0.03n + 0.14$.

Use the sequence to predict the price of a loaf of bread in 2003.

(A) \$1.73 (B) \$1.70 (C) \$1.67 (D) \$1.76

18. The daily growth g of a colony of a certain bacteria growing in a harsh environment can be approximated by the function

$g(k) = 8k^2 - 4k$

where k is the number of bacteria placed in that environment at the beginning of day 1. If the bacteria colony started with 15 bacteria on day 1, approximate the population of the colony after 2 days.

(A) 225 bacteria (B) 3735 bacteria (C) 63 bacteria (D) 3495 bacteria

19. A formal garden has shrubs planted inside a grid consisting of 18 rows. The first row contains 17 shrubs, the second contains 19 shrubs, the third 21, and so on. Write an equation expressing the number of shrubs in a row as a function of the number of the row. Find the number of shrubs in the ninth row.

20. Raul has a stack of 5 index cards. With a pair of scissors, he cuts the stack in half and then places all the resulting card pieces in one stack. If he does this a total of four times, how many card pieces will he have?

Section 9.2: Arithmetic Sequences and Partial Sums

Objective 1: Recognize and write arithmetic sequences

Find a formula for a_n for the arithmetic sequence.

21. $a_1 = 4,\ a_6 = 15$

(A) $a_n = \frac{11}{5}(n-1)+4$ (B) $a_n = 3n-15$ (C) $a_n = \frac{11}{6}(n-1)-15$ (D) $a_n = \frac{19}{5}n+4$

22. $a_1 = 14,\ a_{15} = 21$

(A) $a_n = \frac{3}{2}n+21$ (B) $a_n = \frac{5}{2}n+14$ (C) $a_n = \frac{1}{2}(n-1)+14$ (D) $a_n = \frac{7}{15}(n-1)+21$

23. Determine whether the sequence is arithmetic. If it is, find the common difference.
$\frac{4}{3},\ 3,\ \frac{14}{3},\ \frac{19}{3},\ 8, \ldots$

24. Write the first five terms of the arithmetic sequence. Find the common difference and the *n*th term of the sequence as a function of *n*.
$a_1 = -5;\ a_{k+1} = a_k + 3$

Objective 2: Find an nth partial sum of an arithmetic sequence

25. Find the partial sum. (A) –88 (B) –132 (C) –99 (D) –5

$$\sum_{n=1}^{11}(-2n+4)$$

Find the indicated *n*th partial sum of the arithmetic sequence.

26. $-1,\ 13,\ 27,\ 41,\ \ldots,\ n=20$ (A) 2780 (B) 2220 (C) 3340 (D) 2640

27. $3,\ 9,\ 15,\ 21,\ \ldots,\ n=22$

28. Find the partial sum.

$$\sum_{n=1}^{21}(2n-8)$$

Objective 3: Use arithmetic sequences to model and solve real-life problems

29. In 1992, the average cost of a ticket on a privately-owned airline was \$139. This amount has increased by approximately \$72 yearly. How much should you expect to pay for a ticket on this airline in the year 2009?

(A) \$1224 (B) \$1346 (C) \$1291 (D) \$1363

30. A large asteroid crashed into a moon of another planet, causing several boulders from the moon to be propelled into space toward the planet. Astronomers were able to measure the speed of one of the projectiles. The distance (in feet) that the projectile traveled each second, starting with the first second, was given by the arithmetic progression 22, 66, 110, 154, ... Find the distance that the projectile traveled in the 8th second.

(A) 330 ft (B) 286 ft (C) 352 ft (D) 374 ft

31. The population of a city in 1991 was 754,033. The population of the same city in 1996 was 821,628. Assuming that the change in the number of people is constant over five-year periods, find a formula for the population of the city n five-year periods after 1991 and predict the population of the city in 2006.

32. At a local grocery store, tins of mushrooms are stacked in a triangular formation for display. Each new row has 1 fewer tin than the row beneath it. If there are 8 rows and the top row contains one tin, how many tins of mushrooms are in the display?

Section 9.3: Geometric Sequences and Series

Objective 1: Recognize and write geometric sequences

33. Write the nth term of the sequence as a function of n.

$$\frac{5}{3}, \frac{10}{9}, \frac{20}{27}, \frac{40}{81}, \ldots$$

(A) $a_n = \frac{5}{3}\left(\frac{2}{3}\right)^{n-1}$ (B) $a_n = \frac{5}{3}(2)^{n-1}$ (C) $a_n = \frac{5}{3}\left(\frac{2}{3}\right)^{n+1}$ (D) $a_n = \frac{2}{5}(2)^{n}$

34. Identify the *n*th term of the geometric sequence.

$a_1 = -\frac{1}{3}, r = \frac{1}{5}; n = 7$

(A) $-\frac{1}{9375}$ (B) $-\frac{1}{1{,}171{,}875}$ (C) $-\frac{1}{46{,}875}$ (D) $-\frac{1}{234{,}375}$

35. Determine whether the sequence is geometric. If it is, find the common ratio.

$\frac{3}{4}, \frac{3}{16}, \frac{3}{64}, \frac{3}{256}, \frac{3}{1024}, \ldots$

36. Write the first five terms of the geometric sequence. Determine the common ratio and write the *n*th term of the sequence as a function of *n*.

$a_1 = -4.75, a_{k+1} = 0.95a_k$

Objective 2: Find the sum of a geometric sequence

37. Find the sum of the finite geometric series to three decimal places.

$\sum_{n=3}^{8} 7(0.333)^n$

(A) 0.018 (B) 0.388 (C) 0.5 (D) 0.055

38. Find the sum of the first 7 terms of the geometric series.

$-10 - 2 - \frac{2}{5} - \frac{2}{25} - \ldots$

(A) –13.19 (B) –12.5 (C) –21.28 (D) –52

39. Find the sum of the finite geometric series to three decimal places.

$\sum_{n=0}^{9} \left(\frac{1}{3}\right)^n$

40. Find the sum of the finite geometric series.

$\sum_{n=1}^{6} 4\left(-\frac{1}{3}\right)^{n-1}$

Objective 3: Find the sum of an infinite geometric series

Find the sum of the infinite geometric series.

41. $0.6 + 0.06 + 0.006 + \ldots$ (A) $\frac{2}{3}$ (B) 0.666 (C) 0.018 (D) $\frac{3}{5}$

42. $\sum_{n=1}^{\infty} -4\left(\frac{3}{4}\right)^{n-1}$ (A) −20 (B) 16 (C) −16 (D) Undefined

43. $-3 - \frac{3}{2} - \frac{3}{4} - \frac{3}{8} - \frac{3}{16} - \ldots$

44. $\sum_{n=1}^{\infty} 3\left(-\frac{5}{6}\right)^{n-1}$

Objective 4: Use geometric sequences to model and solve real-life problems

45. A woman made $20,000 during the first year of her new job at city hall. Each year she received a 10% raise. Find her total earnings during the first eight years on the job.

(A) $189,743.42 (B) $229,189.35 (C) $271,589.54 (D) $228,717.76

46. A rubber ball dropped on a hard surface takes a sequence of bounces, each one $\frac{4}{5}$ as high as the preceding one. If this ball is dropped from a height of 15 feet, how far will it have traveled when it hits the surface the fifth time?

(A) $50\frac{53}{125}$ ft (B) $100\frac{106}{125}$ ft (C) $184\frac{19}{125}$ ft (D) $85\frac{106}{125}$ ft

47. In a financial arrangement, you are promised $900 the first day and each day after that you will receive 75% of the previous day's amount. When one day's amount drops below $1, you stop getting paid from that day on. What day is the first day you would receive no payment and what is your total income? Use a formula for the *n*th partial sum of a geometric sequence.

48. There are initially 2500 bacteria in a culture. A scientist is testing a new antibiotic which causes the number of bacteria to halve each hour. The number of bacteria after t hours can be found using the formula

$$N = 2500\left(\frac{1}{2}\right)^t$$

How many bacteria will be present after 4 hours? Round your answer to the nearest whole number.

Section 9.4: Mathematical Induction

Objective 1: Use mathematical induction to prove a statement

49. Find P_{k+1} for the given P_k.

$P_k = k^2$

(A) P_{k+1} (B) $(k+1)^2$ (C) $k(2k-1)$ (D) $k^2 + k$

50. An incomplete mathematical induction proof for all integers n with the given values is shown below. Determine the next step in the proof.

$2^{n+1} > (n+1)^2,\ n \geq 4$

Proof:

1. For $n = 4,\ 2^{4+1} = 2^5 = 32 > 25 = 5^2$
2. Suppose $2^{k+1} > (k+1)^2$
3. Then $2^{[(k+1)+1]} = 2 \cdot 2^{k+1} > ?$

(A) $[(k+1)+1]^2$ (B) 2^k (C) $(n+1)^2$ (D) $2(k+1)^2$

Use mathematical induction to prove the formula or property for all integers n with the given values.

51. $11 + 15 + 19 + \ldots + (4n+7) = 2n^2 + 9n,\ n \geq 1$

52. $\frac{1}{3} + \frac{1}{15} + \frac{1}{35} + \ldots + \frac{1}{(2n-1)(2n+1)} = \frac{n}{2n+1},\ n \geq 1$

Objective 2: Find the sums of powers of integers

Find the sum using the formulas for the sums of powers of integers.

53. $\sum_{n=1}^{8}\left(n^2+n\right)$ (A) 72 (B) 240 (C) 64 (D) 1296

54. $\sum_{n=1}^{15}\left(-2-n^3+2n\right)$ (A) $-50{,}625$ (B) $-14{,}190$ (C) 3389 (D) $-14{,}414$

55. $\sum_{n=1}^{11}\left(n^2-n\right)$

56. $\sum_{n=1}^{6}\left(-3n^3+2n^2-\frac{3}{2}\right)$

Objective 3: Recognize patterns and write the nth term of a sequence

57. Select the formula that represents the nth term of the sequence as a function of n.

$\frac{9}{2}, \frac{81}{10}, \frac{729}{50}, \frac{6561}{250}, \ldots$

(A) $a_n=\frac{9}{2}\left(\frac{9}{5}\right)^{n-1}$ (B) $a_n=\frac{9}{2}\left(\frac{9}{5}\right)^{n+1}$ (C) $a_n=\frac{2}{5}\left(\frac{9}{2}\right)^{n}$ (D) $a_n=\frac{9}{2}\left(\frac{9}{2}\right)^{n-1}$

Write the first five terms of the sequence.

58. $a_0=-1$

$a_1=-4$

$a_n=-a_{n-1}-2a_{n-2}$

(A) $-1, -4, 6, 2, -14$ (B) $-1, -4, 10, 6, -10$ (C) $-1, -4, 1, 7, -19$ (D) $-1, -4, 5, 0, -18$

59. $a_1=\frac{1}{4}$

$a_n=-a_{n-1}\left(\frac{n-1}{n}\right)$

60. Write the nth term of the sequence as a function of n.

$$-\frac{1}{4}, \frac{1}{8}, -\frac{1}{12}, \frac{1}{16}, \ldots$$

Objective 4: Find finite differences of a sequence

61. Find the first five terms and the first and second differences of the sequence.

$a_1 = -4$

$a_n = a_{n-1} - 3$

(A) $-4, -7, -10, -13, -16$
First differences: $-3, -3, -3, -3$
Second differences: $0, 0, 0$

(B) $-4, -8, -12, -16, -20$
First differences: $-4, -4, -4, -4$
Second differences: $0, 0, 0$

(C) $-4, -7, -10, -13, -16$
First differences: $-12, -9, -6, -3$
Second differences: $3, 3, 3$

(D) $-4, -8, -12, -16, -20$
First differences: $-16, -12, -8, -4$
Second differences: $4, 4, 4$

62. Find the first five terms and the first and second differences of the sequence. Determine whether the sequence has a linear model, a quadratic model, or neither.

$a_1 = 4;\ a_n = a_{n-1} - 3$

(A) $4, 0, -4, -8, -12$
First differences: $-4, -4, -4, -4$
Second differences: $0, 0, 0$
Linear

(B) $4, 0, -4, -8, -12$
First differences: $-12, -8, -4, 0$
Second differences: $4, 4, 4$
Quadratic

(C) $4, 1, -2, -5, -8$
First differences: $-3, -3, -3, -3$
Second differences: $0, 0, 0$
Linear

(D) $4, 1, -2, -5, -8$
First differences: $-9, -6, -3, 0$
Second differences: $3, 3, 3$
Quadratic

63. Find the first five terms and the first and second differences of the sequence.

$a_1 = 2;\ a_n = -4(a_{n-1})$

64. The first five terms of a sequence are shown below. Find the first and second differences of the sequence. Determine whether the sequence has a linear model, a quadratic model, or neither.

n	0	1	2	3	4	5
a_n	3	4	19	60	139	268

Section 9.5: The Binomial Theorem

Objective 1: Use the Binomial Theorem to calculate binomial coefficients

65. Use the Binomial Theorem to find the binomial coefficient.

$$\binom{8}{5}$$

(A) 40 (B) 6720 (C) 336 (D) 56

66. Solve the equation for n: (A) 6 (B) 7 (C) 8 (D) 11

$$3\left({}_{n}C_{6}\right) = {}_{n+1}C_{6}$$

67. Use the Binomial Theorem to find the binomial coefficient.

$$\binom{4}{3}$$

68. Solve the equation for n:

$$2\left({}_{n}C_{4}\right) = {}_{n+1}C_{5}$$

Objective 2: Use Pascal's Triangle to calculate binomial coefficients

69. Use Pascal's Triangle to find the binomial coefficient.

$$\binom{14}{11}$$

(A) 364 (B) 242 (C) 14,529,715,200 (D) 2184

Use Pascal's Triangle to find the binomial coefficients.

70. ${}_{11}C_{5}$ and ${}_{11}C_{6}$ (A) 330; 462 (B) 210; 252 (C) 462; 462 (D) 495; 792

71. $\binom{9}{6}$ and $\binom{9}{7}$

72. Use Pascal's Triangle to find the binomial coefficient.

$${}_{13}C_{12}$$

Objective 3: Use binomial coefficients to write binomial expansions

73. Expand and simplify the binomial using Pascal's Triangle to determine the coefficients.
$(a-2b)^3$

(A) $a^3+3a^2b+3ab^2+b^3$
(B) $a^3-6a^2b+12ab^2-8b^3$
(C) $a^3-4a^2b+12ab^2-8b^3$
(D) $a^3-3a^2b-3ab^2-b^3$

Use the Binomial Theorem to expand and simplify the expression.

74. $(v-r)^5$

(A) $v^5-5v^4r-20v^3r^2+60v^2r^3+120vr^4-120r^5$
(B) $v^5-5v^4r+10v^3r^2-10v^2r^3+5vr^4-r^5$
(C) v^5-r^5
(D) $v^5-7v^4r+12v^3r^2-12v^2r^3+7vr^4-r^5$

75. $(x-y)^8$

76. $(g+3h)^3$

Section 9.6: Counting Principles

Objective 1: Solve simple counting problems

77. Suppose you are choosing a wall color from among 3 different paint colors and an accent color from among 5 different paint colors. How many different wall color and accent color combinations are possible?

(A) 15
(B) 14
(C) 8
(D) 18

78. Two friends are partners in a game where, in each round, each member of a team can score from 0 through 14 points. How many ways can the two friends score a total of 14 points in round 4?

(A) 56
(B) 18
(C) 13
(D) 15

79. Don must choose 3 colored toothpicks, one by one, out of a box of 3 toothpicks where each of the following colors are represented: red, yellow, and green.
How many ways can Don pull the toothpicks out of the container if the second toothpick is yellow?

80. Two twelve-sided dice are rolled. If double-number rolls can only be counted once, how many ways are there to arrive at a total of 11?

Objective 2: Use the Fundamental Counting Principal to solve counting problems

81. Account numbers for Century Oil Company consist of eleven digits. If the first digit cannot be a 0, how many account numbers are possible?

(A) 80,000,000,000 (B) 90,000,000,000 (C) 800,000,000,000 (D) 10,000,000,000

82. A custom car dealership offers a variety of options for the interior of a certain car. You can choose from leather, cloth, or vinyl interior in one of 3 different colors. How many choices are there for interior type and color?

(A) 6 (B) 9 (C) 27 (D) 11

83. A cafe serves a variety of stuffed potatoes. You can choose from russet, yellow, or white potatoes with any one of 7 different fillings. How many different varieties of stuffed potatoes can you choose from?

84. A professional baseball coach is scouting players in the minor leagues to pull up to the major leagues. He must find one player each to play shortstop, first base, and center field. He can choose from 4 shortstops, 4 first basemen, and 4 center fielders. How many ways can the coach choose the players he needs?

Objective 3: Use permutations to solve counting problems

85. In how many different ways can two red cards be drawn from a standard deck of cards? (Note: Order of selection is important.)

(A) 2652 (B) 650 (C) 312 (D) 156

86. How many different arrangements can be made using all of the letters in the word MATH?

(A) 4 (B) 24 (C) 16 (D) 36

87. How many different ways can 10 different runners finish in first, second, and third places in a race?

88. How many distinguishable permutations can be made with the letters in the word COMMITTEE?

Objective 4: Use combinations to solve counting problems

89. A college has nine instructors qualified to teach a special computer lab course which requires two instructors to be present. How many different pairs of teachers could teach the class?

(A) 18 (B) 56 (C) 72 (D) 36

90. Two cards are drawn, without replacement, from a standard deck of 52 cards. How many sets of two cards are possible?

(A) 786 (B) 2386 (C) 1326 (D) 296

91. Laura, Rachel, Leila, Hiro, Barry, and Jane are in the math club. The club advisor will assign students to 5-person teams at the next Math Team competition. How many different 5-person teams can be formed from these six students?

92. Curtiss is at a basketball game. The refreshment stand at the basketball game sells hot dogs, tacos, fries, cheeseburgers, hamburgers, popcorn, and pizza to eat and lemonade, coffee, cola, and root beer to drink. If he must order 5 items to eat and 2 items to drink, without ordering more than 1 of any particular item, how many ways can Curtiss order these items?

Section 9.7: Probability

Objective 1: Find the probabilities of events

93. Select the probability that you will encounter only one green light out of five traffic signal lights. Assume red, green, and yellow are equally likely occurrences.

(A) $\frac{4}{243}$ (B) $\frac{80}{243}$ (C) $\frac{1}{5}$ (D) $\frac{1}{243}$

94. In a certain lottery, a participant must choose six numbers from 1 to 50. If these six numbers match the six numbers drawn by the lottery organizers, including the order in which the numbers were drawn, the participant wins or shares the lottery jackpot. What is the probability of winning the lottery if you only purchase one ticket?

(A) $\frac{1}{15{,}890{,}700}$ (B) $\frac{1}{115{,}775{,}100}$ (C) $\frac{1}{11{,}441{,}304{,}000}$ (D) $\frac{1}{228{,}826{,}080}$

95. A box contains 3 green, 2 yellow, and 6 purple balls. Find the probability of obtaining a purple ball in a single random draw.

96. On a spin of the spinner below, find the probability of getting either a lowercase letter or a vowel.

Objective 2: Find the probabilities of mutually exclusive events

97. Two spinners are each divided into four equal sections. The sections on each spinner are colored red, blue, green, and orange. The experiment is to spin both spinners at the same time and record the colors. Use the list of events to determine if D and E are mutually exclusive. If the events are mutually exclusive, find the probability of D or E occurring.

Events
A: Both spinners land on red.
B: Neither spinner lands on green.
C: Exactly one spinner lands on blue.
D: One spinner lands on red and the other lands on orange.
E: Neither spinner lands on orange or blue.

(A) Mutually exclusive; $\frac{7}{16}$

(B) Mutually exclusive; $\frac{3}{8}$

(C) Mutually exclusive; $\frac{5}{8}$

(D) Not mutually exclusive

98. Suppose you mix up the cards below and choose one without looking. What is the probability of selecting neither L nor A?

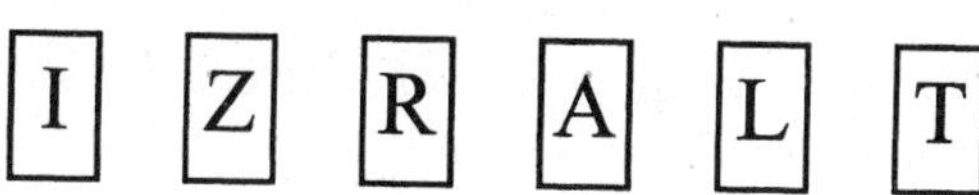

(A) $\frac{2}{3}$

(B) $\frac{1}{3}$

(C) 2

(D) 1

99. A spinner is numbered from 1 through 9 with each number equally likely to occur. What is the probability of obtaining a number less than 3 or greater than 6 with a single spin?

100. Suppose two fair dice are rolled. What is the probability that a sum of 2 or 9 turns up?

Objective 3: Find the probabilities of independent events

101. A drawer contains 6 red socks, 4 white socks, and 8 blue socks. Without looking, you draw out a sock, return it, and draw out a second sock. What is the probability that the first sock is blue and the second sock is white?

(A) $\frac{8}{81}$ (B) $\frac{7}{12}$ (C) $\frac{7}{18}$ (D) $\frac{1}{324}$

102. A coin is tossed and a die is rolled. What is the probability that the coin shows heads and the die shows 2?

(A) $\frac{1}{4}$ (B) $\frac{1}{12}$ (C) $\frac{2}{3}$ (D) $\frac{1}{6}$

103. What is the probability of randomly drawing a card with the number 3 on it and the spinner landing on the number 2? Assume that the sections of the spinner are equal in size.

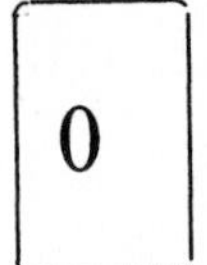

1 2 3 4 5 6 7

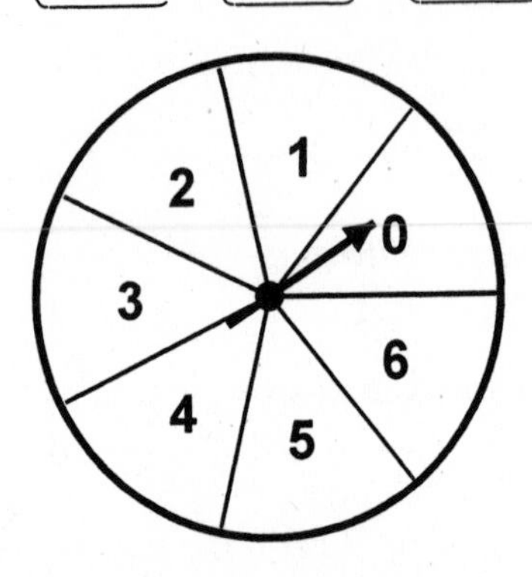

104. Two urns each contain blue balls and white balls. The first urn contains 5 blue balls and 3 white balls, and the second urn contains 3 blue balls and 6 white balls. A ball is drawn randomly from each urn. What is the probability that both balls are blue?

Objective 4: Find the probability of the complement of an event

105. You work at a T-shirt printing business. Of 4400 T-shirts shipped, 396 are printed improperly. If you randomly choose a T-shirt out of the shipment, what is the chance that it is printed correctly?

(A) 9% (B) 91% (C) 3.96% (D) 4.96%

106. Twelve balls numbered from 1 to 12 are placed in an urn. One ball is selected at random. Find the probability that it is *not* number 6.

(A) $\frac{1}{2}$ (B) $\frac{11}{12}$ (C) $\frac{1}{12}$ (D) $\frac{5}{6}$

107. The probability of getting an A in Mr. Allen's class in any semester is 27%. What is the probability of *not* getting an A?

108. Statistics from the past ten years predict that the chance of *not* having rain on any single day in March in Springfield is 65%. If Ja-Yeon's birthday is March 17, what is the chance of it raining on Ja-Yeon's birthday?

Answer Key for Chapter 9 Sequences, Series, and Probability

Section 9.1: Sequences and Series

Objective 1: Use sequence notation to write the terms of a sequence

[1] (B)

[2] (D)

[3] 168

[4] 23, 15, 7, -1, -9

Objective 2: Use factorial notation

[5] (C)

[6] (A)

[7] 56

[8] $(n+3)$

Objective 3: Use summation notation to write sums

[9] (C)

[10] (A)

[11] Answers may vary. Sample answer: $\sum_{i=1}^{6} \frac{i^2}{2}$

[12] $\sum_{i=5}^{9} \frac{(-1)^i x^i}{i+4}$

Objective 4: Find the sum of an infinite series

[13] (C)

[14] (D)

[15] $\frac{14}{5}$

[16] 1

Objective 5: Use sequences and series to model and solve real-life problems

[17] (A)

[18] (D)

[19] Answers may vary. Sample answer: $a_n = 17 + 2(n-1)$; 33

[20] 80

Section 9.2: Arithmetic Sequences and Partial Sums

Objective 1: Recognize and write arithmetic sequences

[21] (A)

[22] (C)

[23] Arithmetic sequence, $d = \frac{5}{3}$

[24] $-5, -2, 1, 4, 7$; $d = 3$; $a_n = 3n - 8$

Objective 2: Find an nth partial sum of an arithmetic sequence

[25] (A)

[26] (D)

[27] 1452

[28] 294

Objective 3: Use arithmetic sequences to model and solve real-life problems

[29] (D)

[30] (A)

[31] $P(n) = 754{,}033 + 67{,}595n$; 956,818

[32] 36

Section 9.3: Geometric Sequences and Series

Objective 1: Recognize and write geometric sequences

[33] (A)

[34] (C)

[35] Geometric sequence, $r = \frac{1}{4}$

[36] $-4.75, -4.5125, -4.28688, -4.07253, -3.8689$; $r = 0.95$; $a_n = -4.75(0.95)^{n-1}$

Objective 2: Find the sum of a geometric sequence

[37] (B)

[38] (B)

[39] 1.5

[40] $4-\frac{4}{3}+\frac{4}{9}-\frac{4}{27}+\frac{4}{81}-\frac{4}{243}=\frac{728}{243}$

Objective 3: Find the sum of an infinite geometric series

[41] (A)

[42] (C)

[43] -6

[44] $\frac{18}{11}$

Objective 4: Use geometric sequences to model and solve real-life problems

[45] (D)

[46] (D)

[47] 25th day; $3596.39 total income

[48] 156

Section 9.4: Mathematical Induction

Objective 1: Use mathematical induction to prove a statement

[49] (B)

[50] (D)

1. For $n=1$, $4\cdot 1+7=11$ and $2\cdot 1^2+9\cdot 1=11$

2. Suppose $11+15+19+...+(4k+7)=2k^2+9k$

3. Then, $11+15+19+...+(4k+7)+4(k+1)+7=2k^2+9k+4(k+1)+7$

$$=2k^2+13k+11$$
$$=2k^2+4k+2+9k+9$$
$$=2(k^2+2k+1)+9k+9$$
$$=2(k+1)^2+9(k+1)$$

[51]

Answers may vary. Sample answer:

1. For $n=1$, $\dfrac{1}{(2\cdot 1-1)(2\cdot 1+1)}=\dfrac{1}{(2-1)(2+1)}=\dfrac{1}{1\cdot 3}=\dfrac{1}{3}=\dfrac{1}{2+1}=\dfrac{1}{2\cdot 1+1}$

2. Suppose $\dfrac{1}{3}+\dfrac{1}{15}+\dfrac{1}{35}+...+\dfrac{1}{(2k-1)(2k+1)}=\dfrac{k}{2k+1}$

3. Then $\dfrac{1}{3}+\dfrac{1}{15}+\dfrac{1}{35}+...+\dfrac{1}{(2k-1)(2k+1)}+\dfrac{1}{[2(k+1)-1][2(k+1)+1]}$

$$=\frac{k}{2k+1}+\frac{1}{[2(k+1)-1][2(k+1)+1]}$$
$$=\frac{k}{2k+1}+\frac{1}{(2k+1)(2k+3)}$$
$$=\frac{k(2k+3)}{(2k+1)(2k+3)}+\frac{1}{(2k+1)(2k+3)}$$
$$=\frac{k(2k+3)+1}{(2k+1)(2k+3)}$$
$$=\frac{2k^2+3k+1}{(2k+1)(2k+3)}$$
$$=\frac{(2k+1)(k+1)}{(2k+1)(2k+3)}$$
$$=\frac{k+1}{2k+3}$$
$$=\frac{k+1}{2(k+1)+1}$$

[52]

Objective 2: Find the sums of powers of integers

[53] (B)

[54] (B)

[55] 440

[56] –1150

Objective 3: Recognize patterns and write the nth term of a sequence

[57] (A)

[58] (A)

[59] $\frac{1}{4}, -\frac{1}{8}, \frac{1}{12}, -\frac{1}{16}, \frac{1}{20}$

[60] $a_n = \frac{(-1)^n}{4n}$

Objective 4: Find finite differences of a sequence

[61] (A)

[62] (C)

[63] 2, –8, 32, –128, 512
First differences: –10, 40, –160, 640
Second differences: 50, –200, 800

n	a_n	First Differences	Second Differences
0	3		
		1	
1	4		14
		15	
2	19		26
		41	
3	60		38
		79	
4	139		50
		129	
5	268		

[64] Neither

Section 9.5: The Binomial Theorem

Objective 1: Use the Binomial Theorem to calculate binomial coefficients

[65] (D)

[66] (C)

[67] 4

[68] 9

Objective 2: Use Pascal's Triangle to calculate binomial coefficients

[69] (A)

[70] (C)

[71] 84; 36

[72] 13

Objective 3: Use binomial coefficients to write binomial expansions

[73] (B)

[74] (B)

[75] $x^8 - 8x^7y + 28x^6y^2 - 56x^5y^3 + 70x^4y^4 - 56x^3y^5 + 28x^2y^6 - 8xy^7 + y^8$

[76] $g^3 + 9g^2h + 27gh^2 + 27h^3$

Section 9.6: Counting Principles

Objective 1: Solve simple counting problems

[77] (A)

[78] (D)

[79] 2

[80] 10

Objective 2: Use the Fundamental Counting Principal to solve counting problems

[81] (B)

[82] (B)

[83] 21

[84] 64

Objective 3: Use permutations to solve counting problems

[85] (B)

[86] (B)

[87] 720

[88] 45,360

Objective 4: Use combinations to solve counting problems

[89] (D)

[90] (C)

[91] 6

[92] 126

Section 9.7: Probability

Objective 1: Find the probabilities of events

[93] (B)

[94] (C)

[95] $\frac{6}{11}$

[96] $\frac{7}{10}$

Objective 2: Find the probabilities of mutually exclusive events

[97] (B)

[98] (A)

[99] $\frac{5}{9}$

[100] $\frac{5}{36}$

Objective 3: Find the probabilities of independent events

[101] (A)

[102] (B)

[103] $\frac{1}{56}$

[104] $\frac{5}{24}$

Objective 4: Find the probability of the complement of an event

[105] (B)

[106] (B)

[107] 73%

[108] 35%

Chapter 10 Topics in Analytic Geometry

Section 10.1: Lines

Objective 1: Find the inclination of a line

1. Find, in radians and degrees, the inclination θ of the line passing through the points.

$(-3, -2), (1, 4)$

(A) 2.27 radians, 120.5° (B) 2.16 radians, 123.7°

(C) 0.87 radian, 59.5° (D) 0.98 radian, 56.3°

2. Find, in radians and degrees, the inclination θ of the line.

$2x + 6y + 40 = 0$

(A) 0.21 radian, 15.2° (B) 2.93 radians, 164.8°

(C) 0.32 radian, 18.4° (D) 2.82 radians, 161.6°

3. Find, in radians and degrees, the inclination θ of the line with a slope of m.

$$m = \frac{5}{13}$$

4. Find, in radians and degrees, the inclination θ of the line passing through the points.

$(6, 8), (-5, 7)$

Objective 2: Find the angle between two lines

Find, in radians and degrees, the angle θ between the lines.

5. $-3x+4y+7=0$

$6x-12y+8=0$

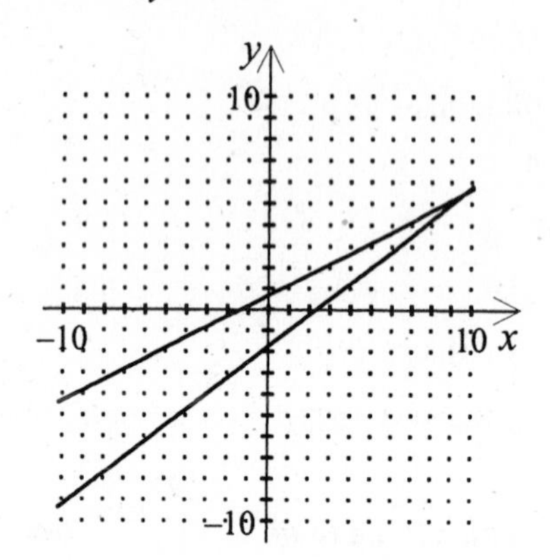

(A) 0.18 radian, 10.3°

(B) 0.29 radian, 13.5°

(C) 2.85 radians, 166.5°

(D) 2.96 radians, 169.7°

6. $x-0.4y-1.2=0$

$0.8x-0.9y+0.7=0$

(A) 2.68 radians, 153.4°

(B) 0.57 radian, 23.4°

(C) 0.46 radian, 26.6°

(D) 2.57 radians, 156.6°

7. $-1.2x+0.2y-0.4=0$

$1.1x+0.5y+1.0=0$

Find, in radians and degrees, the angle θ between the lines.

8. $-8x+7y+5=0$

$-6x-10y+1=0$

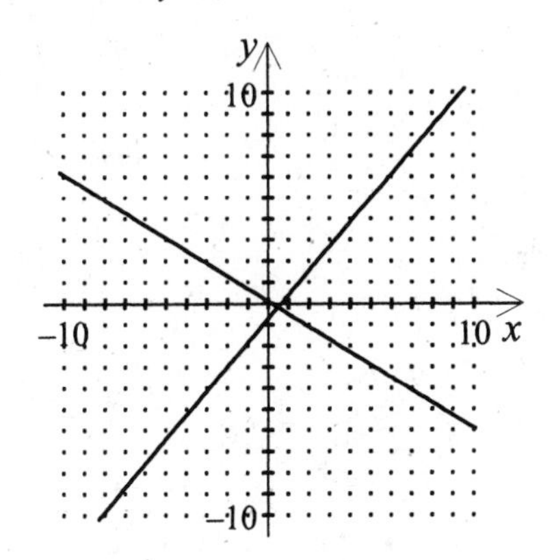

Objective 3: Find the distance between a point and a line

9. Find the distance between the parallel lines.

$2x-y=8$

$2x-y=-2$

(A) 2.24 (B) 8.31 (C) 4.47 (D) 3.13

10. Find the distance between the point and the line.

$(0,\ 0)\quad 5x+6y=-1$

(A) 0.03 (B) 7.87 (C) 12 (D) 0.13

11. Find the distance from $(1,-6)$ to the line containing $(-1,2)$ and $(-5,-2)$, to the nearest hundredth.

12. Find (a) the altitude from vertex B of the triangle to side AC, and (b) the area of the triangle.

$A=(13,\ -12),\ B=(5,\ -11),\ C=(7,\ -8)$

Section 10.2: Introduction to Conics: Parabolas

Objective 1: Recognize a conic as the intersection of a plane and a double-napped cone

13. Identify the diagram that represents an ellipse.

(A)

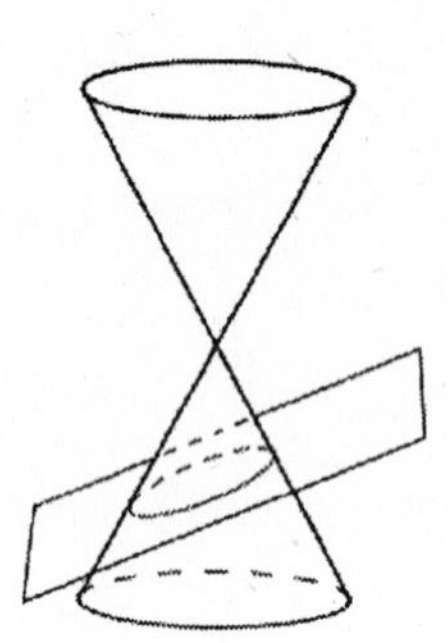

(B)

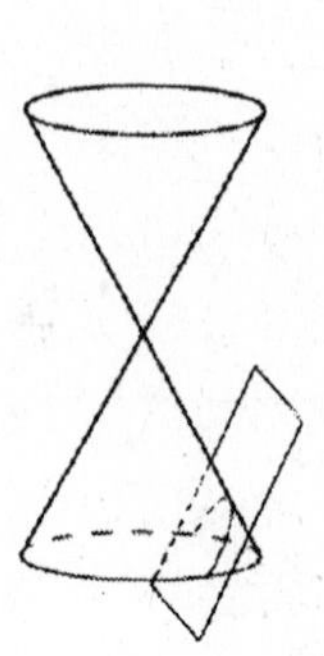

(C)

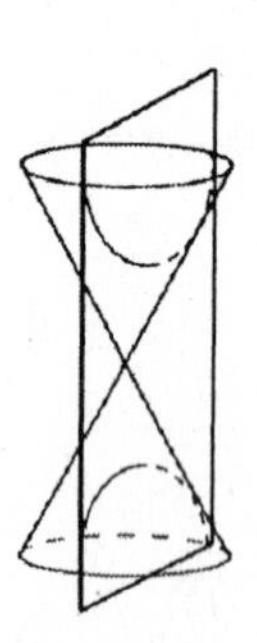

(D)

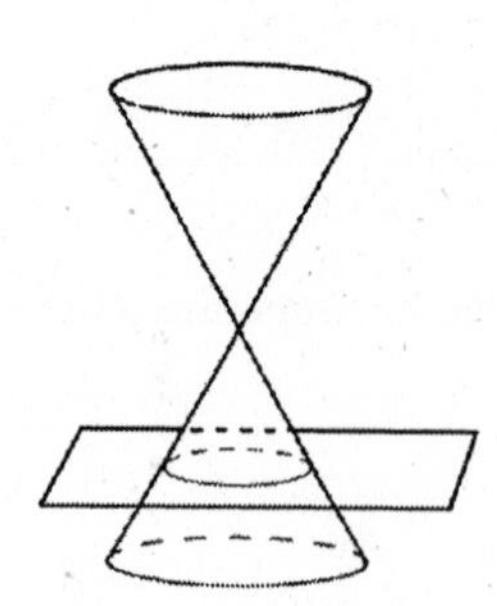

14. Identify the conic.

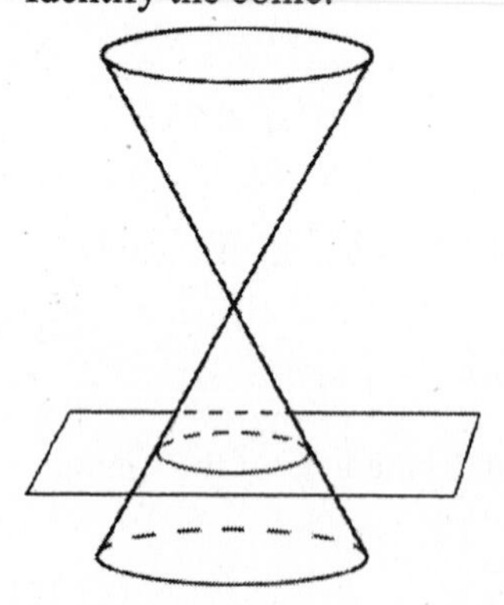

(A) Hyperbola (B) Ellipse (C) Circle (D) Parabola

15. Draw a plane on the following diagram so that the intersection of the plane with the double-napped cone is a circle.

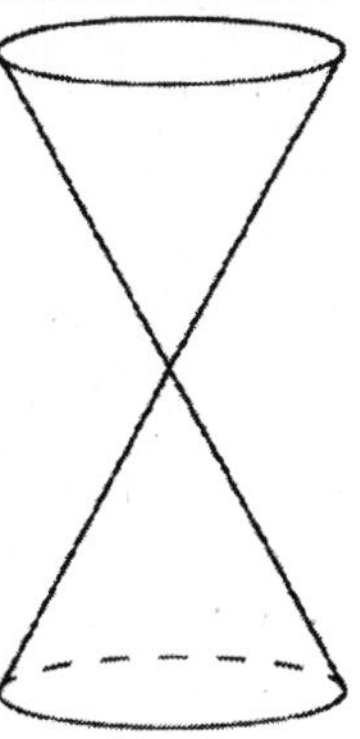

16. Draw a plane on the picture of the double-napped cone so that the intersection of the plane and the cone matches the conic section depicted on the left.

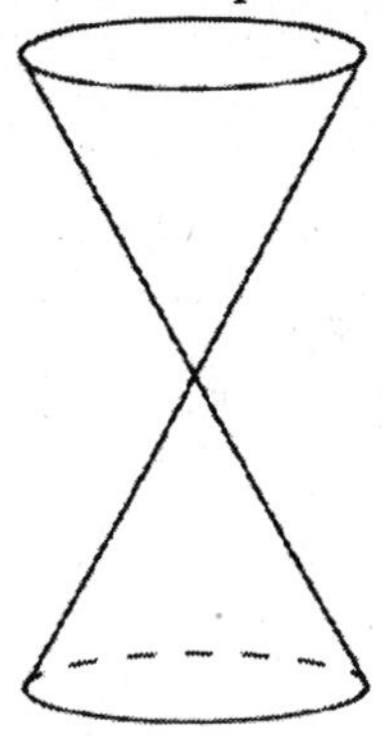

Objective 2: Write the standard form of the equation of a parabola

Find the standard form of the equation of the parabola.

17.

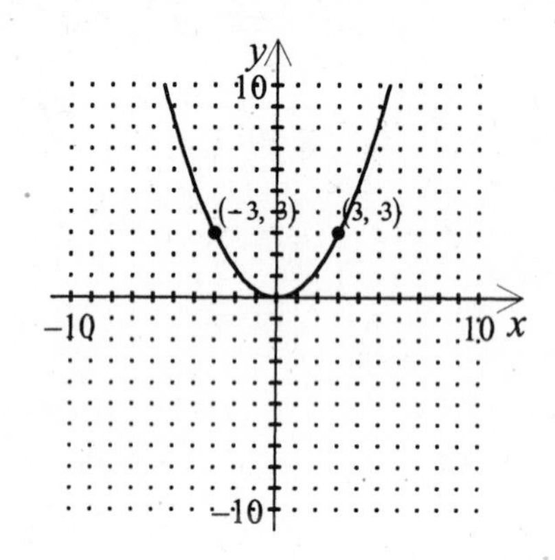

(A) $x^2 = -3y$ (B) $x^2 = 3y$ (C) $y^2 = 3x$ (D) none of these

18. Focus: $(2, 3)$; Vertex: $(2, 1)$

(A) $(x-2)^2 = 8(y-1)$ (B) $(y-2)^2 = 8(x-1)$

(C) $(x+2)^2 = 8(y+1)$ (D) $(y+2)^2 = 8(x+1)$

19.

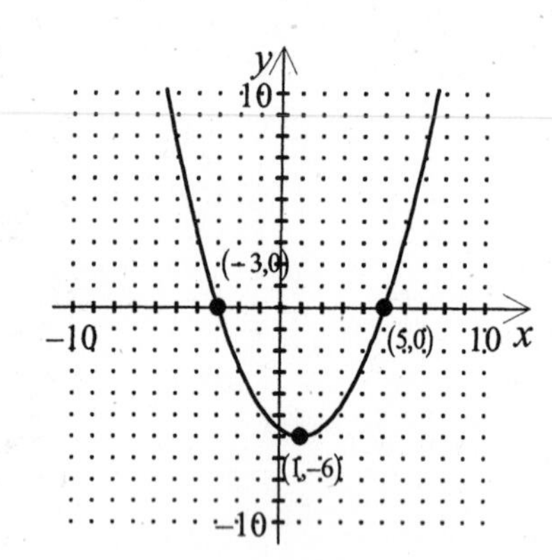

20. Vertex: $(0, 0)$; Focus: at $(-4, 0)$

Objective 3: Use the reflective property of parabolas to solve real-life problems

21. In a factory, a parabolic mirror to be used in a searchlight was placed on the floor. It measured 40 centimeters tall and 90 centimeters wide. Find an equation of the parabola with its vertex at the origin.

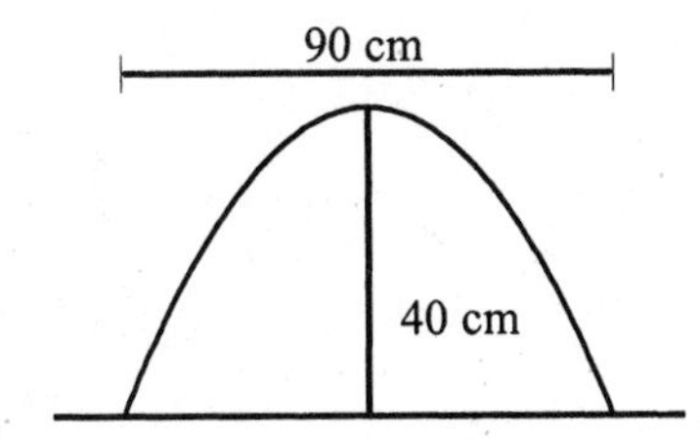

(A) $y = \frac{8}{405}x^2$ (B) $y = -\frac{8}{405}x^2$ (C) $y = -\frac{2}{405}x^2$ (D) $y = \frac{2}{405}x^2$

22. A parachute maker has been experimenting with different parabolic shapes. A particular parachute that is 20.6 feet wide is 8.1 feet higher in the center than it is on the edge (see figure).

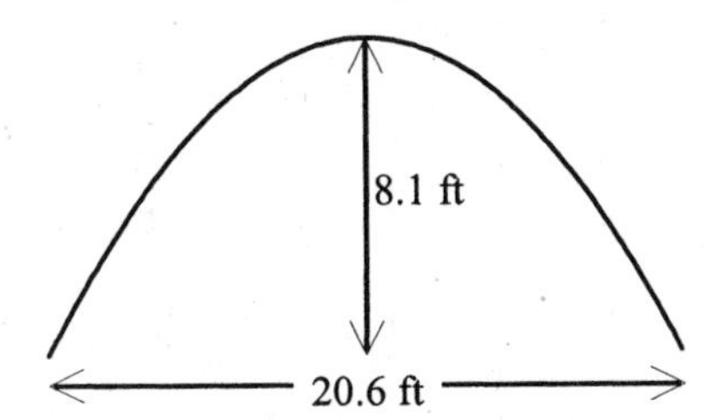

(a) Find an equation of the parabola that models the curve of the parachute. (Assume that the origin is at the center of the parachute.)
(b) How far from the center of the parachute is the parachute surface 3.3 feet lower than in the middle? (Round the answer to the nearest hundredth.)

(A) (a) $y = \frac{-8240}{6561}x^2$
(b) 5.92 ft

(B) (a) $y = \frac{-3240}{10{,}609}x^2$
(b) 5.92 ft

(C) (a) $y = \frac{-810}{10{,}609}x^2$
(b) 6.57 ft

(D) (a) $y = \frac{-405}{21{,}218}x^2$
(b) 6.57 ft

23. Each cable of a suspension bridge is suspended in the shape of a parabola between two towers that are 150 meters apart and whose tops are 23 meters above the roadway. The cables are 0.4 meter above the roadway midway between the towers.
(a) Find an equation for the parabolic shape of the cable. (Assume the origin is on the road, midway between the towers.)
(b) Find the height of the cable 4 meters from the center of the bridge. (Round the answer to the nearest thousandth.)

24. A parabolic television dish antenna is 20 feet across at its opening and 8 feet deep. If the receiver is placed at the focus, how far is it from the vertex?

Section 10.3: Ellipses

Objective 1: Write the standard form of the equation of an ellipse

Find the standard form of the equation of the specified ellipse.

25. Vertices: $(0,\ \pm 8)$; Foci: $(\pm 5,\ 0)$

(A) $\frac{x^2}{89}+\frac{y^2}{64}=5$ (B) $\frac{x^2}{89}+\frac{y^2}{64}=1$ (C) $\frac{x^2}{64}+\frac{y^2}{89}=1$ (D) $\frac{x^2}{64}+\frac{y^2}{89}=5$

26. Foci: $(-4,\ -5)$, $(-8,\ -5)$; Major axis of length 6

(A) $\frac{(x+5)^2}{5}+\frac{(y+6)^2}{9}=1$ (B) $\frac{(x+5)^2}{9}+\frac{(y+6)^2}{5}=1$

(C) $\frac{(x+6)^2}{9}+\frac{(y+5)^2}{5}=1$ (D) $\frac{(x+6)^2}{5}+\frac{(y+5)^2}{9}=1$

27. Center: $(-2,\ 4)$; Vertex: $(7,\ 4)$; Minor axis of length 6

28. Center: $(0,\ 0)$; Foci: $\left(0,\ \pm 2\sqrt{15}\right)$, Vertices: $(0,\ \pm 8)$

Objective 2: Use properties of ellipses to model and solve real-life problems

29. Rachel rides her bike through a tunnel shaped like the top half of an ellipse. The tunnel is 11 meters wide and 3 meters high. On her bike, the top of Rachel's helmet is 1.8 meters above the ground. If she were to ride through the tunnel 4.1 meters from the center, would her helmet miss the ceiling? If so, by how much?

(A) Yes, she would clear the ceiling by 0.2 m.

(B) Yes, she would clear the ceiling by 0.4 m.

(C) Yes, she would clear the ceiling by 0.3 m. (D) No, she would hit the ceiling.

30. The opening under a bridge is in the shape of a semielliptical arch and has a span of 106 feet. The height of the arch, at a distance of 28 feet from the center, is 37 feet. Find the height of the arch at its center.

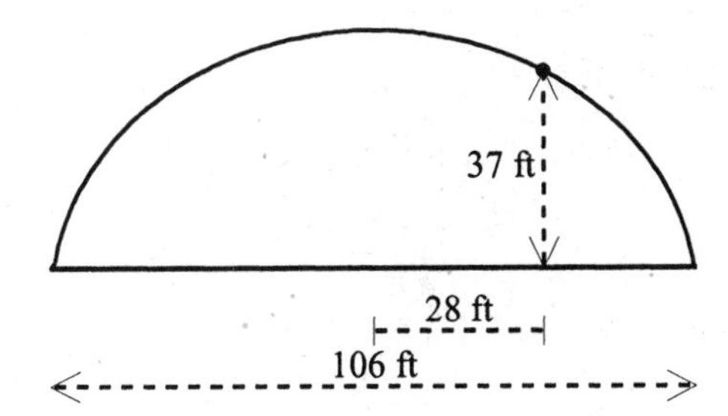

(A) $43\frac{26}{45}$ ft (B) $44\frac{1}{2}$ ft (C) $43\frac{2}{3}$ ft (D) $44\frac{26}{45}$ ft

31. A satellite orbits around the moon in an elliptical path. Assuming the moon to be the center of a rectangular coordinate system, find the equation of the elliptical path of a satellite whose x-intercepts are $\pm 90{,}000$ km and whose y-intercepts are $\pm 73{,}000$ km.

32. A skating park has a track shaped like an ellipse. If the length of the track is 52 m and the width of the track is 36 m, find the equation of the ellipse.

Objective 3: Find the eccentricity of an ellipse

Find the eccentricity of the ellipse.

33. $16x^2 + 4y^2 + 96x - 32y + 144 = 0$ (A) $\frac{\sqrt{3}}{2}$ (B) $\frac{\sqrt{11}}{5}$ (C) $\frac{\sqrt{5}}{2}$ (D) $\frac{2\sqrt{5}}{5}$

34. $\frac{x^2}{25} + \frac{y^2}{9} = 1$ (A) $\frac{4}{5}$ (B) $\frac{\sqrt{70}}{2}$ (C) $\frac{2\sqrt{5}}{9}$ (D) $\frac{3\sqrt{3}}{4}$

35. $49x^2 - 588x + 36y^2 - 504y + 1764 = 0$

36. $\frac{(x+2)^2}{49} + \frac{(y-2)^2}{25} = 1$

Section 10.4: Hyperbolas

Objective 1: Write the standard form of the equation of a hyperbola

Find the standard form of the equation of the specified hyperbola.

37.

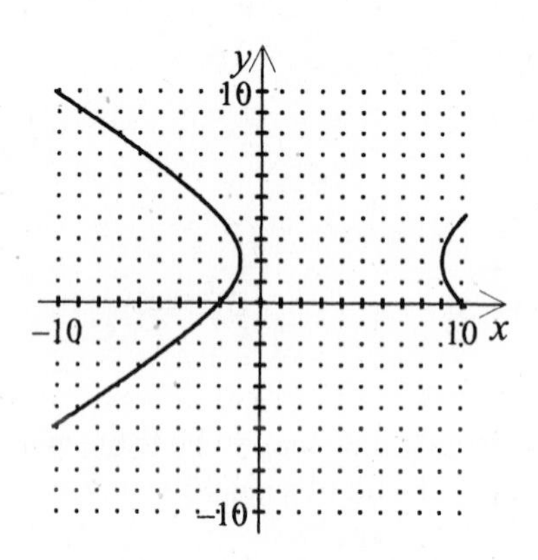

(A) $\dfrac{(x-4)^2}{25}-\dfrac{(y-2)^2}{9}=1$

(B) $\dfrac{(x+2)^2}{9}-\dfrac{(y+4)^2}{25}=1$

(C) $\dfrac{(x+4)^2}{9}-\dfrac{(y+2)^2}{25}=1$

(D) $\dfrac{(x-2)^2}{25}-\dfrac{(y-4)^2}{9}=1$

38. Center: $(5,\ 2)$
Focus: $(25,\ 2)$
Vertex: $(17,\ 2)$

(A) $\dfrac{(x-5)^2}{144}+\dfrac{(y-2)^2}{256}=1$

(B) $\dfrac{(x-5)^2}{144}-\dfrac{(y-2)^2}{256}=1$

(C) $\dfrac{(x+5)^2}{144}-\dfrac{(y+2)^2}{256}=1$

(D) $\dfrac{(y-2)^2}{144}-\dfrac{(x-5)^2}{256}=1$

39. $36x^2+288x-49y^2+98y-1237=0$

40. Vertices: $(\pm 2,\ 0)$

Asymptotes: $y=\pm\dfrac{5}{2}x$

Objective 2: Find the asymptotes of a hyperbola

41. Find the equations of the asymptotes of the hyperbola.
$-16x^2+49y^2+64x+294y-407=0$

(A) Asymptotes: $y=\frac{4}{7}x-\frac{29}{7},\ y=-\frac{4}{7}x-\frac{13}{7}$ (B) Asymptotes: $y=\pm\frac{7}{4}x$

(C) Asymptotes: $y=\pm\frac{4}{7}x$ (D) Asymptotes: $y=\frac{7}{4}x-\frac{13}{2},\ y=-\frac{7}{4}x+\frac{1}{2}$

42. Find the vertices, foci, and the equations of the asymptotes of the hyperbola.
$\frac{x^2}{36}-\frac{y^2}{25}=1$

(A) Vertices: $(0,\ \pm 6)$; Foci: $\left(0,\ \pm\sqrt{61}\right)$; Asymptotes: $y=\pm\frac{5}{6}x$

(B) Vertices: $(\pm 6,\ 0)$; Foci: $\left(\pm\sqrt{61},\ 0\right)$; Asymptotes: $y=\pm\frac{5}{6}x$

(C) Vertices: $\left(0,\ \pm\frac{36}{25}\right)$; Foci: $\left(\pm\sqrt{61},\ 0\right)$; Asymptotes: $x=\pm\frac{6}{5}y$

(D) Vertices: $\left(\pm\frac{36}{25},\ 0\right)$; Foci: $\left(0,\ \pm\sqrt{61}\right)$; Asymptotes: $x=\pm\frac{6}{5}y$

Find the equations of the asymptotes of the hyperbola.

43. $\frac{9(x+2)^2}{4}-\frac{4(y+5)^2}{9}=1$

44. $9y^2-4x^2-36=0$

Objective 3: Use properties of hyperbolas to solve real-life problems

45. You and a friend live 2 miles apart and are talking on the phone. You hear a crack of thunder and 2 seconds later your friend hears the crack. Find an equation that gives the possible places where the lightning could have occurred. (Use miles as the unit of distance. The speed of sound is about 1100 ft/sec.)

(A) $\dfrac{x^2}{\frac{551}{576}}+\dfrac{y^2}{\frac{25}{576}}=1$ (B) $\dfrac{x^2}{\frac{25}{576}}+\dfrac{y^2}{\frac{551}{576}}=1$ (C) $\dfrac{x^2}{\frac{25}{576}}-\dfrac{y^2}{\frac{551}{576}}=1$ (D) $\dfrac{x^2}{\frac{551}{576}}-\dfrac{y^2}{\frac{25}{576}}=1$

46. The focus of a hyperbolic mirror has coordinates $(30,\ 0)$. Find the vertex of the mirror if $(30,\ 24)$ is one of the points on the mirror and the other focus has coordinates $(-30,\ 0)$.

(A) $(6,\ 0)$ (B) $(0,\ 18)$ (C) $(54,\ 0)$ (D) $(18,\ 0)$

47. An explosion is monitored by three scientific stations. Station A is located at $(2500,\ 0)$, Station B at $(-2500,\ 0)$, and Station C at $(-2500,\ -6600)$. Station A records the explosion 4 seconds after Station B. Station C records the explosion 6 seconds after Station B. If the coordinates are measured in feet and sound travels at 1100 feet per second, determine the coordinates of the explosion.

48. Two LORAN stations are positioned 296 miles apart along a straight shore. The stations transmit synchronized pulses which travel at the speed of light (186,000 miles per second) in order to find the coordinates of ships in the area. The station closest to a ship being tracked by the LORAN stations is called the master station.
(a) A ship records a time difference of 0.000516 second between the LORAN signals. Determine where the ship would reach shore if it were to follow the indicated hyperbolic path.
(b) If the ship were to enter a harbor located between the two stations 8 miles from the master station, what time difference should it be looking for?
(c) If the ship is 80 miles offshore when the desired time difference is obtained, what is the exact location of the ship?

Objective 4: Classify a conic from its general equation

49. Identify the equation of a circle.

(A) $5y^2-9y-12x^2-3x+6=0$ (B) $-3x^2-9x-3y^2+6y+1=0$

(C) $6y^2-9x+y=8$ (D) $6x^2-9x+12y^2+y=-1$

50. Identify the equations that represent parabolas.

(i) $3x^2 + 7y^2 + 3x + 11y - 8 = 0$

(ii) $5x^2 + 2y^2 + 5x + 3y - 8 = 0$

(iii) $5x^2 + 4y^2 + 5x + 15y + 25 = 0$

(A) iii only (B) i, ii, and iii (C) ii and iii only (D) None of these

Classify the graph of the equation as a circle, a parabola, an ellipse, or a hyperbola.

51. $3x^2 + 3y^2 - 3x - 6y - 7 = 0$

52. $4x^2 + 4y^2 + 9x - 10 = 0$

Section 10.5: Rotation of Conics

Objective 1: Rotate the coordinate axes to eliminate the xy-term in the equation of a conic

53. The $x'y'$-coordinate system has been rotated θ degrees from the *xy*-coordinate system. The coordinates of a point on the *xy*-coordinate system are given. Find the coordinates of the point on the rotated coordinate system.

$\theta = 78°$, $(8, 7)$

(A) $x' = -5.18$, $y' = 9.28$ (B) $x' = 9.49$, $y' = 5.39$

(C) $x' = -6.37$, $y' = 8.51$ (D) $x' = 8.51$, $y' = 6.37$

54. Rotate the axes to eliminate the *xy*-term.

$6x^2 - 12xy + 6y^2 + 58x + 8y + 105 = 0$

(A) $12x^2 + 12y^2 - 25\sqrt{2}y + 105 = 0$ (B) $12x^2 + 12y^2 + 33\sqrt{2}x + 105 = 0$

(C) $12y^2 + 33\sqrt{2}x - 25\sqrt{2}y + 105 = 0$ (D) $12x^2 + 33\sqrt{2}x - 25\sqrt{2}y + 105 = 0$

55. Rotate the axes to eliminate the *xy*-term. Sketch the graph of the resulting equation, showing both sets of axes.

$xy - 3 = 0$

56. Graph the conic. Determine the angle θ through which the axes are rotated.

$21x^2 + 24xy - 11y^2 = 375$

Objective 2: Use the discriminant to classify a conic

Find the discriminant and classify the graph.

57. $8x^2 + 8xy + 3y^2 - 4x - 2y = 0$

(A) –32; ellipse (B) 160; hyperbola (C) 160; parabola (D) –32; hyperbola

58. $5x^2 + 7xy + 4y^2 - 3 = 0$

(A) –129; hyperbola (B) –31; ellipse (C) –129; parabola (D) –31; hyperbola

Use the discriminant to classify the graph, use the Quadratic Formula to solve for *y*, and then use a graphing utility to graph the equation.

59. $x^2 + xy + 2y^2 + 4x - 2y = 0$

60. $3x^2 + 12xy - 13y^2 + 4x + 3y - 73 = 0$

Section 10.6: Parametric Equations

Objective 1: Evaluate a set of parametric equations for a given value of the parameter

Evaluate the set of parametric equations for the given value of the parameter.

61. $x = 3t$, $y = 8t^2 + 9t$, $t = -2$

(A) $x = -2, y = 270$ (B) $x = -6, y = 14$ (C) $x = -2, y = 14$ (D) $x = -6, y = 270$

62. $x = 8 + 3t$, $y = 3t^2$, $t = -2$

(A) $x = 2, y = 12$ (B) $x = -22, y = -12$ (C) $x = 2, y = -6$ (D) $x = -22, y = 12$

63. $x = -1 - 4\cos\theta$, $y = -3 + 4\sin\theta$, $\theta = \dfrac{4\pi}{3}$

64. $x = 3t^2$, $y = -4\ln t$, $t = 1$

Objective 2: Sketch the curve that is represented by a set of parametric equations

65. Sketch the curve represented by the parametric equations (indicate the direction of the curve).

$x = -\tan t$

$y = \sec t$

(A)

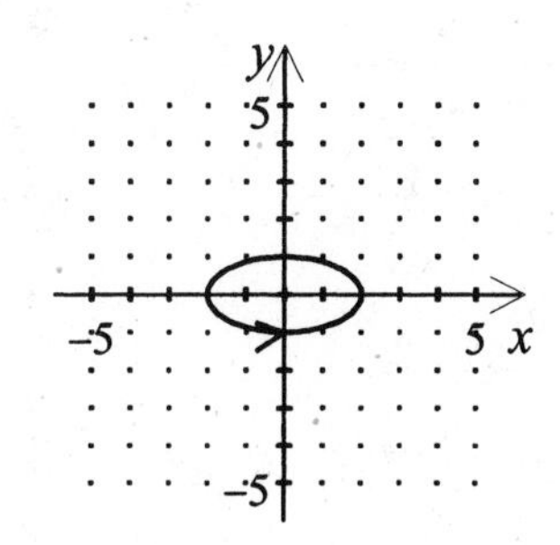

(B)

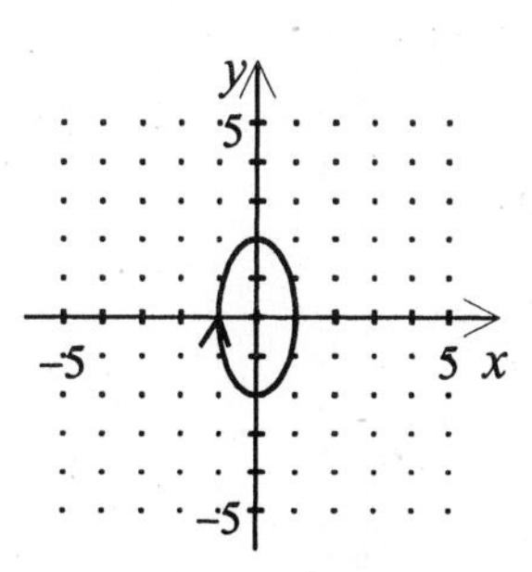

(C)

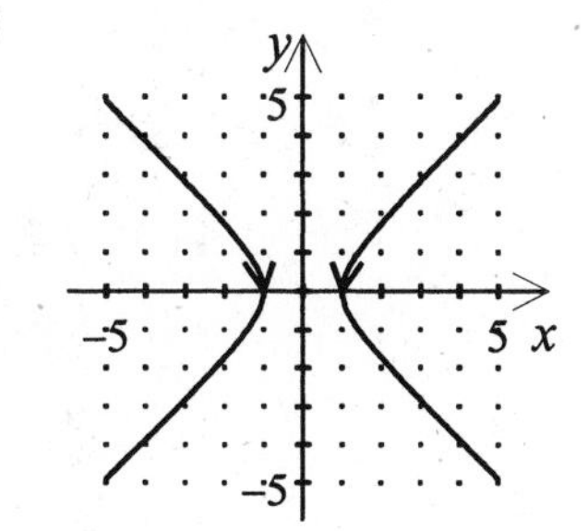

(D)

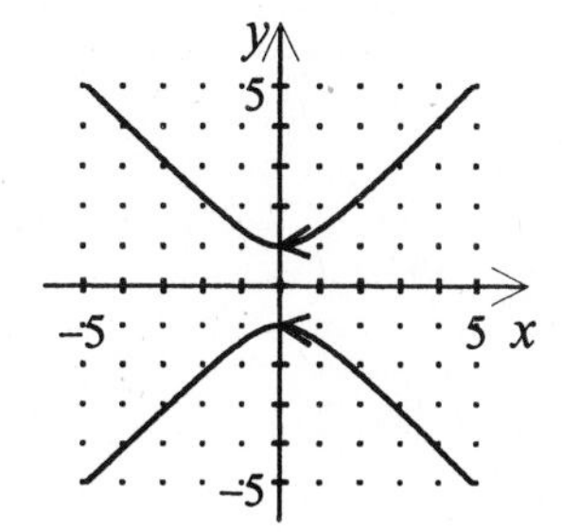

Sketch the curve represented by the parametric equations.

66. $x = t$

$y = t^2 + 3t - 1$

(A) $y = x^2 + 3x - 1$

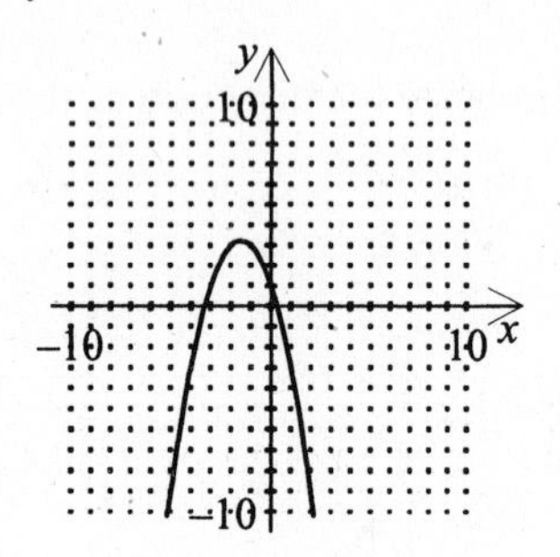

(B) $y = x^2 + 3x - 1$

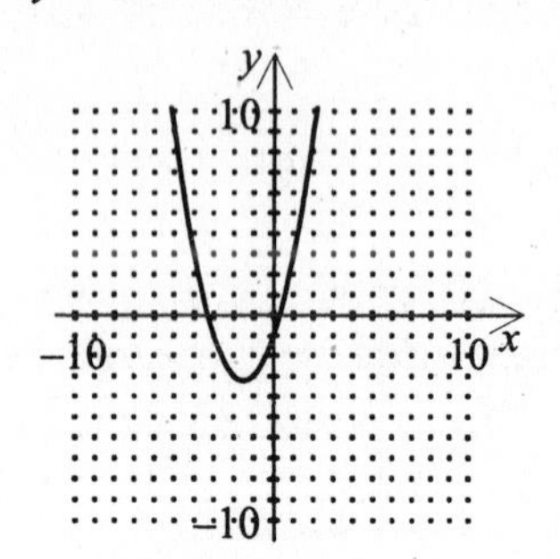

(C) $y = \sqrt{x^2 + 3x - 1}$

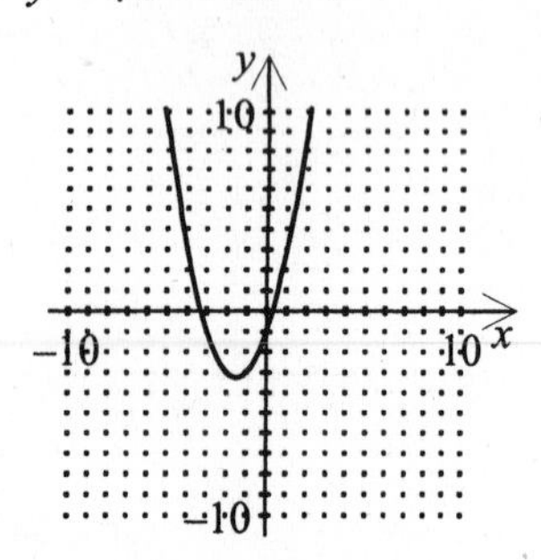

(D) $y = \sqrt{x^2 + 3x - 1}$

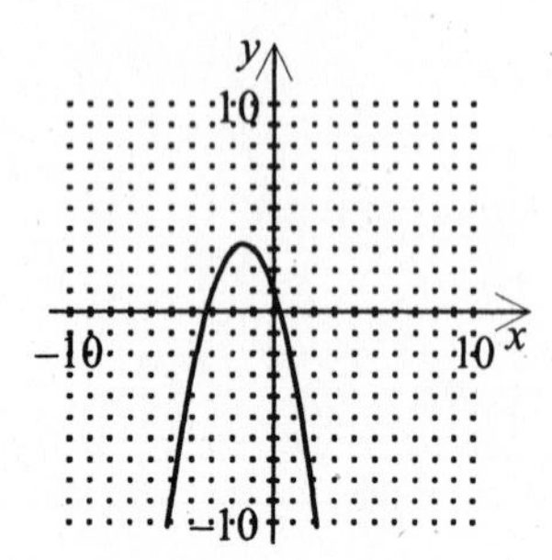

(66.)

Sketch the curve represented by the parametric equations.

67. $x = 5\sin 2t$

$y = 3\cos 2t$

68. Sketch the curve represented by the parametric equations (indicate the direction of the curve).

$x = -\tan t$

$y = \sec t$

Objective 3: Rewrite a set of parametric equations as a single rectangular equation

Eliminate the parameter and obtain the standard form of the rectangular equation.

69. $x = 6t - 6$

$y = -3t - 2$

(A) $y = \frac{1}{2}x + 5$ (B) $y = -4x + \frac{1}{5}$ (C) $y = -\frac{1}{2}x - 5$ (D) $y = -x - \frac{10}{3}$

70. $x = t$

$y = 2t^2 + 2t + 4$

(A) $y = 2x^2 + 2x + 4$ (B) $y = \sqrt{2x^2 + 2x + 4}$

(C) $y = \sqrt{2x^2 + 2x + 4}$ (D) $y = 2x^2 + 2x + 4$

71. $x = 4\sin 2t$

$y = 5\cos 2t$

72. $x = 8t^2 - 6$

$y = 9t + 5$

Objective 4: Find a set of parametric equations for a graph

Identify the set of parametric equations for the given rectangular equation.

73. $x^2 + y^2 = 64$

(A) $x = 8\cot t$, $y = 8\tan t$
(B) $x = 8t$, $y = 8t$
(C) $x = \sqrt{t}$, $y = 8$
(D) $x = 8\cos t$, $y = 8\sin t$

74. $y = x^2 - 2x$

(A) $x = 0,\ y = x^2 - 2x$
(B) $x = x^3,\ y = x^2 - 2x$
(C) $x = t^2 - 2t,\ y = t$
(D) $x = t,\ y = t^2 - 2t$

75. Find a set of parametric equations for the given rectangular equation using $t = \frac{x}{4}$.

$$y = -\frac{1}{16}x^2 - \frac{1}{4}x + 8$$

76. Find a set of parametric equations for the given rectangular equation using $t = x$.

$$y = x^3 + x - 2$$

Section 10.7: Polar Coordinates

Objective 1: Plot points on the polar coordinate system

77. Identify the graph of the given point.

$$\left(-5, \frac{5\pi}{3}\right)$$

(A)

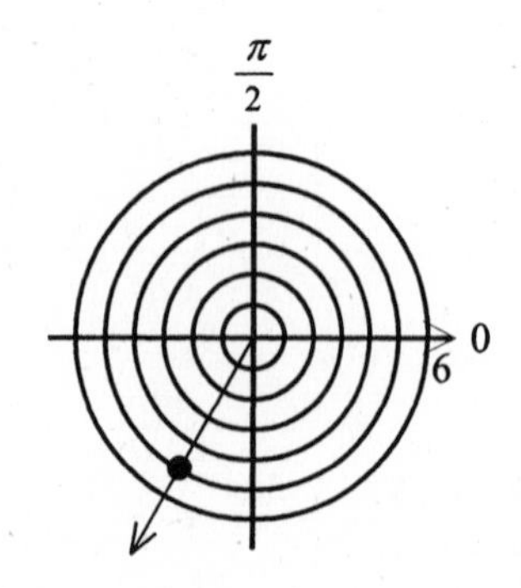

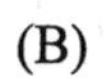

(B)

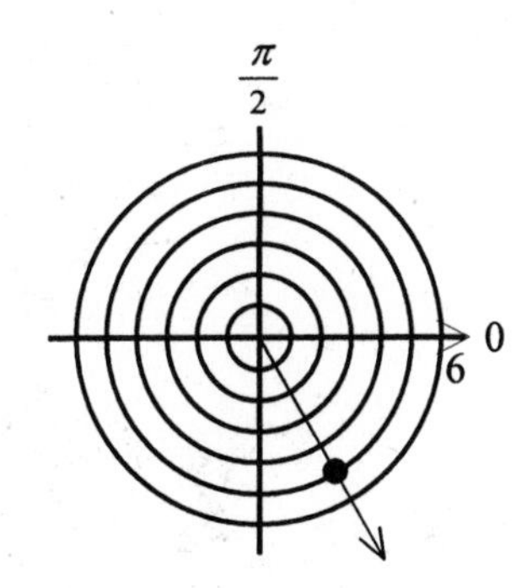

(C)

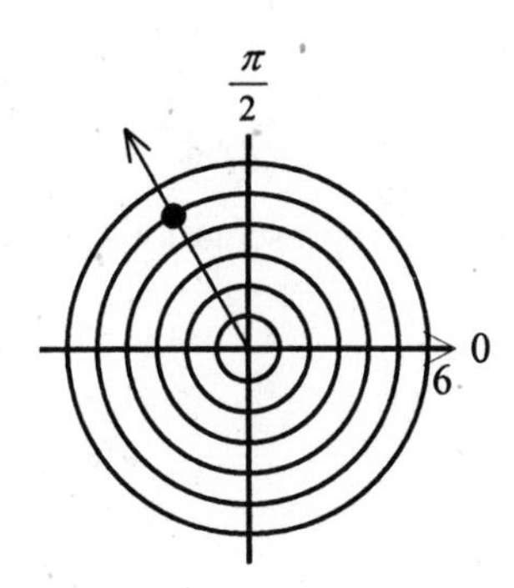

(D)

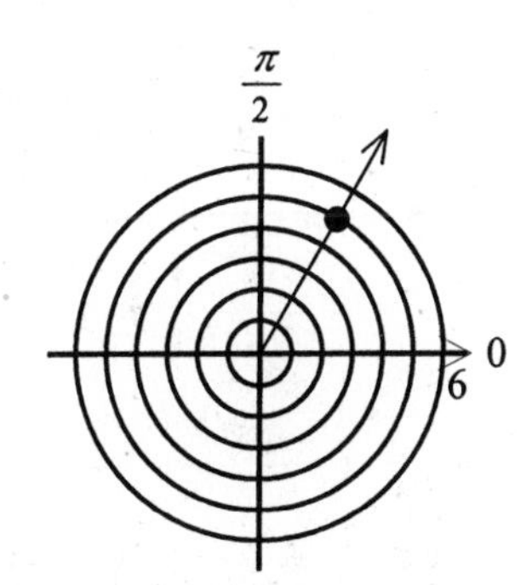

(77.)

78. Identify the point graphed.

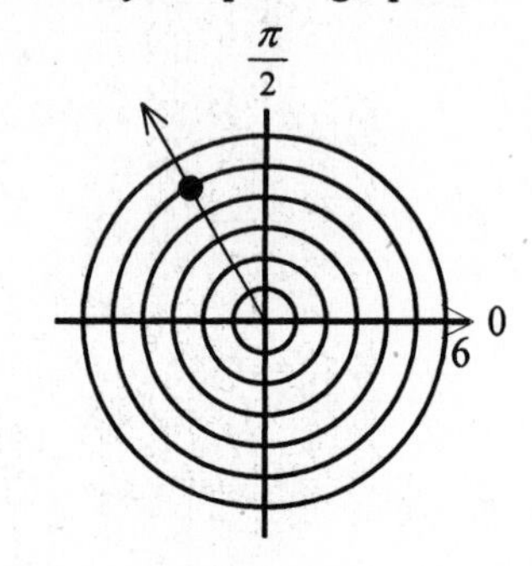

(A) $\left(5, -\frac{\pi}{6}\right)$ (B) $\left(-5, -\frac{\pi}{6}\right)$ (C) $\left(5, -\frac{\pi}{3}\right)$ (D) $\left(-5, -\frac{\pi}{3}\right)$

Plot the point given in polar coordinates and find two additional polar representations.

79. $\left(-5, -\frac{\pi}{4}\right)$

80. $(5, 2.1)$

Objective 2: Convert points from rectangular to polar form and vice versa

81. A point in rectangular coordinates is given. Convert the point to polar coordinates.
$(4, -4)$

(A) $\left(4\sqrt{2}, \frac{7\pi}{4}\right)$ (B) $\left(4\sqrt{2}, \frac{5\pi}{4}\right)$ (C) $\left(4\sqrt{2}, \frac{9\pi}{4}\right)$ (D) $\left(-4\sqrt{2}, \frac{7\pi}{4}\right)$

82. A point in polar coordinates is given. Convert the point to rectangular coordinates.
$(-7, -0.31)$

(A) $(0.952, -0.305)$ (B) $(-7.000, 0.038)$

(C) $(-6.666, 2.135)$ (D) $(-2.135, -6.666)$

83. A point in rectangular coordinates is given. Convert the point to polar coordinates.
$\left(-3, 3\sqrt{3}\right)$

84. A point in polar coordinates is given. Convert the point to rectangular coordinates.
$\left(7, \frac{3\pi}{2}\right)$

Objective 3: Convert equations from rectangular to polar form and vice versa

85. Convert the polar equation to rectangular form.
$r = 6\cos\theta$

(A) $(x+3)^2 + y^2 = 9$ (B) $y = x$ (C) $(x-3)^2 + y^2 = 9$ (D) $x^2 + (y+3)^2 = 9$

Convert the rectangular equation to polar form.

86. $(x-7)^2 + y^2 = 49$

(A) $\theta = -\frac{\pi}{4}$ (B) $r = 14\cos\theta$ (C) $r = 14\sin\theta$ (D) $r = \pm 7$

87. $x^2 + (y+4)^2 = 16$

88. Convert the polar equation to rectangular form.
$r = \frac{1}{2 - 2\sin\theta}$

Section 10.8: Graphs of Polar Equations

Objective 1: Graph polar equations by point plotting

Identify the graph of the polar equation.

89. $r = -3\sin 3\theta,\ 0 \le \theta \le 2\pi$

(A)

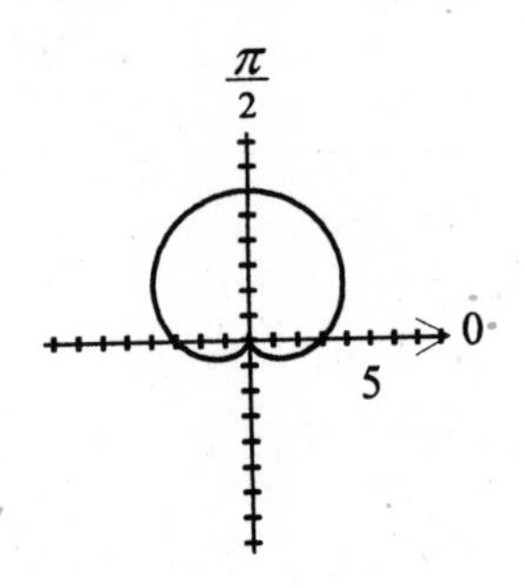

(B)

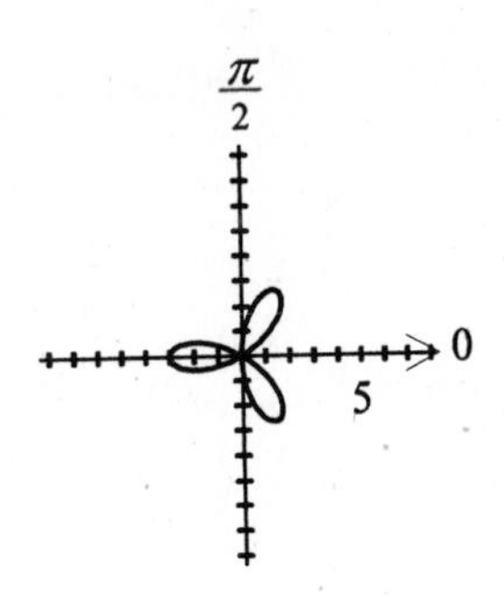

(C)

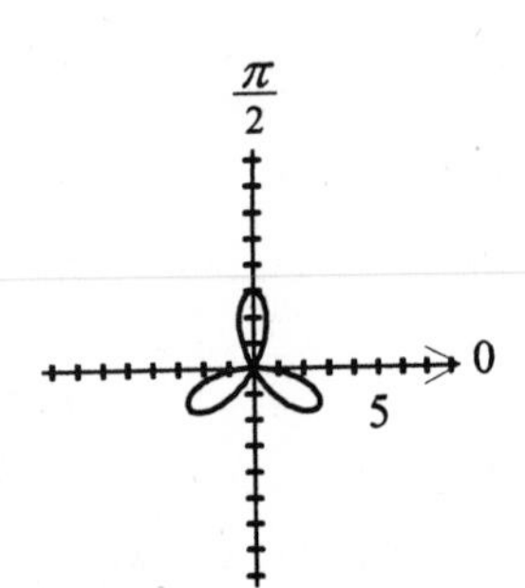

(D)

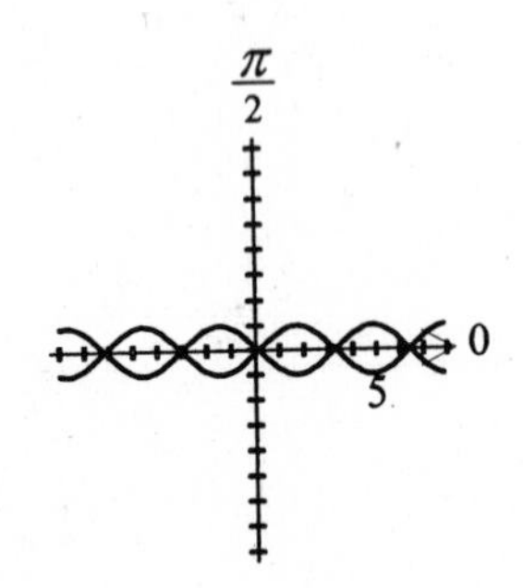

(89.)

Identify the graph of the polar equation.

90. $3 = r\cos\left(\theta + 40°\right)$

(A)

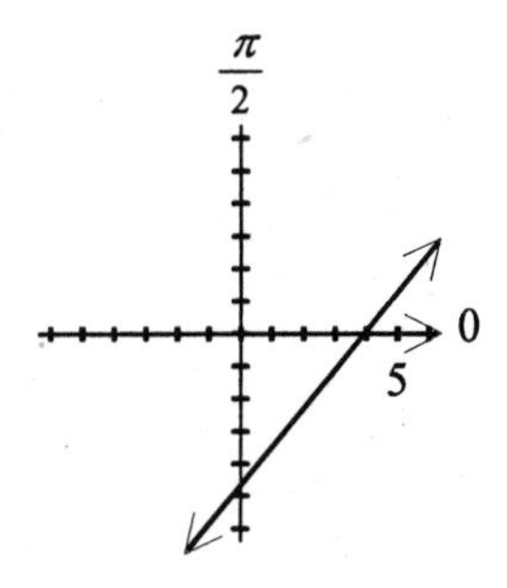

(B)

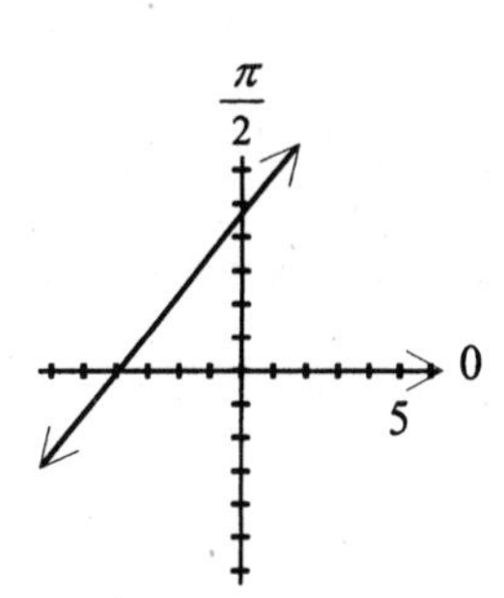

(C)

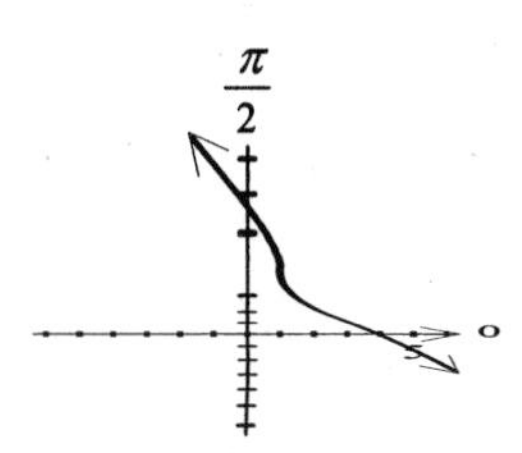

(D)

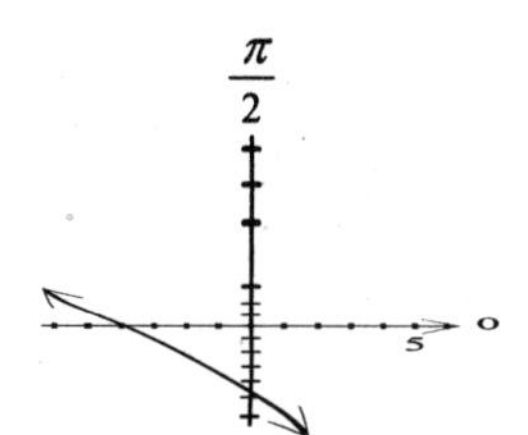

Sketch the graph of the polar equation.

91. $r = \dfrac{16}{5 + 3\cos\theta}$

92. $\theta = \dfrac{\pi}{2}$

Objective 2: Use symmetry to sketch graphs of polar equations

93. Test for symmetry with respect to $\theta = \frac{\pi}{2}$, the polar axis, and the pole.

$r = \frac{1}{1 + \cos\theta}$

(A) The line does not have symmetry.

(B) The graph is symmetric with respect to the polar axis.

(C) The graph is symmetric with respect to the pole.

(D) The graph is symmetric with respect to $\theta = \frac{\pi}{2}$.

94. Test for symmetry with respect to $\theta = \frac{\pi}{2}$, the polar axis, and the pole.

$r = -2 - 2\cos\theta$

(A) The graph is symmetric with respect to the line $\theta = \frac{\pi}{2}$.

(B) The graph is symmetric with respect to the polar axis.

(C) The graph is symmetric with respect to the polar axis and the pole.

(D) The graph is symmetric with respect to the line $\theta = \frac{\pi}{2}$ and the pole.

95. Test for symmetry with respect to $\theta = \frac{\pi}{2}$, the polar axis, and the pole.

$r = \frac{1}{2 - 3\sin\theta}$

96. Test for symmetry with respect to $\theta = \frac{\pi}{2}$, the polar axis, and the pole.

$r = 2\cos 2\theta$

Objective 3: Use zeros and maximum r-values to sketch graphs of polar equations

97. Find the maximum value of $|r|$ and any zeros of r.

$r = 4(1 - 2\sin\theta)$

(A) 12; $\theta = \frac{7\pi}{6}, \frac{11\pi}{6}$ (B) 12; $\theta = \frac{\pi}{6}, \frac{5\pi}{6}$ (C) 4; $\theta = \frac{5\pi}{6}, \frac{7\pi}{6}$ (D) 4; $\theta = \frac{\pi}{6}, \frac{11\pi}{6}$

98. Sketch the graph of the polar equation $r = \dfrac{16}{5 - 3\cos\theta}$.

(A)

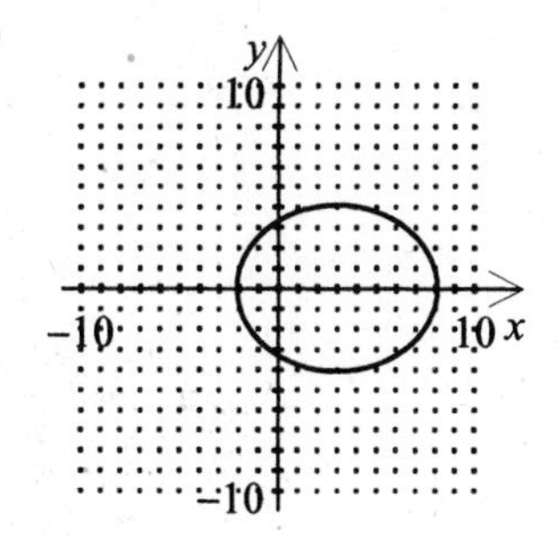

(B)

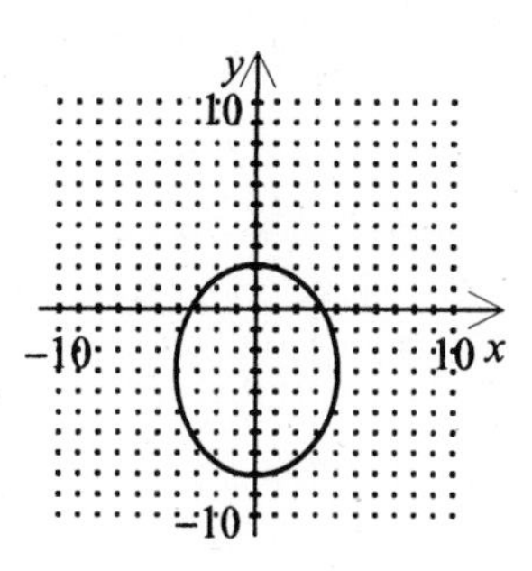

(C)

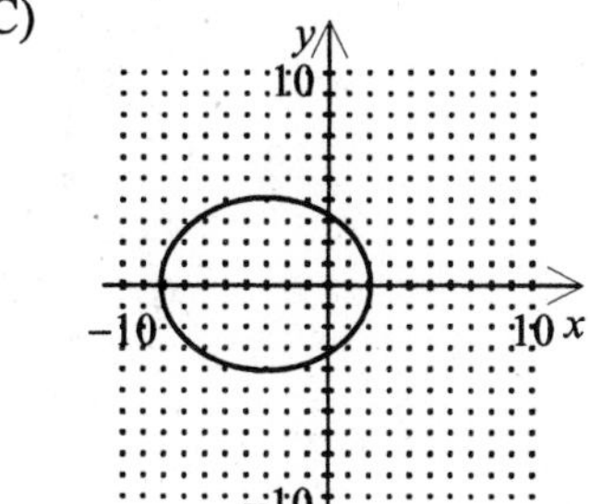

(D)

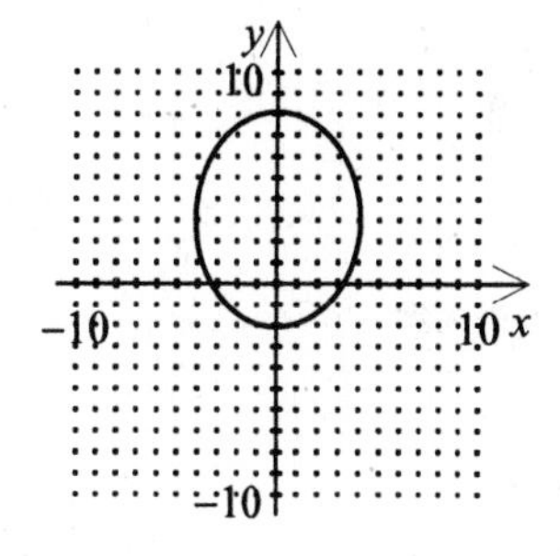

99. Find the maximum value of $|r|$ and any zeros of r.

$r = 4\cos 4\theta$

100. Find the maximum value of $|r|$ and any zeros of r.

$r = 10\left(\sqrt{2} - 2\sin\theta\right)$

Objective 4: Recognize special polar graphs

101. Identify the type of polar graph.

$r = 2 + 4\cos\theta$

(A)

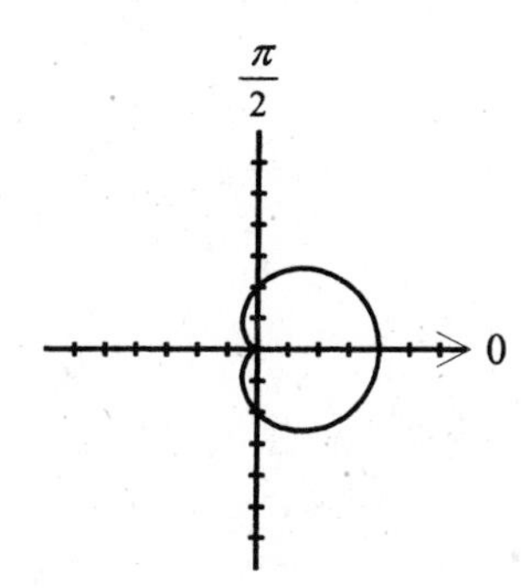

(B)

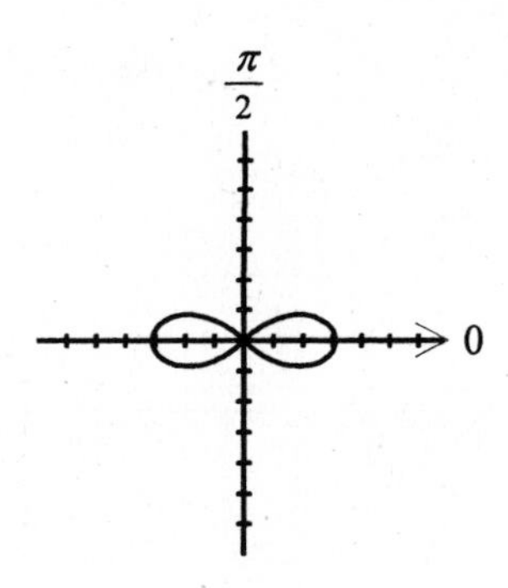

(C)

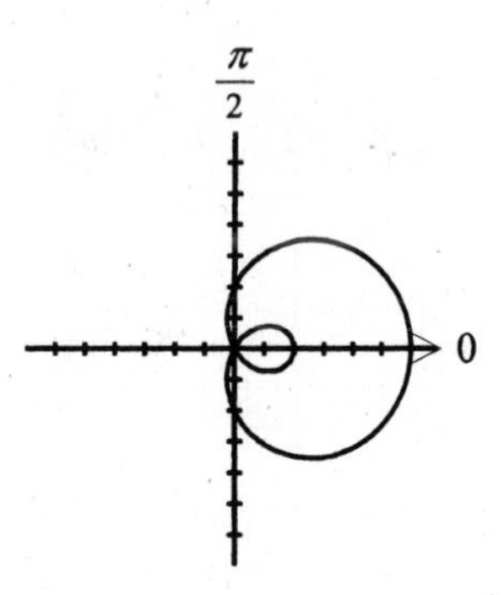

(D)

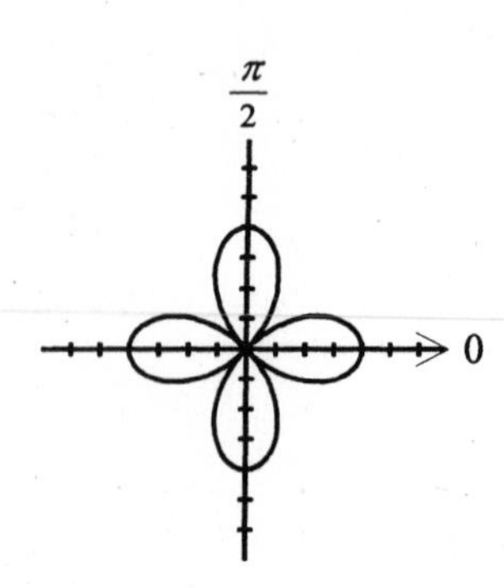

(101.)

102. Identify the graph of a cardioid.

(A)

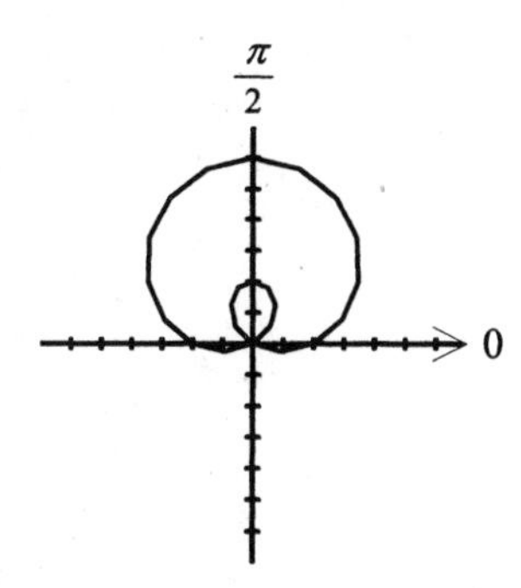

(B)

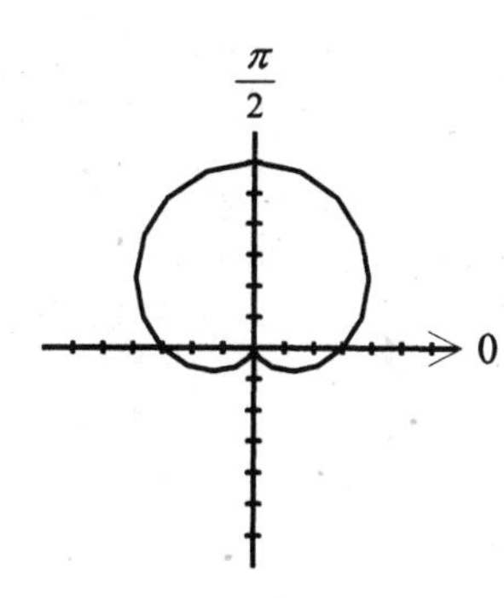

(C)

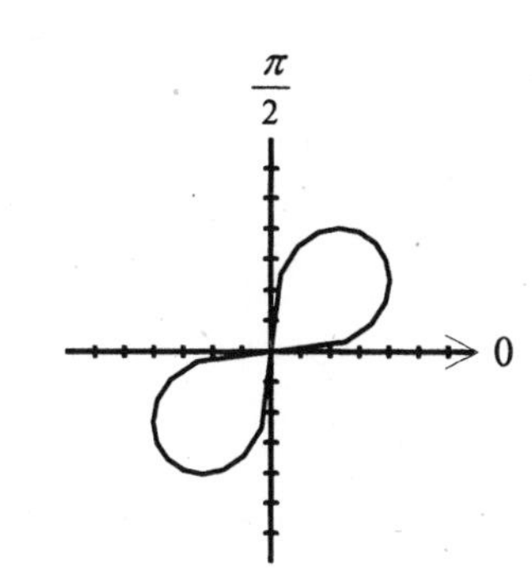

(D)

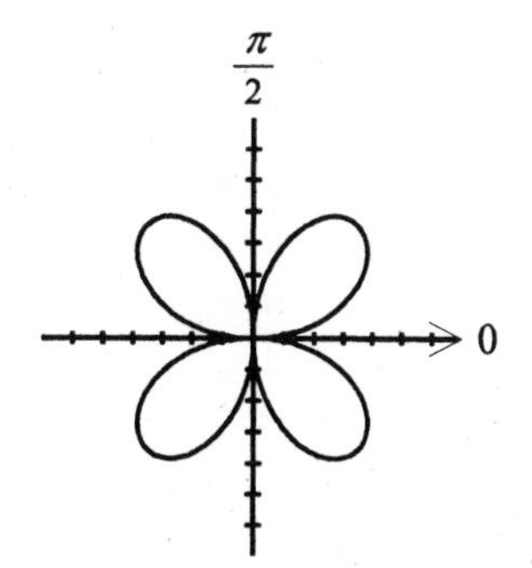

103. Identify the graph for the equation $r = \cos 4\theta$, as a cardioid, a rose, a circle, a lemniscate, or a limaçon.

104. Identify the type of polar graph.
$r = 5\sin 3\theta$

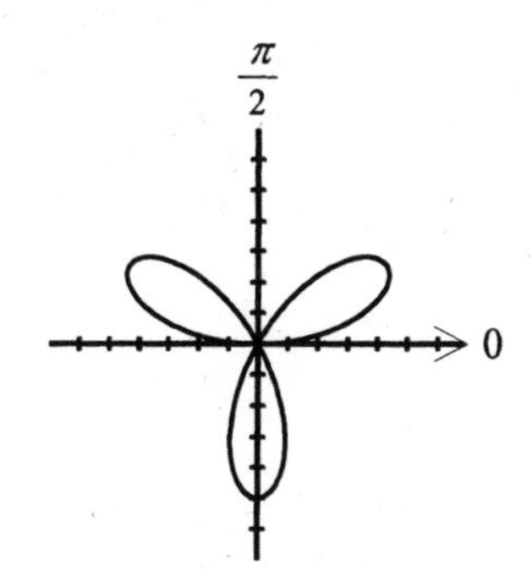

Section 10.9: Polar Equations of Conics

Objective 1: Define conics in terms of eccentricity

105. Graph.

$$r = \frac{3}{2 + 4\sin\theta}$$

(A)

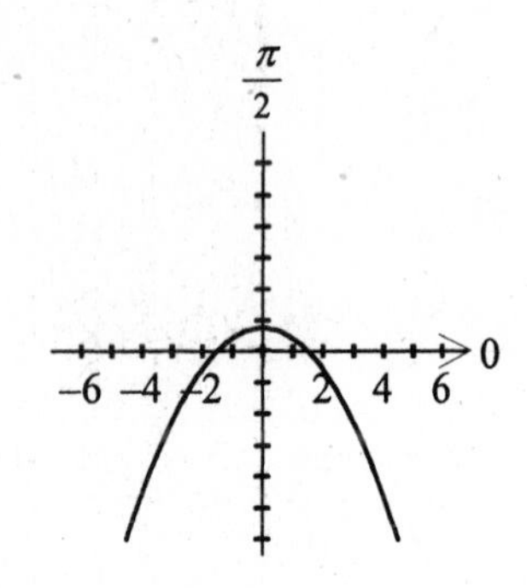

(B)

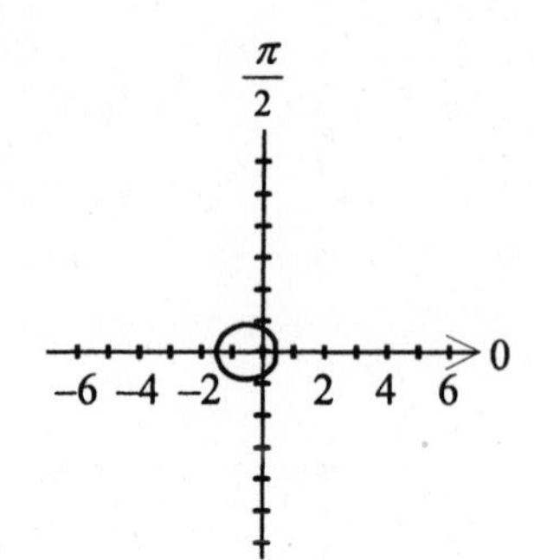

(C)

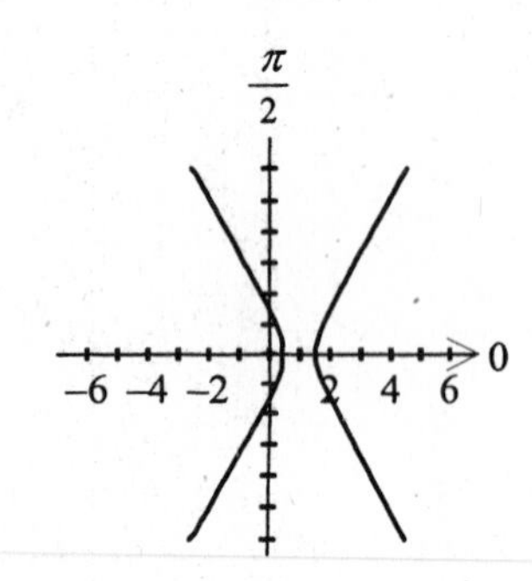

(D)

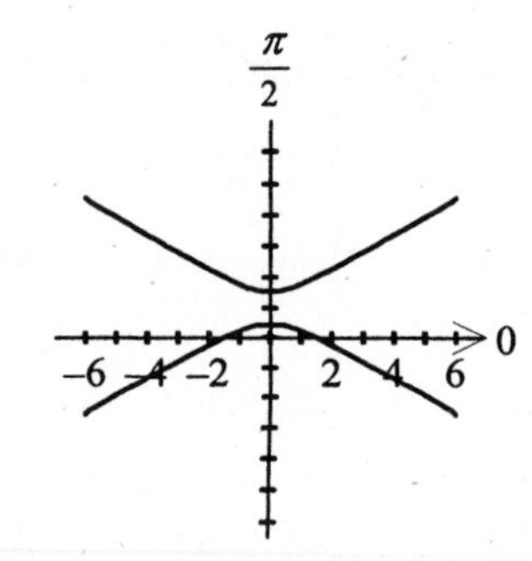

106. Identify the equation of an ellipse.

(A) $r = \frac{5}{2 - 2\cos\theta}$ (B) $r = \frac{5}{3 - 2\cos\theta}$ (C) $r = \frac{5}{2 - 3\sin\theta}$ (D) $r = \frac{5}{2 - 3\cos\theta}$

107. Use a graphing utility to graph the polar equation. Identify the graph.

$$r = \frac{4}{3 + \sin\theta}$$

108. Use a graphing utility to graph the polar equation for $e = \frac{1}{2}$.

$$r = \frac{4e}{1 + e\sin\theta}$$

Objective 2: Write equations of conics in polar form

109. Which of the following is an equation of the hyperbola with vertices at $(-9,\ 0)$ and $(5,\ \pi)$?

(A) $r = \dfrac{45}{2+7\sin\theta}$ (B) $r = \dfrac{45}{2-7\sin\theta}$ (C) $r = \dfrac{45}{2+7\cos\theta}$ (D) $r = \dfrac{45}{2-7\cos\theta}$

110. Which of the following is an equation of the parabola with vertex at $\left(\dfrac{3}{2},\ \pi\right)$?

(A) $r = \dfrac{3}{1-\sin\theta}$ (B) $r = \dfrac{3}{1+\sin\theta}$ (C) $r = \dfrac{3}{1+\cos\theta}$ (D) $r = \dfrac{3}{1-\cos\theta}$

111. Find the polar equation of a hyperbola with its focus at the pole.
Eccentricity: $e = \dfrac{8}{5}$; Directrix: $y = -\dfrac{1}{8}$

112. Find the polar equation of a hyperbola with vertices at $\left(-9,\ \dfrac{\pi}{2}\right)$ and $\left(1,\ \dfrac{3\pi}{2}\right)$.

Objective 3: Use equations of conics in polar form to model real-life problems

113. In a distant galaxy, a planet travels in an elliptical orbit with the planet's sun at one focus. The length of the major axis is $2a$, where $a = 4.89 \times 10^9$ miles, and $e = 0.0148$.
(a) Find the polar equation of the planet's orbit if the focus is at the pole.
(b) Find the minimum distance from the sun to the planet using the equation $r = a(1-e)$.

(A) (a) $r = \dfrac{4.82 \times 10^9}{1-0.0148\cos\theta}$; (b) 4.82×10^9 miles

(B) (a) $r = \dfrac{4.89 \times 10^9}{1-0.0148\cos\theta}$; (b) 7.24×10^7 miles

(C) (a) $r = \dfrac{4.89 \times 10^9}{1-0.0148\cos\theta}$; (b) 4.82×10^9 miles

(D) (a) $r = \dfrac{4.82 \times 10^9}{1-0.0148\cos\theta}$; (b) 7.24×10^7 miles

114. A comet with a parabolic orbit with the sun as its focus is 620,000 miles from the center of the sun at its closest point.
(a) Find the polar equation of the parabola with its focus at the pole and vertex at $(620{,}000,\ 0)$.
(b) Find the distance between the center of the sun and the comet when $\theta = 30°$.

(A) (a) $\dfrac{1{,}240{,}000}{1+\cos\theta}$; (b) 332,000 miles

(B) (a) $\dfrac{1{,}240{,}000}{1+\cos\theta}$; (b) 665,000 miles

(C) (a) $\dfrac{620{,}000}{1-\cos\theta}$; (b) 332,000 miles

(D) (a) $\dfrac{620{,}000}{1-\cos\theta}$; (b) 665,000 miles

115. A comet with a hyperbolic orbit has the sun as a focus.
(a) Find the polar equation of the orbit with the focus at the pole, directrix at $x = 1{,}120{,}000$, and an eccentricity of 1.14.
(b) How close does the comet come to the surface of the sun? (Assume all measurements are in miles and the diameter of the sun is 864,000 miles.)

116. A satellite travels in an elliptical orbit with a planet as a focus. It is 172 miles above the surface of the planet at its lowest point, and 124,000 miles above the surface of the planet at its highest point.
(a) Write a polar equation of the orbit with the planet at the pole and one of the vertices at $(127000,\ 0)$. (Assume the planet has a radius of 3000 miles.)
(b) Find the distance between the surface of the planet and the satellite when $\theta = \dfrac{\pi}{3}$.

Answer Key for Chapter 10 Topics in Analytic Geometry

Section 10.1: Lines

Objective 1: Find the inclination of a line

[1] (D)

[2] (D)

[3] 0.37 radians, 21°

[4] 0.09 radian, 5.2°

Objective 2: Find the angle between two lines

[5] (A)

[6] (C)

[7] 0.59 radians, 33.9°

[8] 1.39 radians, 79.8°

Objective 3: Find the distance between a point and a line

[9] (C)

[10] (D)

[11] 7.07

[12] (a) $\sqrt{13}$; (b) 13

Section 10.2: Introduction to Conics: Parabolas

Objective 1: Recognize a conic as the intersection of a plane and a double-napped cone

[13] (A)

[14] (C)

Answers may vary. Sample answer:

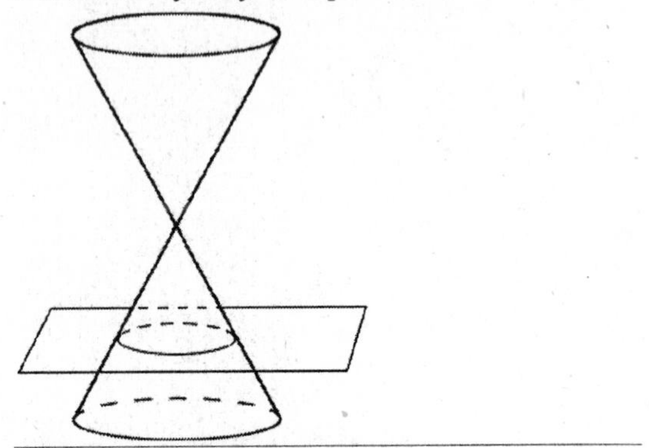

[15]

Answers may vary. Sample answer:

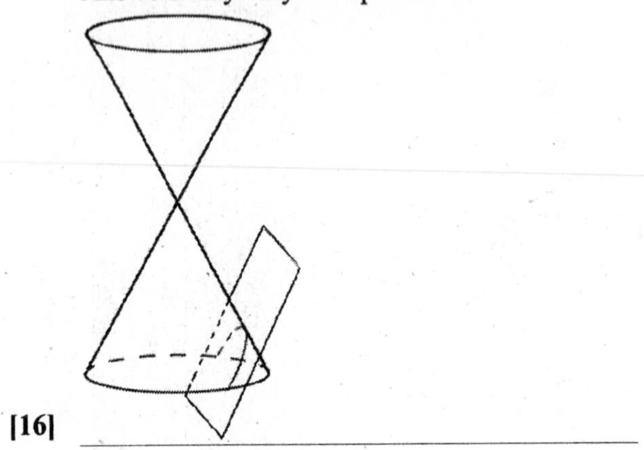

[16]

Objective 2: Write the standard form of the equation of a parabola

[17] (B)

[18] (A)

[19] $(x-1)^2 = \frac{8}{3}(y+6)$

[20] $y^2 = -16x$

Objective 3: Use the reflective property of parabolas to solve real-life problems

[21] (B)

[22] (C)

[23] (a) $y = \frac{113}{28{,}125}x^2 + 0.4$; (b) 0.464 meters

[24] 3.13 ft

Section 10.3: Ellipses

Objective 1: Write the standard form of the equation of an ellipse

[25] (B)

[26] (C)

[27] $\frac{(x+2)^2}{81} + \frac{(y-4)^2}{9} = 1$

[28] $\frac{x^2}{4} + \frac{y^2}{64} = 1$

Objective 2: Use properties of ellipses to model and solve real-life problems

[29] (A)

[30] (A)

[31] $\dfrac{x^2}{8.10\times10^9}+\dfrac{y^2}{5.33\times10^9}=1$

[32] $\dfrac{x^2}{676}+\dfrac{y^2}{324}=1$

Objective 3: Find the eccentricity of an ellipse

[33] (A)

[34] (A)

[35] $\dfrac{\sqrt{13}}{7}$

[36] $\dfrac{2\sqrt{6}}{7}$

Section 10.4: Hyperbolas

Objective 1: Write the standard form of the equation of a hyperbola

[37] (A)

[38] (B)

[39] $\dfrac{(x+4)^2}{49}-\dfrac{(y-1)^2}{36}=1$

[40] $\dfrac{x^2}{4}-\dfrac{y^2}{25}=1$

Objective 2: Find the asymptotes of a hyperbola

[41] (A)

[42] (B)

[43] Asymptotes: $y = \pm \dfrac{9}{4}(x+2) - 5$

[44] Asymptotes: $y = \pm \dfrac{2}{3}x$

Objective 3: Use properties of hyperbolas to solve real-life problems

[45] (C)

[46] (D)

[47] $(-2500,\ 640.909)$

[48] (a) The ship would reach shore 100 miles from the master station.
(b) 0.00151 second
(c) The ship is at $(272.1,\ 80)$.

Objective 4: Classify a conic from its general equation

[49] (B)

[50] (D)

[51] Circle

[52] Circle

Section 10.5: Rotation of Conics

Objective 1: Rotate the coordinate axes to eliminate the xy-term in the equation of a conic

[53] (A)

[54] (C)

[55] $\frac{(x')^2}{6} - \frac{(y')^2}{6} = 1$

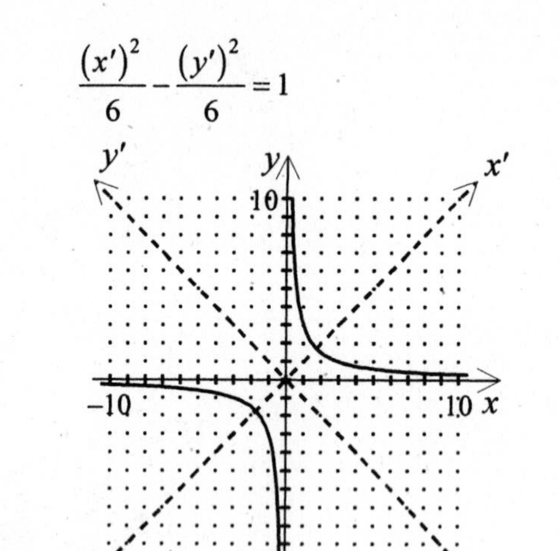

[56] $\theta = 18.43°$

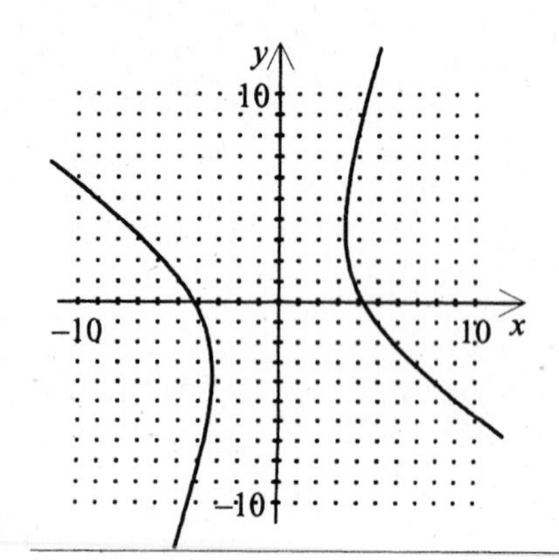

Objective 2: Use the discriminant to classify a conic

[57] (A)

[58] (B)

ellipse

$$y = \frac{-(x-2) \pm \sqrt{(x-2)^2 - 4(2)(x^2+4x)}}{2(2)}$$

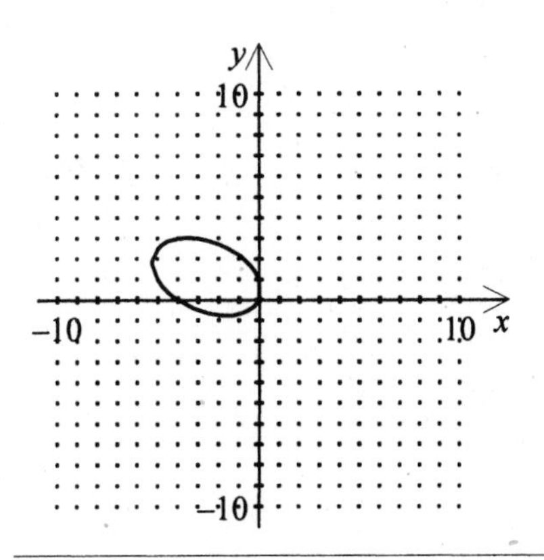

[59]

Hyperbola

$$y = \frac{-(12x+3) \pm \sqrt{(12x+3)^2 - 4(-13)(3x^2+4x-73)}}{2(-13)}$$

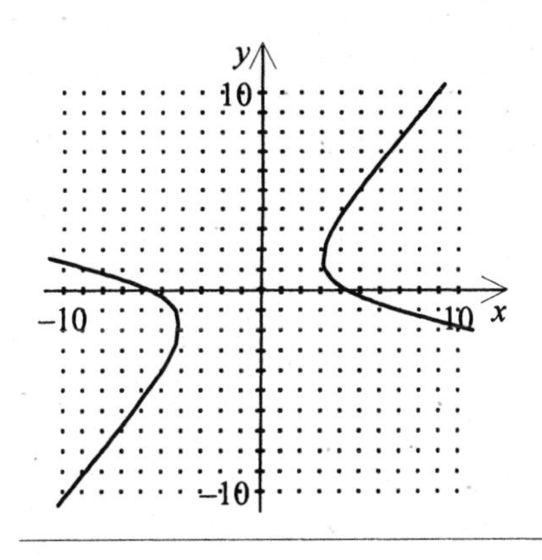

[60]

Section 10.6: Parametric Equations

Objective 1: Evaluate a set of parametric equations for a given value of the parameter

[61] (B)

[62] (A)

[63] $x = 1,\ y = -3 - 2\sqrt{3}$

[64] $x = 3, y = 0$

Objective 2: Sketch the curve that is represented by a set of parametric equations

[65] (D)

[66] (B)

$$\frac{x^2}{25} + \frac{y^2}{9} = 1$$

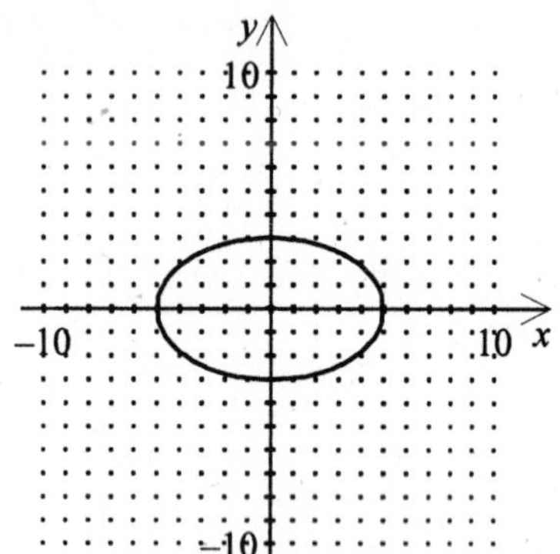

[67]

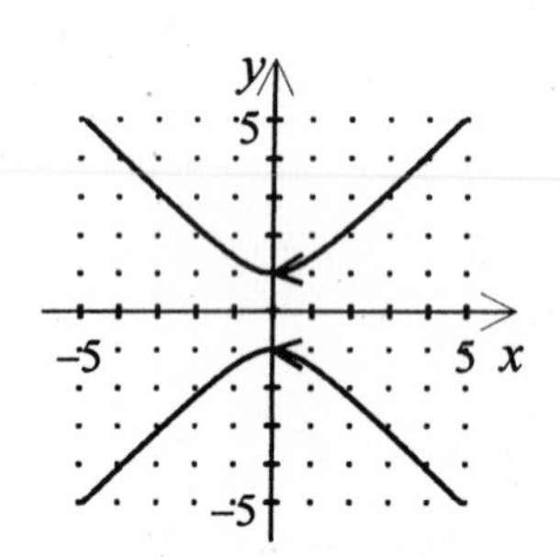

[68]

Objective 3: Rewrite a set of parametric equations as a single rectangular equation

[69] (C)

[70] (D)

[71] $\frac{x^2}{16} + \frac{y^2}{25} = 1$

[72] $x = 8\left(\frac{y-5}{9}\right)^2 - 6$

Objective 4: Find a set of parametric equations for a graph

[73] (D)

[74] (D)

[75] $x = 4t$ and $y = -t^2 - t + 8$

[76] $x = t$ and $y = t^3 + t - 2$

Section 10.7: Polar Coordinates

Objective 1: Plot points on the polar coordinate system

[77] (C)

[78] (D)

$\left(-5, \frac{7\pi}{4}\right), \left(5, -\frac{5\pi}{4}\right)$

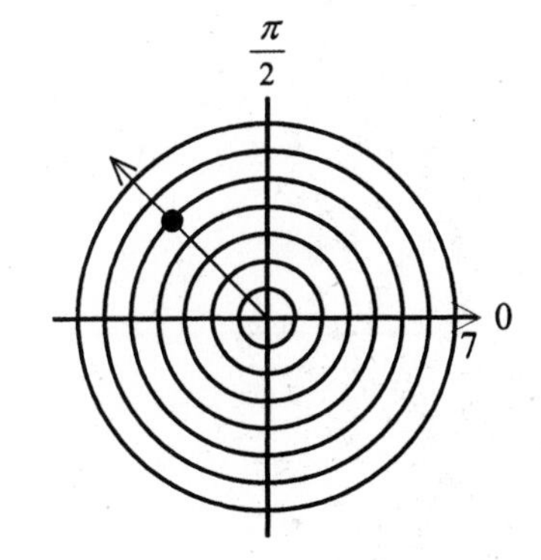

[79]

$(5,\ 8.38)$, $(-5,\ -1.04)$

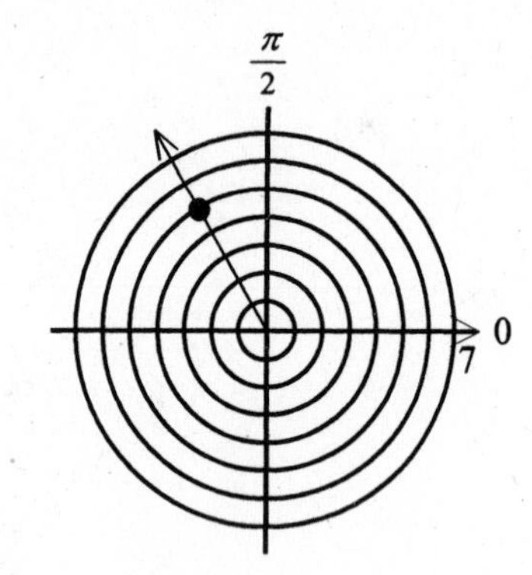

[80] ______________________

Objective 2: Convert points from rectangular to polar form and vice versa

[81] (A)

[82] (C)

[83] $\left(6,\ \frac{2\pi}{3}\right)$

[84] $(0, -7)$

Objective 3: Convert equations from rectangular to polar form and vice versa

[85] (C)

[86] (B)

[87] $r = -8 \sin \theta$

[88] $4x^2 - 4y - 1 = 0$

Section 10.8: Graphs of Polar Equations

Objective 1: Graph polar equations by point plotting

[89] (C)

[90] (A)

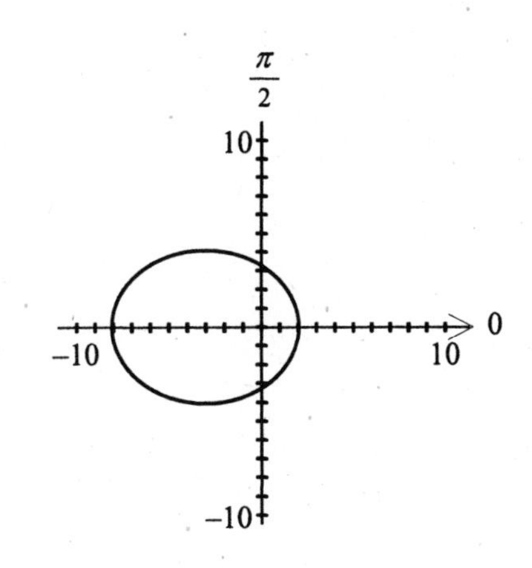

[91]

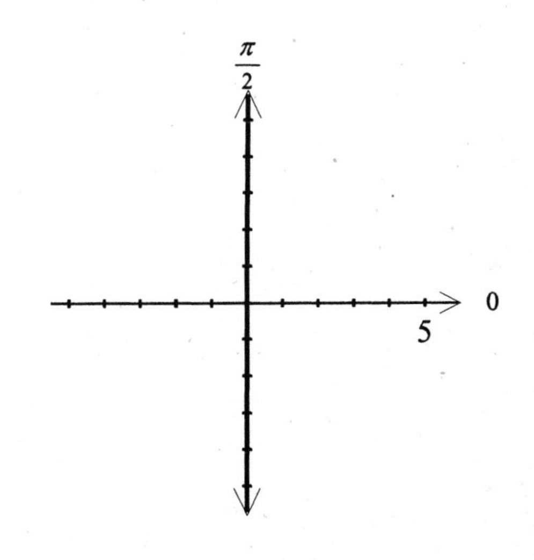

[92]

Objective 2: Use symmetry to sketch graphs of polar equations

[93] (B)

[94] (B)

[95] The graph is symmetric with respect to $\theta = \frac{\pi}{2}$.

[96] The line $\theta = \frac{\pi}{2}$: The graph is symmetric with respect to the line $\theta = \frac{\pi}{2}$.
Polar axis: The graph is symmetric with respect to the polar axis.
The pole: The graph is symmetric with respect to the pole.

Objective 3: Use zeros and maximum r-values to sketch graphs of polar equations

[97] (B)

[98] (A)

[99] $4;\ \theta = \frac{\pi}{8}, \frac{3\pi}{8}, \frac{5\pi}{8}, \frac{7\pi}{8}, \frac{9\pi}{8}, \frac{11\pi}{8}, \frac{13\pi}{8}, \frac{15\pi}{8}$

[100] $10\sqrt{2} + 20;\ \theta = \frac{\pi}{4}, \frac{3\pi}{4}$

Objective 4: Recognize special polar graphs

[101] (C)

[102] (B)

[103] Rose

[104] Rose

Section 10.9: Polar Equations of Conics

Objective 1: Define conics in terms of eccentricity

[105] (D)

[106] (B)

Ellipse

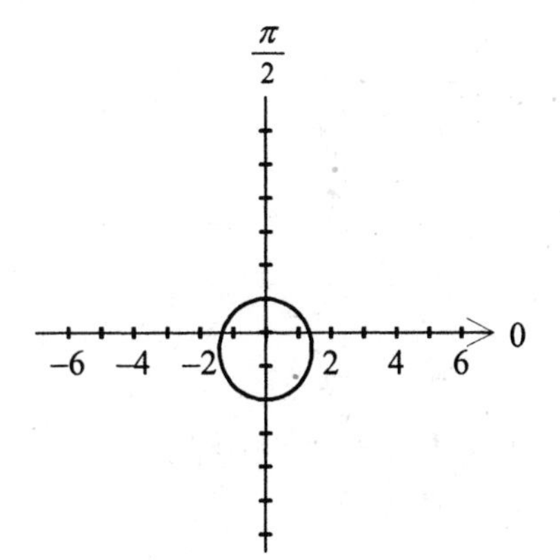

[107] ______

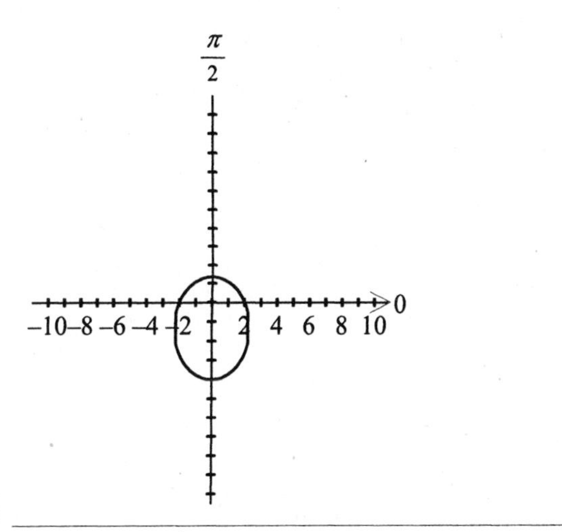

[108] ______

Objective 2: Write equations of conics in polar form

[109] (D)

[110] (D)

[111] $r = \dfrac{1}{5 - 8\sin\theta}$

[112] $r = \dfrac{9}{4 - 5\sin\theta}$

Objective 3: Use equations of conics in polar form to model real-life problems

[113] (C)

[114] (B)

[115] (a) $r = \dfrac{1{,}276{,}800}{1 + 1.14\cos\theta}$; (b) 165,000 miles

[116] (a) $r = \dfrac{6189.41}{1 - 0.951264\cos\theta}$
(b) 8803.56 miles

Appendix A Review of Fundamental Concepts of Algebra

Section A.1: Real Numbers and Their Properties

Objective 1: Represent and classify real numbers

1. Identify which of the following numbers are natural numbers.

$5,\ -\frac{7}{20},\ -7,\ 10\pi,\ 15,\ 0.1,\ \sqrt{7},\ -\frac{7}{25}$

(A) $10\pi,\ 15$ (B) $5,\ \sqrt{7}$ (C) $5,\ 15$ (D) $\sqrt{7},\ 15,\ -\frac{7}{25},\ 5$

2. Write the rational number as the ratio of two integers.

$0.\overline{96}$

(A) $\frac{48}{5}$ (B) $\frac{48}{25}$ (C) $\frac{32}{33}$ (D) $\frac{24}{25}$

3. Identify the numbers that are *not* integers.

$24,\ \sqrt{2},\ 2.3,\ 0,\ -4.2222,\ \frac{4}{7},\ -9$

4. Use a calculator to find the decimal form of the rational number. If it is a nonterminating decimal, write the repeating pattern.

$\frac{4}{11}$

Objective 2: Order real numbers and use inequalities

5. Use inequality notation to describe the set.
Julia scored at least 22 points.

(A) $k > 22$ (B) $k \le 22$ (C) $k < 22$ (D) $k \ge 22$

6. Identify the verbal description of the interval.

$[-10,\ 3)$

(A) All real numbers greater than -10 and less than or equal to 3

(B) All real numbers greater than -10 and less than 3

(C) All real numbers greater than or equal to -10 and less than or equal to 3

(D) All real numbers greater than or equal to -10 and less than 3

7. Place the correct symbol (<, >, or =) between the pair of real numbers.

$-\frac{5}{6} \bigcirc -\frac{3}{4}$

8. Verbally describe the subset of real numbers represented by the inequality. Then sketch the subset on the real number line. State whether the interval is bounded or unbounded.

$x < 4$

Objective 3: Find the absolute values of real numbers and find the distance between two real numbers

9. Caleb is on a mountain 11,246 feet above sea level. Leslie is in a submarine 3379 feet below sea level. Which of the following can be used to find the difference between Caleb's elevation and Leslie's elevation?

(A) $|11{,}246|-|-3379|$ (B) $|-3379-11{,}246|$

(C) $|11{,}246-3379|$ (D) $|3379-11{,}246|$

10. Identify the expression that corresponds to the distance shown by the bar.

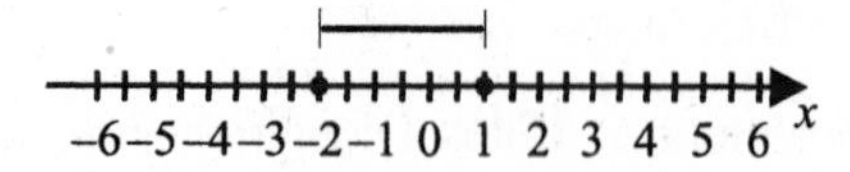

(A) $|2+1|$ (B) $-2+1$ (C) $-2-1$ (D) $|-2-1|$

11. Order the values from smallest to largest.

$|-15|, 11, |-7|, 2$

12. Evaluate the expression.

$|-3| - |-11|$

Objective 4: Evaluate algebraic expressions

Evaluate the expression for the given value of x.

13. $(5 \cdot x)\left(\frac{2}{7}\right)$ for $x = -3$ (A) $\frac{10}{7}$ (B) $-\frac{7}{30}$ (C) $-\frac{10}{21}$ (D) $-\frac{30}{7}$

Evaluate the expression for the given value of x.

14. $\dfrac{x^2-4}{8-x^3}$ for $x=15$ (A) $-\dfrac{221}{343}$ (B) $\dfrac{1}{13}$ (C) $-\dfrac{17}{259}$ (D) None of these

15. $-x^2+x+3$ for $x=5$

16. $4x^2+2x-5$ for $x=-5$

Objective 5: Use the basic rules and properties of algebra

17. Identify which of the following is an example of the Commutative Property of Multiplication.

(A) $7\cdot(8\cdot5)=(7\cdot8)\cdot5$ (B) $7+8=8+7$ (C) $7\cdot8=8\cdot7$ (D) $5\cdot(7+8)=5\cdot7+5\cdot8$

18. Identify the rule of algebra illustrated by the equation.
$85+0=85$

(A) Commutative Property of Addition (B) Distributive Property

(C) Associative Property of Addition (D) Additive Identity Property

19. Give an example of the Multiplicative Inverse Property.

20. Perform the operation. (Write the fractional answer in simplest form.)
$\dfrac{3}{2}\div\dfrac{2}{9}$

Section A.2: Exponents and Radicals

Objective 1: Use properties of exponents

21. Simplify the expression. (A) $9x$ (B) $4x^{10}$ (C) $\dfrac{4x^{10}}{9}$ (D) $\dfrac{9}{4x^{10}}$
$\left(\dfrac{3x^{-2}}{2x^3}\right)^{-2}$

22. Evaluate the expression. (A) $-\dfrac{216}{35}$ (B) $\dfrac{37}{216}$ (C) $-\dfrac{35}{216}$ (D) $\dfrac{216}{37}$
$6^{-3}-6^{-1}$

23. Evaluate each expression.

(a) $\dfrac{1}{9^{-2}}$ (b) $(-6)^{-3}$ (c) $\sqrt[3]{-27}$ (d) 4^0

24. Simplify the expression.

$$\frac{\left(2^4x^{-5}y\right)^{-1}}{\left(2^5x^2y^3\right)^{-3}}$$

Objective 2: Use scientific notation to represent real numbers

25. Identify the number written in decimal form.

1.14×10^7

(A) 11,400,000 (B) 114,000,000 (C) 1,140,000 (D) 0.000000114

26. Find the product written in scientific notation.

$\left(9.4 \times 10^{-4}\right)\left(2.1 \times 10^1\right)$

(A) 1.974×10^{-2} (B) 1.974×10^{-3} (C) 11.5×10^{-3} (D) 11.5×10^{-4}

Write the number in scientific notation.

27. A virus takes up a volume of approximately 0.000000000000029 cubic centimeter.

28. In 1995, Cambodia had a population of about 10,720,000 people.

Objective 3: Use properties of radicals

29. Perform the operation and simplify the expression.

$\left(\sqrt[3]{x^5}\right)^3$

(A) $x^{3/5}$ (B) $x^{9/5}$ (C) x^5 (D) x^{15}

Evaluate the expression without using a calculator.

30. $\sqrt[4]{\dfrac{1}{16}}$ (A) $\dfrac{1}{4}$ (B) $\dfrac{1}{2}$ (C) $\dfrac{1}{32}$ (D) $\dfrac{1}{8}$

Evaluate the expression without using a calculator.

31. $\sqrt[3]{9} \cdot \sqrt[3]{3}$

32. Use a calculator to approximate the number. (Round your answer to three decimal places.)
$\sqrt[6]{72}$

Objective 4: Simplify and combine radicals

33. Simplify by removing all possible factors from each radical.
$\sqrt{20}$

(A) $10\sqrt{2}$ (B) $5\sqrt{2}$ (C) $2\sqrt{5}$ (D) $2\sqrt{10}$

Simplify the expression.

34. $2\sqrt{6} - \sqrt{81} + 4\sqrt{96}$ (A) $9\sqrt{6}$ (B) $18\sqrt{6} - 9 + 4\sqrt{96}$ (C) $5\sqrt{183}$ (D) $18\sqrt{6} - 9$

35. $\sqrt{63} + \sqrt{28}$

36. $8\sqrt{2} - \sqrt{81} + 8\sqrt{18}$

Objective 5: Rationalize denominators and numerators

37. Rationalize the denominator of the expression. Then simplify the answer.
$\frac{4}{7 - \sqrt{6}}$

(A) $\frac{4\sqrt{6}}{7\sqrt{6} - 6}$ (B) $\frac{28 + \sqrt{6}}{43}$ (C) $\frac{28 + 4\sqrt{6}}{43}$ (D) $\frac{16}{55}$

Rationalize the numerator of the expression. Then simplify the answer.

38. $\frac{2 - \sqrt{2}}{4}$ (A) $\frac{2\sqrt{2} - 2}{4\sqrt{2}}$ (B) $\frac{3}{8}$ (C) $\frac{2}{8 + \sqrt{2}}$ (D) $\frac{1}{4 + 2\sqrt{2}}$

Rationalize the numerator of the expression. Then simplify the answer.

39. $\dfrac{12\sqrt{6}}{5}$

40. Rationalize the denominator of the expression. Then simplify the answer.

$\dfrac{\sqrt{x}}{\sqrt{x}+\sqrt{3}}$

Objective 6: Use properties of rational exponents

41. Perform the operation and simplify.

$x^{-1/3} \cdot x^{-3/5}$

(A) $\dfrac{1}{x^{14/15}}$ (B) $\dfrac{1}{x^{1/5}}$ (C) $x^{14/15}$ (D) $x^{1/5}$

42. Write the expression as a single radical. Then simplify your answer.

$\sqrt{\sqrt[5]{2x}}$

(A) $\sqrt{10x^2}$ (B) $\sqrt{2x^{10}}$ (C) $\sqrt[10]{2x}$ (D) $\sqrt{10x}$

43. Use a calculator to approximate the number. (Round your answer to three decimal places.)

$(1.9)^{-1.1} + 3\sqrt{12}$

44. Evaluate the expression without using a calculator.

$(-4)^{1/2}$

Section A.3: Polynomials and Factoring

Objective 1: Write polynomials in standard form

45. Identify the polynomial written in standard form.

$\dfrac{3x^2+5x-8}{2}$

(A) $\dfrac{3}{2}x^2+\dfrac{5}{2}x-4$ (B) $\dfrac{3}{2}x^2+5x-8$ (C) $6x^2+10x-16$ (D) Not a polynomial

46. Identify the polynomial written in standard form.

(A) $4-4x^2-3x^4-5x$ (B) $-4x^2-3x^4-5x+4$

(C) $-5x+4-3x^4-4x^2$ (D) $-3x^4-4x^2-5x+4$

47. Write a polynomial that fits the description. (There are many correct answers.)
A second-degree polynomial with leading coefficient 7

48. Write the polynomial in standard form.
$-5x^2-2x+2x^3-3$

Objective 2: Add, subtract, and multiply polynomials

Perform the operations and identify the result written in standard form.

49. $4a^4\left(3a^4+6a^3-5a+4\right)$

(A) $12a^8+3a^7-5a^5+4a^4$ (B) $12a^8+24a^7-20a^5+16a^4$

(C) $7a^8+10a^7-a^5+8a^4$ (D) $12a^{16}+24a^{12}-4a^4$

50. $(5x+4)(4x+4)$

(A) $20x^2+4x-17$ (B) $20x^2+36x-17$ (C) $20x^2+36x+16$ (D) $20x^2-35x+16$

51. $\left(5k^2+3\right)-(3k+6)-\left(9k^2-5k\right)$

52. $3x(x-3)+(x-4)(x-5)$

Objective 3: Use special products to multiply polynomials

Multiply or find the special product.

53. $(4x-6y)^2$

(A) $16x^2-4xy+36y^2$ (B) $16x^2-48xy+36y^2$

(C) $16x^2+36y^2$ (D) $16x^2-24xy+36y^2$

Multiply or find the special product.

54. $\left(-2x-y^3\right)^3$

(A) $-8x^3-4x^2y^3-2xy^6-y^9$

(B) $-8x^3-12x^2y^3-6xy^6-y^9$

(C) $-x^3-6x^2y^3+12xy^6-3y^9$

(D) $-8x^3-y^9$

55. $\left(4x-\frac{3}{5}y\right)\left(4x+\frac{3}{5}y\right)$

56. $\left[(x-6)+5\right]\left[(x-6)-5\right]$

Objective 4: Remove common factors from polynomials

Factor out the common factor.

57. $\frac{3}{5}x^7-\frac{3}{5}x^6+\frac{3}{5}x^5-\frac{4}{5}x^3$

(A) $x^3\left(\frac{3}{5}x^4-\frac{3}{5}x^3+\frac{3}{5}x^2-\frac{4}{5}\right)$

(B) $\frac{1}{5}x^3\left(3x^4+3x^3+3x^2+4\right)$

(C) $x^3\left(3x^4-3x^3+3x^2-4\right)$

(D) $\frac{1}{5}x^3\left(3x^4-3x^3+3x^2-4\right)$

58. $16x^5-40x^8$

(A) $8x^5\left(2-5x^3\right)$

(B) $8\left(2x^5-5x^8\right)$

(C) $8x^4\left(2x-5x^7\right)$

(D) $x^5\left(16-40x^3\right)$

59. $5x(x+2)+2(x+2)$

60. $15x^5-9x^4+12x^3+9x^2$

Objective 5: Factor special polynomial forms

Completely factor the expression.

61. $25n^2 + 60n + 36$

(A) $(5n+6)(5n-6)$ (B) $(5n+6)^2$ (C) $(5n-36)(5n+1)$ (D) $(5n-6)^2$

62. $81x^2 - 16y^2$

(A) $(9x-4y)^2$ (B) $(9x+4y)(9x+4y)$

(C) $(9x+4y)(9x-4y)$ (D) $(9x-4y)(9x-4y)$

63. $-9x^4 + 81x^2$

64. $x^3 - 8$

Objective 6: Factor trinomials as the product of two binomials

Factor the trinomial.

65. $(f+g)y^2 + 6y(f+g) - 7(f+g)$

(A) $(f+g)(y-7)(y+1)$ (B) $(f+g)(y+7)(y-1)$

(C) $(y+7)(y-1)$ (D) $-(f+g)(y^2-6y-7)$

66. $2(x-6)^2 + 11(x-6) + 12$

(A) $(2x-2)(3x-9)$ (B) $(x-2)(2x-9)$ (C) $(x-2)(3x-9)$ (D) $(2x-2)(2x-9)$

67. $30x^2 + 21x - 36$

68. $4v^5 + 12v^4 - 40v^3$

Objective 7: Factor polynomials by grouping

Factor by grouping.

69. $2x^3 + 7x^2 - 6x - 21$

(A) $(x-3)(2x-7)$ (B) $(x^2-3)(2x+7)$ (C) $(x^2-3)(2x-7)$ (D) $(x-3)(2x+7)$

70. $3x^7 - 6x^5 + 2x^4 - 4x^2$

(A) $x^2(3x^3-2)(x^2+2)(x+1)$ (B) $x^2(3x^3-2)(x^2+2)$

(C) $x^2(3x^3+2)(x^2-2)$ (D) $x^2(3x^3+2)(x^2-2)(x+1)$

71. $4x^8 + 7x^7 - 12x - 21$

72. $3x^7 - 15x^5 + 2x^3 - 10x$

Section A.4: Rational Expressions

Objective 1: Find domains of algebraic expressions

Find the domain of the expression.

73. $\dfrac{-7}{6+x}$

(A) All real numbers x such that $x \neq -6$

(B) All real numbers x such that $x \neq 6$

(C) All real numbers x such that $x \neq 0$

(D) All real numbers x such that $x \neq -7$

74. $3x^2 + 2$

(A) All real numbers x such that $x \geq 0$ (B) All real numbers

(C) All real numbers x such that $x \leq 0$ (D) All real numbers x such that $x \geq 3$

Find the domain of the expression.

75. $\dfrac{x^2+7x+10}{x^2-2x-48}$

76. $\sqrt{x+49}$

Objective 2: Simplify rational expressions

Write the rational expression in simplest form.

77. $\dfrac{-x}{x-x^2}$

(A) $\dfrac{1}{x-1},\ x \neq 0, x \neq 1$

(B) $-\dfrac{1}{x+1},\ x \neq 0, x \neq -1$

(C) $\dfrac{1}{x+1},\ x \neq 0, x \neq -1$

(D) $-\dfrac{1}{x-1},\ x \neq 0, x \neq 1$

78. $\dfrac{x^2+x-30}{5x^2-125}$

(A) $\dfrac{x+5}{5(x+6)},\ x \neq \pm 6$

(B) $\dfrac{x+6}{5(x+5)},\ x \neq \pm 5$

(C) $\dfrac{x-6}{5(x-5)},\ x \neq \pm 5$

(D) $\dfrac{x-5}{5(x-6)},\ x \neq \pm 6$

79. $\dfrac{x^2+2x-24}{4-x}$

80. $\dfrac{x^2+7x-8}{x^2+19x+88}$

Objective 3: Add, subtract, multiply, and divide rational expressions

Perform the operation(s) and simplify.

81. $\dfrac{x}{x^2-1}-\dfrac{1}{1-x^2}$ (A) $\dfrac{1}{x+1}$ (B) $\dfrac{x+1}{x-1}$ (C) $\dfrac{1}{x-1}$ (D) $\dfrac{x-1}{x+1}$

82. $\dfrac{3y^2+17y+10}{3y^2-17y+10}\cdot\dfrac{3y^2-2y}{25-y^2}$

(A) $\dfrac{y(3y+2)}{(5-y)(y-5)}$ (B) $\dfrac{(3y-2)(3y+2)}{(5-y)(y-5)}$ (C) $\dfrac{y(3y-2)}{(5-y)(y-5)}$ (D) $\dfrac{y}{(5-y)(y-5)}$

83. $\dfrac{x-1}{x^2-4}-\dfrac{x-4}{x^2-2x-8}$

84. $\dfrac{2x^2-x-1}{2x^2+5x+2}\div\dfrac{x^2+4x-5}{2x^2+14x+20}$

Objective 4: Simplify complex fractions

Simplify the complex fraction.

85. $\dfrac{\dfrac{2}{x^2+2x-8}+\dfrac{1}{x^2-7x+10}}{\dfrac{3}{x^2-x-20}+\dfrac{1}{x^2-3x+2}}$

(A) $\dfrac{4x^2-3x+2}{3(2x-1)(x+4)}$ (B) $\dfrac{3(2x-1)(x+4)}{4x^2-3x+2}$ (C) $\dfrac{3(x-1)(x-2)}{2(x+1)(2x-7)}$ (D) $\dfrac{2(x+1)(2x-7)}{3(x-1)(x-2)}$

86. $\dfrac{\left(36-x^2\right)^{1/2}+5x^2\left(36-x^2\right)^{-1/2}}{36-x^2}$

(A) $\dfrac{36+4x^2}{\left(36-x^2\right)^{3/2}}$ (B) $\dfrac{41x^2}{\left(36-x^2\right)^{3/2}}$ (C) $\dfrac{36+3x^2}{\left(36-x^2\right)^{3/2}}$ (D) None of these

Simplify the complex fraction.

87. $\dfrac{\dfrac{x^2+2x+1}{15x}}{\dfrac{x+1}{3x}}$

88. $\dfrac{\dfrac{2}{x}+\dfrac{1}{2x}}{\dfrac{1}{3x}-\dfrac{3}{4x}}$

Section A.5: Solving Equations

Objective 1: Identify different types of equations

89. Which of the following is an identity?

(A) $7(x-3)=x-21$ (B) $7(x-3)=7x-3$

(C) $7(x-3)=21-7x$ (D) $7(x-3)=7x-21$

90. Which of the following is a conditional equation?

(A) $\dfrac{5}{x}\cdot\left(-\dfrac{9}{x}\right)=-\dfrac{45}{x^2}$ (B) $\dfrac{x}{5}\div\dfrac{x}{9}=\dfrac{9}{5}$ (C) $\dfrac{5}{x}-\dfrac{9}{x}=-\dfrac{4}{x}$ (D) $\dfrac{5}{x}-\dfrac{9}{x}=-4$

91. Determine whether the equation is an identity or a conditional equation.
$(5x+3)^2=3(x+6)$

92. Determine whether the equation is a conditional or an identity.
$x^2+10x+23=(x+5)^2-2$

Objective 2: Solve linear equations in one variable

Solve the equation and check your solution.

93. $1=2(x-1)+2-x$ (A) 1 (B) 5 (C) 4 (D) 0

Solve the equation and check your solution.

94. $4(x+3)+1=5(x+2)+5$ (A) -3 (B) -2 (C) 9 (D) 26

95. Find the solution to the equation.
$5(x-5)+5x=4x-3$

96. Solve the equation and check your solution.
$1\frac{1}{2}j-27=9$

Objective 3: Solve quadratic equations by factoring, extracting square roots, completing the square, and using the Quadratic Formula

97. Use the quadratic formula to solve the equation.
$2x^2-1=7x$

(A) $\frac{7\pm\sqrt{57}}{4}$ (B) $\frac{-7\pm\sqrt{41}}{4}$ (C) $\frac{-7\pm\sqrt{57}}{4}$ (D) $\frac{7\pm\sqrt{41}}{4}$

98. Solve the equation by factoring. (A) $-8, 1$ (B) $-8, -1$ (C) $-1, 8$ (D) $1, 8$
$x^2+9x+8=0$

99. Solve the equation by extracting square roots.
$(x-2)^2=1$

100. Use the quadratic formula to solve the equation.
$2x^2+8x+2=0$

Objective 4: Solve polynomial equations of degree three or greater

Find all real solutions of the equation. Check your solutions in the original equation.

101. $21x^4-56x^2+35=0$

(A) $\pm\sqrt{\frac{5}{3}}, \pm 1$ (B) $\pm\sqrt{\frac{1}{3}}, \pm\sqrt{\frac{5}{6}}$ (C) $\pm 1, \pm 2$ (D) $\pm\sqrt{\frac{5}{7}}, \pm\sqrt{\frac{7}{3}}$

Find all real solutions of the equation. Check your solutions in the original equation.

102. $2x^4 - 3x^3 - 9x^2 + 8x + 12$

(A) $(2x-3)(x+2)(x-1)(x+2)$ (B) $(2x-3)(x-2)(x-1)(x-2)$

(C) $(2x+3)(x-2)(x+1)(x-2)$ (D) $(2x+3)(x+2)(x+1)(x+2)$

103. $x^3 + 4x^2 - 49x - 196 = 0$

104. $343x^3 + 64 = 0$

Objective 5: Solve equations involving radicals

Find all real solutions of the equation. Check your solutions in the original equation.

105. $\sqrt{x-5} + 7 = -5$ (A) 7 (B) 149 (C) –29 (D) No solution

106. $(x-6)^{5/4} = 243$ (A) –75 or 87 (B) –75 (C) 87 (D) 87 or 75

107. $3x(x+75)^{1/2} + 6(x+75)^{3/2} = 0$

108. $\sqrt{3x-5} = \sqrt{2x+2}$

Objective 6: Solve equations involving absolute values

Find all real solutions of the equation. Check your solutions in the original equation.

109. $|x^2 - 2| = 2x + 6$ (A) –3 (B) –2, 4 (C) 1, 9 (D) 2, 6

110. $4|x-3| - 16 = -20$ (A) 3 (B) –4, –2 (C) 2, 4 (D) No solution

111. $|x-2| = x^2 - 4$

Find all real solutions of the equation. Check your solutions in the original equation.

112. $|6x-4|=6$

Section A.6: Solving Inequalities

Objective 1: Represent solutions of linear inequalities in one variable

113. Which is the graph of the inequality? $3<x<6$

(A)

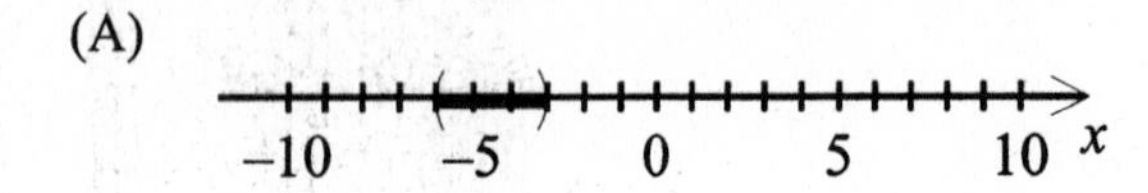

(B)

−10 −5 0 5 10 x

(C)

−10 −5 0 5 10 x

(D)

−10 −5 0 5 10 x

114. Which of the following intervals represents $-6 \le x < 6$?

(A) $(-6, 6)$ (B) $[-6, 6]$ (C) $[-6, 6)$ (D) $(-6, 6]$

Graph.

115. $-3 < x < 4$

116. $x > -4$

Objective 2: Solve linear inequalities in one variable

Solve the inequality.

117. $-2 \le -2x - 4 \le 4$

(A) $8 \le x \le 2$ (B) $-4 \le x \le -1$ (C) $-1 \le x \le -4$ (D) $2 \le x \le 8$

118. $\frac{7}{6} - \frac{1}{3}x + \frac{5}{6} \ge 6x - \frac{4}{3}$

(A) $x \le \frac{10}{17}$ (B) $x \le \frac{10}{19}$ (C) $x \ge \frac{10}{19}$ (D) None of these

119. $-1 \le \frac{5-2x}{5} \le 7$

120. $16x - 8x + 13 > 9x - (10 - 7x)$

Objective 3: Solve inequalities involving absolute values

121. Solve the inequality.
$|3-5x| - 3 < 3$

(A) $x > -\frac{3}{5}, x < -\frac{3}{5}$ (B) $-\frac{3}{5} < x < \frac{9}{5}$ (C) $x > \frac{9}{5}, x < -\frac{3}{5}$ (D) $-\frac{3}{5} < x < -\frac{3}{5}$

122. Which is the graph of the solution for the inequality?
$|8x+4| > 12$

(A) −10 −5 0 5 10 x

(B) −10 −5 0 5 10 x

(C) −10 −5 0 5 10 x

(D) −10 −5 0 5 10 x

123. Graph the solution for the inequality.
$|2x-3| \le 3$

124. Solve: $\left|\dfrac{x+5}{7}\right| - 8 \ge -7$

Objective 4: Solve polynomial and rational inequalities

125. Solve the inequality and give the solution in interval notation.
$x^2 + 8x - 8 > 0$

(A) $\left(-4 - 2\sqrt{6},\ -4 + 2\sqrt{6}\right)$

(B) $\left(-\infty,\ -4 - 2\sqrt{6}\right) \cup \left(-4 + 2\sqrt{6},\ \infty\right)$

(C) $\left(-\infty,\ -4 - 2\sqrt{6}\right] \cup \left[-4 + 2\sqrt{6},\ \infty\right)$

(D) $\left[-4 - 2\sqrt{6},\ -4 + 2\sqrt{6}\right]$

126. Solve the inequality and graph the solution on the real number line.
$x^2 - x > 12$

(A) $x < -3,\ x > 4$

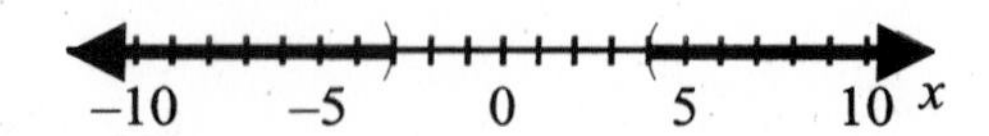

(B) $-4 < x < 3$

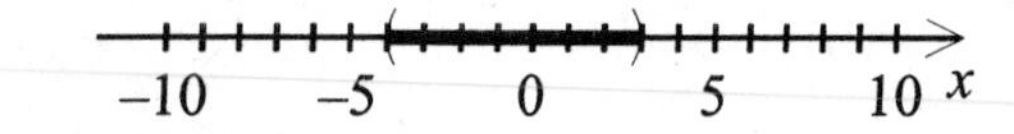

(C) $-3 < x < 4$

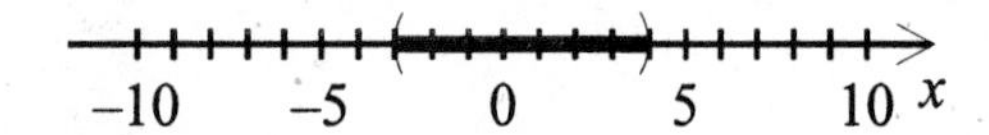

(D) $x < -4,\ x > 3$

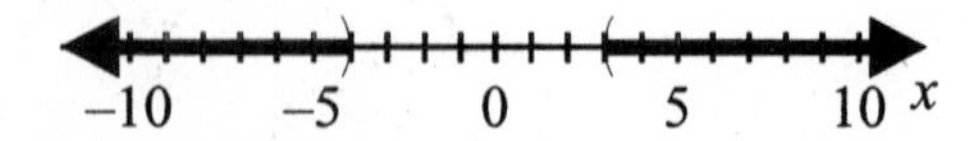

127. Solve the inequality.
$\dfrac{x+2}{x-5} \le 0$

128. Solve the inequality and give the solution in interval notation.

$$\frac{(x-9)(x+6)}{x-4} \geq 0$$

Section A.7: Errors and the Algebra of Calculus

Objective 1: Avoid common algebraic errors

129. Identify the missing factor which, if placed in the parentheses, would make the equation true.

$$(3-5x)^{6/5} + 5(3-5x)^{1/5} = (3-5x)^{1/5}(\quad)$$

(A) $(3-5x)^{6/5}+5$ (B) $-5x-8$ (C) $-5x+8$ (D) $(3-5x)^{6}+5$

130. Which equation is correct?

(A) $-5+\frac{x}{3x} = -5+\frac{1}{2x}$ (B) $\frac{-6+3x}{-6} = 1+3x$

(C) $-6+\frac{x}{3x} = -6+\frac{1}{2x}$ (D) $\frac{-5-5x}{-5} = 1+x$

131. Correct the error.

$$(4x+16)^2 = 4(x+4)^2 \quad \times$$

132. If there is an error in the equality, correct the error. If no error exists, write "True."

$$\sqrt{x-49} = \sqrt{x} - 7$$

Objective 2: Recognize and use algebraic techniques that are common in calculus

Insert the required factor in the parentheses.

133. $\frac{3x+4}{\left(x^2+2x+7\right)^3} = (\quad)\frac{1}{\left(x^2+2x+7\right)^3}(9x+12)$

(A) $\frac{1}{3}$ (B) x^2+2x+7 (C) 3 (D) None of these

Insert the required factor in the parentheses.

134. $\dfrac{16x^2}{25} - \dfrac{81x^2}{49} = \dfrac{x^2}{(\ )} - \dfrac{x^2}{(\ )}$

(A) $\dfrac{49}{81}, \dfrac{25}{16}$ (B) $\dfrac{1}{16}, \dfrac{1}{81}$ (C) $\dfrac{25}{16}, \dfrac{49}{81}$ (D) $\dfrac{1}{25}, \dfrac{1}{49}$

135. Simplify the expression.

$$\frac{(x+7)^{3/4}(x+5)^{-2/3} - (x+5)^{1/3}(x+7)^{-1/4}}{\left[(x+7)^{3/4}\right]^2}$$

136. Write the fraction as a sum of two or more terms.

$$\frac{9x^6 - 7x^5 - x^2 + 7}{x^3}$$

Section A.8: Graphical Representation of Data

Objective 1: Plot points in the Cartesian plane

137. Identify the graph of the points $A(-3,\ 1)$, $B(3,\ -2)$, and $C(1,\ -2)$.

(A)

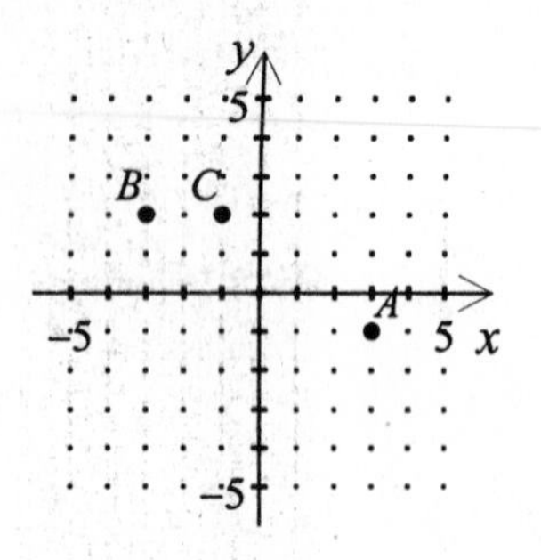

(B)

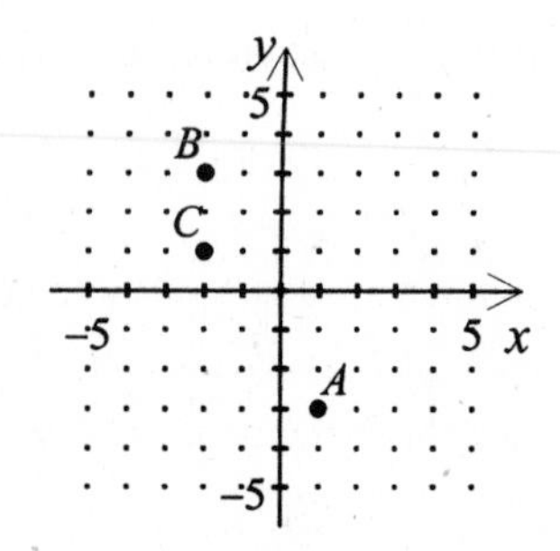

(C)

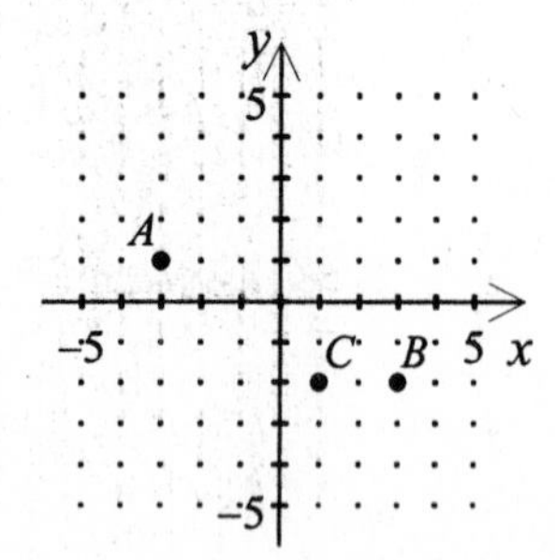

(D)

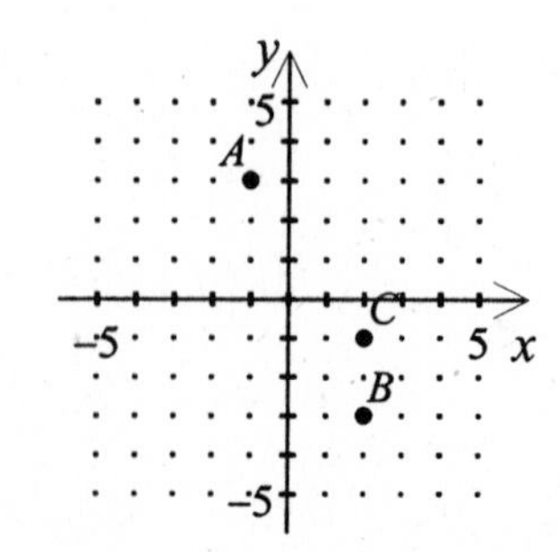

138. Determine the quadrant in which (x, y) is located so that the conditions are satisfied.

$x > 0$ and $y > 0$

(A) Quadrant I (B) Quadrant II (C) Quadrant III (D) Quadrant IV

139. Name the coordinates of point A and the quadrant in which A is located.

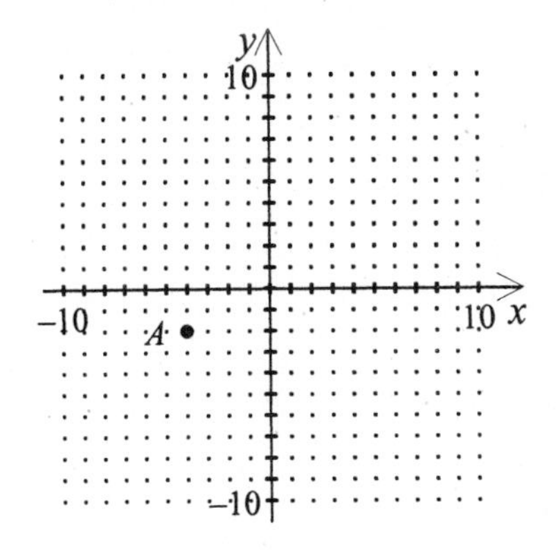

140. Find the coordinates of the point that is located 3 units to the right of the y-axis and 2 units above the x-axis.

Objective 2: Use the Distance Formula to find the distance between two points

141. Find the length of the hypotenuse of the triangle.

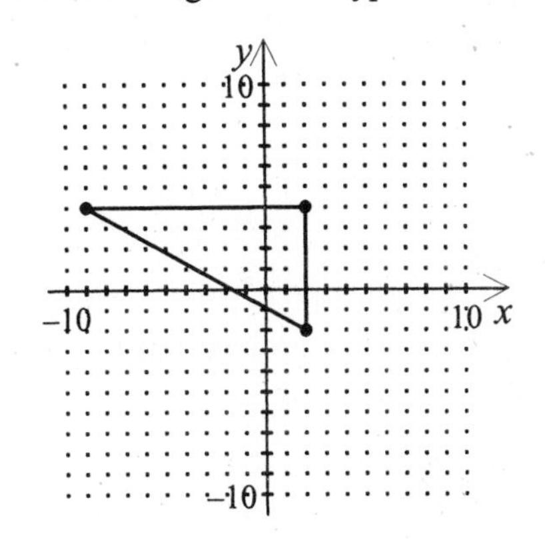

(A) 9.22 (B) 11.18 (C) 12.53 (D) 7.28

Find the distance between the points.

142. $\left(0, \sqrt{10}\right)$ and $\left(-\sqrt{2}, 0\right)$ (A) 12 (B) $\sqrt{11}$ (C) $2\sqrt{3}$ (D) $2\sqrt{26}$

Find the distance between the points.

143. $(6, 4)$ and $(-7, 4)$

144. Verify that the triangle with vertices $S(-7, 0)$, $T(-1, 0)$, and $U(-4, -4)$ is an isosceles triangle.

Objective 3: Use the Midpoint Formula to find the midpoint of a line segment

145. Find the midpoint of the line segment connecting $(5, -5)$ and $(-10, -18)$.

(A) $(-5, -23)$ (B) $\left(-\frac{5}{2}, -\frac{23}{2}\right)$ (C) $\left(\frac{15}{2}, \frac{13}{2}\right)$ (D) $(5, 23)$

146. Find the midpoint of $\overline{AB}$.

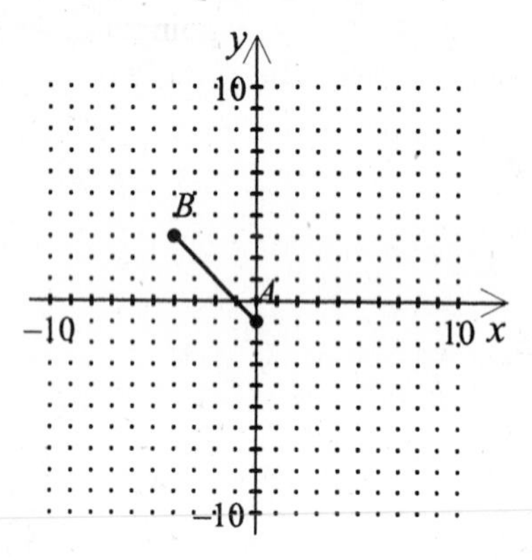

(A) $(1, -2)$ (B) $(-2, 1)$ (C) $(2, -1)$ (D) $(-1, 2)$

147. $M(-2, 1)$ is the midpoint of $\overline{RS}$. If S has coordinates $(5, 5)$, find the coordinates of R.

148. Find the midpoint of the line segment connecting the two points. Then show that the midpoint is the same distance from each point.

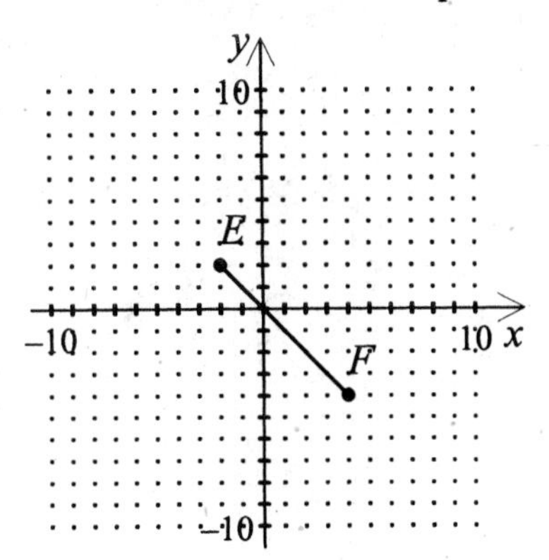

Objective 4: Use a coordinate plane to model and solve real-life problems

149. A highway map of Ohio has a coordinate grid superimposed on top of the state. Springfield is at point $(2, -4)$ and Cleveland is at point $(5, 3)$. The Springfield History Club is going to Cleveland to see the Rock and Roll Hall of Fame. The map shows a highway rest area halfway between the cities. What are the coordinates of the rest area? What is the distance between Springfield and Cleveland? (one unit = 27.31 miles)

(A) Rest area = $\left(-\frac{1}{2}, \frac{7}{2}\right)$
Cleveland = 173 miles

(B) Rest area = $\left(\frac{3}{2}, \frac{7}{2}\right)$
Cleveland = 189 miles

(C) Rest area = $\left(\frac{7}{2}, -\frac{1}{2}\right)$
Cleveland = 208 miles

(D) Rest area = $\left(-\frac{3}{2}, -\frac{7}{2}\right)$
Cleveland = 193 miles

150. On a street map, the side of each square represents 1 mile. The local Q-Mart is located at $(1, 2)$, and the Circle Cineplex is located at $(16, 10)$. A diagonal street runs directly between the two locations. Approximately how far is it from Q-Mart to Circle Cineplex?

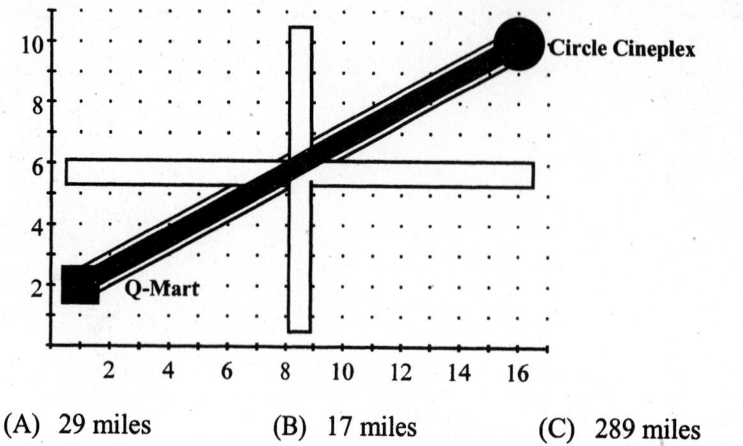

(A) 29 miles (B) 17 miles (C) 289 miles (D) 23 miles

151. Juanita made a sketch of a circular pool on a graph grid. On the graph the diameter of the pool has endpoints at $(6, 1)$ and $(-8, -5)$. What are the coordinates of the center of the pool?

152. You live in Carson City, Nevada, which has approximate latitude-longitude coordinates of $(39\ N,\ 120\ W)$ and your friend lives in Ottawa, Ohio, with coordinates of $(41\ N,\ 84\ W)$. You plan to meet halfway between the two cities. Find the coordinates of the halfway point.

Answer Key for Appendix A Review of Fundamental Concepts of Algebra

Section A.1: Real Numbers and Their Properties

Objective 1: Represent and classify real numbers

[1] (C)

[2] (C)

[3] $\sqrt{2},\ 2.3,\ -4.2222,\ \frac{4}{7}$

[4] $0.\overline{36}$

Objective 2: Order real numbers and use inequalities

[5] (D)

[6] (D)

[7] $-\frac{5}{6} < -\frac{3}{4}$

$x < 4$ denotes the set of all real numbers less than 4.

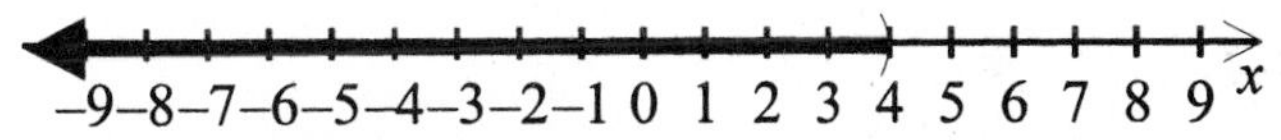

[8] The interval is unbounded.

Objective 3: Find the absolute values of real numbers and find the distance between two real numbers

[9] (B)

[10] (D)

[11] $2,\ |-7|,\ 11,\ |-15|$

[12] −8

Objective 4: Evaluate algebraic expressions

[13] (D)

[14] (C)

[15] -17

[16] 85

Objective 5: Use the basic rules and properties of algebra

[17] (C)

[18] (D)

[19] Answers may vary. Sample answer: $4 \cdot \frac{1}{4} = 1$

[20] $\frac{27}{4}$

Section A.2: Exponents and Radicals

Objective 1: Use properties of exponents

[21] (C)

[22] (C)

[23]
(a) 81
(b) $-\frac{1}{216}$
(c) -3
(d) 1

[24] $2^{11}x^{11}y^{8}$

Objective 2: Use scientific notation to represent real numbers

[25] (A)

[26] (A)

[27] 2.9×10^{-14}

[28] 1.072×10^{7}

Objective 3: Use properties of radicals

[29] (C)

[30] (B)

[31] 3

[32] 2.040

Objective 4: Simplify and combine radicals

[33] (C)

[34] (D)

[35] $5\sqrt{7}$

[36] $32\sqrt{2} - 9$

Objective 5: Rationalize denominators and numerators

[37] (C)

[38] (D)

[39] $\frac{72}{5\sqrt{6}}$

[40] $\frac{x-\sqrt{3x}}{x-3}$

Objective 6: Use properties of rational exponents

[41] (A)

[42] (C)

[43] 10.886

[44] $\sqrt{-4}$ = not a real number

Section A.3: Polynomials and Factoring

Objective 1: Write polynomials in standard form

[45] (A)

[46] (D)

[47] Answers may vary. Sample answer: $7x^2+12x+5$

[48] $2x^3-5x^2-2x-3$

Objective 2: Add, subtract, and multiply polynomials

[49] (B)

[50] (C)

[51] $-4k^2+2k-3$

[52] $4x^2 - 18x + 20$

Objective 3: Use special products to multiply polynomials

[53] (B)

[54] (B)

[55] $16x^2 - \frac{9}{25}y^2$

[56] $x^2 - 12x + 11$

Objective 4: Remove common factors from polynomials

[57] (D)

[58] (A)

[59] $(5x + 2)(x + 2)$

[60] $3x^2\left(5x^3 - 3x^2 + 4x + 3\right)$

Objective 5: Factor special polynomial forms

[61] (B)

[62] (C)

[63] $-9x^2(x + 3)(x - 3)$

[64] $(x - 2)\left(x^2 + 2x + 4\right)$

Objective 6: Factor trinomials as the product of two binomials

[65] (B)

[66] (B)

[67] $3(5x-4)(2x+3)$

[68] $4v^3(v+5)(v-2)$

Objective 7: Factor polynomials by grouping

[69] (B)

[70] (C)

[71] $(x^7-3)(4x+7)$

[72] $x(3x^4+2)(x^2-5)$

Section A.4: Rational Expressions

Objective 1: Find domains of algebraic expressions

[73] (A)

[74] (B)

[75] All real numbers x such that $x \neq 8,\ x \neq -6$

[76] All real numbers x such that $x \geq -49$

Objective 2: Simplify rational expressions

[77] (A)

[78] (B)

[79] $-x-6,\ x \neq 4$

[80] $\dfrac{x-1}{x+11},\ x \neq -11,\ x \neq -8$

Objective 3: Add, subtract, multiply, and divide rational expressions

[81] (C)

[82] (A)

[83] $\dfrac{1}{(x+2)(x-2)}$

[84] 2

Objective 4: Simplify complex fractions

[85] (C)

[86] (A)

[87] $\dfrac{x+1}{5}$

[88] –6

Section A.5: Solving Equations

Objective 1: Identify different types of equations

[89] (D)

[90] (D)

[91] Conditional equation

[92] Identity

Objective 2: Solve linear equations in one variable

[93] (A)

[94] (B)

[95] $\frac{11}{3}$

[96] 24

Objective 3: Solve quadratic equations by factoring, extracting square roots, completing the square, and using the Quadratic Formula

[97] (A)

[98] (B)

[99] $1, 3$

[100] $-2 \pm \sqrt{3}$

Objective 4: Solve polynomial equations of degree three or greater

[101] (A)

[102] (C)

[103] $\pm 7, -4$

[104] $-\frac{4}{7}$

Objective 5: Solve equations involving radicals

[105] (D)

[106] (C)

[107] –50

[108] 7

Objective 6: Solve equations involving absolute values

[109] (B)

[110] (D)

[111] –3, 2

[112] $\frac{5}{3}, -\frac{1}{3}$

Section A.6: Solving Inequalities

Objective 1: Represent solutions of linear inequalities in one variable

[113] (C)

[114] (C)

[115]
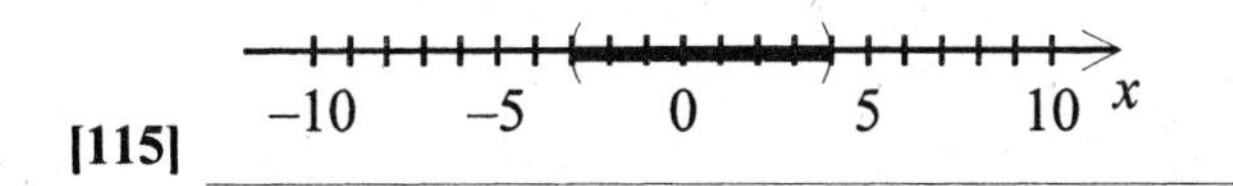

[116]
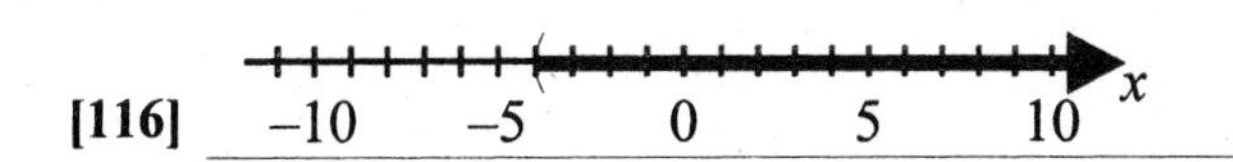

Objective 2: Solve linear inequalities in one variable

[117] (B)

[118] (B)

[119] $-15 \le x \le 5$

[120] $x < \dfrac{23}{8}$

Objective 3: Solve inequalities involving absolute values

[121] (B)

[122] (C)

[123] −10 −5 0 5 10 x

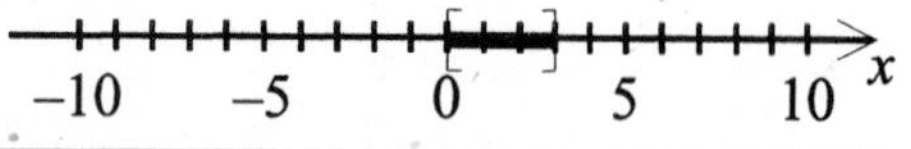

[124] $x \le -12,\ x \ge 2$

Objective 4: Solve polynomial and rational inequalities

[125] (B)

[126] (A)

[127] $-2 \le x < 5$

[128] $[-6, 4) \cup [9, \infty)$

Section A.7: Errors and the Algebra of Calculus

Objective 1: Avoid common algebraic errors

[129] (C)

[130] (D)

[131] $(4x+16)^2 = [4(x+4)]^2 = 16(x+4)^2$

[132] $\sqrt{x-49}$ cannot be simplified.

Objective 2: Recognize and use algebraic techniques that are common in calculus

[133] (A)

[134] (C)

[135] $\dfrac{2}{(x+7)^{7/4}(x+5)^{2/3}}$

[136] $9x^3 - 7x^2 - \dfrac{1}{x} + \dfrac{7}{x^3}$

Section A.8: Graphical Representation of Data

Objective 1: Plot points in the Cartesian plane

[137] (C)

[138] (A)

[139] $(-4, -2)$; Quadrant III

[140] $(3, 2)$

Objective 2: Use the Distance Formula to find the distance between two points

[141] (C)

[142] (C)

[143] 13

[144] Students must show that two sides are equal in length.

Objective 3: Use the Midpoint Formula to find the midpoint of a line segment

[145] (B)

[146] (B)

[147] $(-9, -3)$

[148] $(1, -1)$; $d \approx 4.24$

Objective 4: Use a coordinate plane to model and solve real-life problems

[149] (C)

[150] (B)

[151] $(-1, -2)$

[152] $(40\,N, 102\,W)$